Lecture Notes in Computer Scienc

Commenced Publication in 1973
Founding and Former Series Editors:
Gerhard Goos, Juris Hartmanis, and Jan van Leeuwen

Evangelos Kranakis   Jaroslav Opatrny (Eds.)

Ad-Hoc, Mobile, and Wireless Networks

6th International Conference, ADHOC-NOW 2007
Morelia, Mexico, September 24-26, 2007
Proceeedings

 Springer

Volume Editors

Evangelos Kranakis
Carleton University, School of Computer Science
Herzberg Building, 1125 Colonel By Drive, Ottawa, Ontario K1S 5B6, Canada
E-mail: kranakis@scs.carleton.ca

Jaroslav Opatrny
Concordia University
Department of Computer Science and Software Engineering
1455 de Maisonneuve Blvd West, Montréal, Québec H3G 1M8, Canada
E-mail: opatrny@cs.concordia.ca

Library of Congress Control Number: 2007934039

CR Subject Classification (1998): C.2, D.2, H.4, H.3, I.2.11, K.4.4, K.6.5

LNCS Sublibrary: SL 5 – Computer Communication Networks and
Telecommunications

ISSN   0302-9743
ISBN-10   3-540-74822-9 Springer Berlin Heidelberg New York
ISBN-13   978-3-540-74822-9 Springer Berlin Heidelberg New York

Springer is a part of Springer Science+Business Media

springer.com

© Springer-Verlag Berlin Heidelberg 2007
Printed in Germany

Typesetting: Camera-ready by author, data conversion by Scientific Publishing Services, Chennai, India
Printed on acid-free paper   SPIN: 12120345   06/3180   5 4 3 2 1 0

Preface

The sixth international conference on AD-HOC NetwOrks and Wireless was held in the city of Morelia, Michoacan State, Mexico. It follows the tradition of a multidisciplinary research program on all aspects of ad hoc networks that aims to create a collaborative forum between mathematicians, computer scientists and engineers. Previous Ad-Hoc Networks and Wireless conferences were held in Ottawa, Canada (2006), Cancun, Mexico (2005), Vancouver, Canada (2004), Montreal, Canada (2003), and Toronto, Canada (2002).

This year there were 50 submissions of which 20 were accepted for presentation and inclusion in the conference proceedings. We would like to express thanks to the members of the the Program Committee (see the organization page) for their work with reviewing and selecting the papers for the conference. Many thanks also to Lali Barriere, Stephane Durocher, Angelo Fanelli, Vinay Kolar, Georgios Lioudakis, Jan Manuch, Nathalie Mitton, Gianpiero Monaco, Luca Moscardelli, Alfredo Navarra, Katerina Potika, Christine Stoll for their help in additional refereeing of the submitted papers.

We would like to thank the invited speakers Andrea Werneck Richa, Breno de Medeiros, and Jorge Urrutia for their excellent tutorials and research presentations. Many thanks to Christine Laurendeau for maintaining the conference Web page, Edgar Chavez for coordinating local activities, and Cuauhtemoc Rivera, Chair of local arrangements of ENC 2007 (the annual conference of the Mexican Computer Science Society).

September 2007

Evangelos Kranakis
Jaroslav Opatrny

Conference Organization

Steering Committee

Evangelos Kranakis, Carleton University, Canada
Michel Barbeau, Carleton University, Canada
S.S. Ravi, SUNY Albany, USA
Ioanis Nikolaidis, University of Alberta, Canada
Violet R. Syrotiuk, Arizona State University, USA
Thomas Kunz, Carleton University, Canada

Technical Program Committee

Evangelos Kranakis, Ottawa (TPC Co-chair)
Jaroslav Opatrny, Montreal (TPC Co-chair)

Breno de Medeiros, Talahassee
Nael Abu-Ghazaleh, Binghamton
Euripides Markou, Hamilton
Azzedine Boukerche, Ottawa
Pedro Ruiz Martinez, Murcia
Prosenjit Bose, Ottawa
Pat Morin, Ottawa
Mike Burmester, Talahassee
Lata Narayanan, Montreal
Edgar Chavez, Morelia
Ioanis Nikolaidis, Edmonton
Vinod Choyi, Ottawa
Aris Pagourtzis, Athens
Francesc Comellas, Barcelona
Paolo Penna, Salerno
Jurek Czyzowicz, Gatineau
Giuseppe Persiano, Salerno
Stefan Dobrev, Ottawa
Sunil Shende, Camden
Michele Flammini, L'Aquila
Yannis Stamatiou, Ioannina
Leszek Gasieniec, Liverpool
Ivan Stojmenovic, Ottawa
Christos Kaklamanis, Patras
Jorge Urrutia, Mexico City
Rastislav Kralovic, Bratislava

Peter Widmayer, Zurich
Danny Krizanc, Middletown
Laco Stacho, Vancouver
Marina Papatriantafilou, Gothenborg
Pierre Fraigniaud, Paris
Tao Wan, Nortel

Publicity Chair

Paul Boone, Carleton University

Local Arrangements

Edgar Chavez, Universidad Michoacana, Morelia (coordinator)
Karina Figueroa
Cuauhtemoc Rivera
Erick Sadit Tellez

The Sponsoring Institutions

Universidad Michoacana de San Nicolas de Hidalgo
Sociedad Mexicana de Ciencias de la Computacion

Table of Contents

Protocols

Quality of Service and Performance

Local Routing on Tori

Maia Fraser*

University of Colima, Facultad de Ciencias, C.P. 28045, Colima, Col., México
fraser@ucol.mx

Abstract. We show that face routing, the well-known local routing algorithm for plane graphs introduced by Kranakis and Urrutia, does not in fact succeed for embedded graphs on the torus or higher genus surfaces (contrary to conjecture). We then describe a generalization of face routing, and prove that this algorithm does provide a local routing algorithm for arbitrary graphs embedded in the torus. Finally we discuss extension of this type of algorithm to surfaces of genus g, describing the problems encountered in this setting.

1 Introduction

Kranakis, Urrutia et al., introduced the basic face routing algorithm in 1999 in [1] (see also [2]). It provides a local routing algorithm for graphs embedded in the plane (i.e. surface of genus 0). This means it provides a way of successfully routing messages from one arbitrary point to another in such graphs using only local information (that is sender, destination and current coordinates) plus some small, bounded memory.

The basic method uses a *connecting curve* γ from sender to destination as a reference. In the original treatment this is a straight line, but any other compact connecting curve which can be defined by the given available information will do. Face routing then consists in following faces along γ.

Since only finitely many faces meet γ and all are topologically trivial (i.e. have interior homeomorphic to a disk) this algorithm is guaranteed to stop at the destination in finite time. The full argument is given in Sect. 3.

Local routing algorithms are of special interest for internet and wireless applications since the topology of such networks is constantly changing and no accurate global picture exists. Of particular interest for wireless adhoc networks is the local routing problem for *3-D graphs*, that is graphs embedded in Euclidean 3-space ($\mathbb{R}^3$). This represents for example the problem of routing in a wireless network (of sensors, laptops etc..) in a building. Finding a local routing algorithm for graphs embedded in surfaces of arbitrary genus has been considered as a step towards solving the 3-D problem. It was conjectured [4] that face routing should succeed for graphs embedded in the torus (surface of genus 1) and possibly for graphs embedded in surfaces of arbitrary genus.

---

* The author wishes to thank J. Urrutia, E. Kranakis and J. Opatrny for introducing her to this subject and gratefully acknowledges the many helpful comments E. Kranakis made regarding a preliminary version of this paper.

E. Kranakis and J. Opatrny (Eds.): ADHOC-NOW 2007, LNCS 4686, pp. 1–14, 2007.
© Springer-Verlag Berlin Heidelberg 2007

1.1   Results of the Paper

We show in Sect. 3 that in fact face routing fails on all surfaces of positive genus. Namely, in Prop. 1 we construct an embedding of a graph in a surface of arbitrary positive genus for which face routing fails.

In Sect. 4.2 we propose a modification of face routing, *generalized face routing* (GFR), which we prove succeeds on the torus (Thm. 1). It requires the use of three connecting curves which pairwise combine to give the longitude and meridian of the torus.

In fact, our work suggests that for an algorithm based on face routing to succeed in surfaces of arbitrary positive genus it must have access to a certain number of curves of distinct homotopy classes on the surface (i.e. curves which cannot be deformed one into the other). This dependence on global topological information casts doubt on the suitability of such algorithms for approaching the 3-D local routing problem. In Section 7 we briefly discuss this and other limitations of generalized face routing for surfaces of genus greater than 1.

1.2   Notation and Terminology

Throughout this paper Σ_g will be taken to be a compact oriented surface of genus g (where g is a nonnegative integer) and G an arbitrary connected graph embedded in Σ_g.

We will sometimes refer to directed subgraphs of G, meaning subgraphs of the original graph where each edge involved has been assigned a direction. In particular we will be interested in oriented cycles, that is directed cycles where the directions are compatible, i.e. where each edge's final vertex is the initial vertex of the next edge. We will speak of a *directed edge* when we wish to refer to an edge with a specified direction and simply an *edge* otherwise. Given a directed edge e, we write $-e$ to denote the oppositely directed edge. We define a *face* of G to be the closure of an open region that does not meet the graph but is bounded by edges of the graph.

We assume standard *orientation* conventions so that given a subsurface S with boundary component β, the induced orientation of β is such that when travelling along β in this direction, S lies to the left (i.e. the inner side is the left one). By extension, we will also speak of the *inner side* and the *outer side* of any oriented embedded curve β, even one not necessarily closed. We will sometimes refer to non-closed curves as *curve segments* to make the distinction clear.

Finally, the *complement* of a subset X in Σ_g will be denoted X^c.

2   Border Cycles in Embedded Graphs

It will be useful to define boundary components of faces locally, since standard face routing actually follows the boundary components of faces and not the faces themselves. To this end we define:

Definition 1 (Border Cycle, Clear Collar). *Let σ in G be an oriented cycle and suppose there is some $\epsilon > 0$ such that no edge of G meets N_ϵ, where N_ϵ*

denotes the open region to the inner side and within a distance of ε of σ but not including σ itself. Then σ will be called a (left-handed) border cycle, *and* N_ϵ *a* (left) clear collar *for σ.*

Informally a left-handed border cycle is one that can be traversed while resting the left hand on the surface (in the clear collar), without lifting the hand or touching any edges of the graph. Using the clear collars one may immediately show that every face boundary (with its canonical orientation as boundary of the face) is a border cycle and every border cycle is a boundary component of some face. We now observe the following.

Lemma 1. *Given a directed edge e in G, there is a unique (left-handed) border cycle containing e and this may be constructed locally starting from e.*

This allows us to make the following definitions.

Definition 2 (Border Cycle Determined by a Directed Edge). *We will refer to the unique border cycle containing a directed edge e as the* border cycle determined by the directed edge e.

Definition 3 (Border Cycle Determined by a Curve at a Vertex). *Given a vertex v in G and an embedded segment δ with one endpoint at v such that δ ∩ G = {v}, δ must enter v between two edges of G since it meets G only at v. Suppose a direction is assigned to one of these, e, so that δ lies to its left and let β be the border cycle determined by e. β will be referred to as the* border cycle determined by δ at v. *This is illustrated in Fig. 1.*

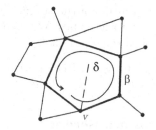

Fig. 1. The left-handed border cycle β determined by δ at v is shown in boldface (the curve δ is dashed). Note the clear collar swept out as shown by the arrow during the construction of β.

In the plane or the sphere all faces have interiors homeomorphic to the disk. In tori or higher genus surfaces this is no longer the case; the faces may have more complicated topology and moreover they may have several boundary components (each of which is a border cycle).

Definition 4 (Trivial Face, Trivial Border Cycle). *A face whose interior is homeomorphic to a disk will be referred to as a* trivial face. *A boundary of such a disk will be referred to as a* trivial border cycle.

The faces of G cover the entire surface Σ_g, but it is useful to record what part of this is accomplished by the trivial faces of G.

Definition 5 (Tiled Region). *Let $\mathcal{R}(G)$ denote the union of G together with all trivial faces of G. This closed region will be called the* tiled region *defined by G.*

We make the following observations concerning $\mathcal{R}(G)$ (the first is immediate, the second follows from basic arguments in algebraic topology).

Lemma 2. *$\mathcal{R}(G)$ is connected and its boundary components, if any, are non-trivial border cycles, given the orientation induced on them as boundaries of $\mathcal{R}(G)^c$ (the complement of $\mathcal{R}(G)$).*

Lemma 3. *When Σ_g is a torus, i.e. $g = 1$, $\mathcal{R}(G)$ is either all of Σ_g or it is an embedded disk or its complement is an embedded annulus.*

By Lemma 3, if there exists a non-trivial border cycle in Σ_1 which is not homotopic to zero (i.e. cannot be continuosly shrunk to a point) then $\mathcal{R}(G)^c$ must be an embedded annulus. Thus there exists exactly one other non-trivial border cycle (the other boundary component of the annulus).

Definition 6 (Mate of a Non-Trivial Border Cycle). *Let β be a nontrivial border cycle in Σ_1 which is not homotopic to zero then the other non-trivial border cycle in Σ_1 will be referred to as the mate of β.*

3   Face Routing

In this section, we define face routing of the standard type in such a way that it can be used on surfaces of arbitrary genus. We then discuss its properties on such surfaces. First, we establish some terminology.

Definition 7 (Connecting Curve). *Given vertices s and d of G, a compact embedded curve γ starting at s and ending at d will be referred to as a connecting curve from s to d.*

Definition 8 (Generic Crossing, Generic Connecting Curve). *A transversal intersection point of a curve with an edge that is not an endpoint of either the edge or the curve is called a generic crossing. A connecting curve γ will be said to be generic (for G) if each intersection point of γ with G is either a generic crossing of γ with some edge of G or else an endpoint of γ occuring at a vertex of G where γ is transversal to all adjacent edges.*

The reason for this terminology is that intersection points generically have such properties, i.e. we can always achieve them[1] by slightly perturbing γ in an arbitrarily small neighbourhood of each of its intersections with G (but keeping its endpoints fixed). To simplify the description of face routing we make one last definition concerning border cycles.

---

[1] Since they are open conditions.

Definition 9 (Border Cycle Determined by a Generic Crossing). *Given a generic crossing of an embedded curve γ with G, we will refer to the border cycle determined by γ at this crossing meaning the border cycle determined by e where e is the edge of G at which the crossing occurs, with direction chosen so that γ crosses over to its inner side.*

We may now define standard face routing:

Algorithm 1 (Face Routing). [2] *Let γ be a generic connecting curve from a vertex s to a vertex d or a subsegment of this connecting curve and let $\mathcal{X} = \gamma(t_0)$ be an intersection of γ with G, which is not d. Then* face routing (along γ to d) starting from $\mathcal{X}$ *is defined as follows:*

1. *Let β_0 be the border cycle determined by γ at $\mathcal{X}$. Set $i := 0$.*
2. *Travel around β_i (stopping if d encountered) to locate the intersection of γ with β_i having highest t-value (say t_*) and then return there.*
3. *Let e be the directed edge of β_i containing $\gamma(t_*)$. Increment i by one and set $t_i := t_*$. Let β_i be the border cycle determined by $-e$. Go to step 2.*

Analogously we define reverse face routing (along γ to s) starting from *an intersection point $\mathcal{Y} = \gamma(t_0)$ of γ with G which is not s (replacing "highest" with "lowest" and d with s in the algorithm description above).*

Definition 10 (Reaching of Face Routing). *In either case - (forward) face routing or reverse face routing - we will refer to the process as* reaching a border cycle (or vertex or edge) $\mathcal{Z}$ *if at some point during the respective algorithm there is a β_i equal to (or resp. containing) $\mathcal{Z}$.*

Definition 11 (Success in General). *For any of the routing algorithms defined in this paper, we will say the algorithm* succeeds *if the destination vertex is contained in one of the border cycles traversed by the algorithm. When routing does not succeed, we say it* fails.

Comment - Extent of γ: Note that the extent of γ, i.e. whether it is a whole connecting curve or just a subsegment thereof, is important in both algorithms, since we maximize (or minimize) t among all crossings of $\gamma(t)$ with a given border cycle. A longer γ may pass through a given border cycle again, thus changing that cycle's successor in the sequence of border cycles visited (see Fig. 3).

Comment - Forward/Reverse Not Inverses. Note that reverse face routing is just (forward) face routing on the same but oppositely oriented connecting curve. However, reverse face routing *does not in general un-do*[3] face routing, namely the sequences of border cycles these procedures define need not be reflections of each other (see Fig. 3).

---

[2] We have omitted the usual checking of adjacent vertices (for d) since this is mainly a performance enhancing feature. Of course, success of the algorithm without checking implies success of the usual algorithm (with checking). And conversely, any example where the algorithm without checking fails can be converted to an example where the usual algorithm fails (by inserting extra vertices).

[3] See definition of *slow face routing* in Section 4.1.

Lemma 4. *Let $\beta_0, \beta_1, \beta_2, \ldots$ be the sequence of border cycles generated by face routing along γ from s to d. Suppose $d \in \beta_n$ for some n then the above sequence is finite and β_n must be its last element. Conversely if the sequence of border cycles generated by face routing is finite and consists of more than just β_0 or we know β_0 is trivial[4], then the final border cycle must include d.*

3.1   Success and Limitations of Face Routing

We are now ready to look at cases when face routing does or does not succeed.

Lemma 5. *Let s and d be vertices of G in Σ_g (note we are in the general genus setting). If γ is any connecting curve from s to d which meets only trivial faces of G (i.e. which remains on the interior of $\mathcal{R}(G)$) then face routing succeeds along γ. We may also allow holes in Σ_g away from G.*

Proof. Let $\beta_0, \beta_1, \beta_2, \ldots$ be the border cycle sequence generated by face routing along γ.

We claim that the triviality of all faces encountered along γ means there can be no repetition in this sequence, i.e. $i \neq j \Rightarrow \beta_i \neq \beta_j$. This implies the sequence must be finite since only finitely many faces meet γ. But now, let β_n be the last element of the sequence and F the face it defines. By Lem. 4, $d \in \beta_n$, so face routing succeeds along γ.

It remains to prove the claim. Assume without loss of generality that γ is already generic. For each β_i in the sequence $\beta_0, \beta_1, \beta_2, \ldots$, let t_i be its maximal t-value. Note that $t_0, t_1, t_2, \ldots$ is a non-decreasing sequence of real numbers. Moreover, γ crosses from the inner side of β_i to the inner side of β_{i+1} at the value t_i for any pair of successive border cycles (β_i, β_{i+1}) in the border cycle sequence. This is because β_i is trivial so the face F determined by t_i has only one boundary component and thus if γ were to cross to the inner side of β_i (and interior of F) at t_i, it would need to intersect β_i again later at a higher t value (since d is in the complement of the interior of F) contradicting t_i being maximal. In particular, we cannot have $t_i = t_{i+1}$.

Now suppose two border cycles in the sequence coincide, say $\beta_i = \beta_j$ with $i < j$. Then $t_{j-1} \leq t_i$ because $\gamma(t_{j-1})$ is a crossing of γ with $\beta_j = \beta_i$ and t_i is the maximal t-value of β_i. However $i \leq j - 1$ so $t_i \leq t_{j-1}$. Thus $t_i = t_k = t_{j-1}$ for all k with $i \leq k \leq j - 1$ (and in fact for all $k \geq i$), so $t_i = t_{i+1}$, which cannot happen. Repetition is thus impossible. $\qquad\square$

An immediate corollary of Lem. 5 is the following.

Corollary 1. *If G is embedded in Σ_g in such a way that all faces are trivial (i.e $\mathcal{R}(G) = \Sigma_g$) then face-routing will succeed along any connecting curve.*

---

[4] We need to exclude the case where the border cycle sequence consists of just one border cycle β_0 and this cycle meets γ just once (at s). This situation, which can only occur in tori and higher genus surfaces, means there is no edge at all of β_0 that crosses γ (except at endpoints) and face routing goes no further, although d need not be on β_0.

This is the situation on the sphere (and plane), so we obtain:

Corollary 2 (Kranakis, Urrutia et al.). *Let s and d be two vertices of an embedded graph G on the sphere (or plane) Σ_0. Then face routing succeeds along any connecting curve γ from s to d.*

The method of proof for Lem. 5 also shows:

Lemma 6. *Let s and d be vertices of G in Σ_g and let γ be a connecting curve from $s = \gamma(0)$ to $d = \gamma(1)$. Suppose γ intersects a (possibly non-trivial) border cycle β at a value $t = t_{last}$ such that $\gamma(t)$ lies in the interior of $\mathcal{R}(G)$ for all $t : t_{last} < t < 1$. Then face routing along γ reaches d from β.*

We now show that on surfaces of positive genus, unlike on the sphere, face routing does not in general succeed.

Proposition 1. *Suppose g is a positive integer. Then there exists an embedded graph in Σ_g and a generic connecting curve γ along which face routing fails.*

Proof. Let G be the graph shown embedded on the torus in Fig. 2. It has seven vertices r, s, t, u, v, w and d, and edges st, sw, tu, uv, vw, wt and sr and rd. Face routing fails along the connecting curve γ.

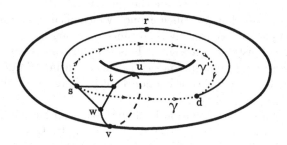

Fig. 2. Embedded graph on the torus for which face routing fails. In this example there are two faces: one is trivial and is bounded by the cycle (swt), the other is non-trivial and has two boundary components, the border cycle $(wvut)$ and the border cycle $(stuvwsrdr)$. Face routing along γ generates the infinite border cycle sequence $\beta_0 = (swt)$, $\beta_1 = (wvut)$, ..., $\beta_{2k} = \beta_0$, $\beta_{2k+1} = \beta_1$, ... and never reaches d.

To obtain an example on a surface Σ_g of genus $g > 1$ choose any curve σ in Σ_1 bounding a disk away from G and from all connecting curves. Then glue a punctured surface of genus $g - 1$ in along σ. The result is a surface Σ_g of genus g with a graph and connecting curves as required.   □

Note that in Figure 2, face routing along the connecting curve γ' does succeed. The curve γ' exits and then re-enters the tiled region across the non-trivial border cycle $(wvut)$, while γ crosses a non-trivial border cycle once only.

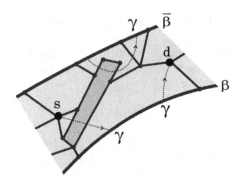

Fig. 3. The Phenomenon of Diversion. The two non-trivial border cycles β and $\bar{\beta}$ are boundary components of the tiled region $\mathcal{R}(G)$ (shaded). The connecting curve γ between s and d has 3 components in $\mathcal{R}(G)$. Face routing starts at s on one of these but then at the darkly shaded face it hops to a different component of γ. However, this component meets $\bar{\beta}$ inopportunely; at the highest t value along $\bar{\beta}$, the curve γ exits the tiled region $\mathcal{R}(G)$. Face routing then hops forever back and forth between $\bar{\beta}$ and the trivial border cycle just inside $\mathcal{R}(G)$ at this maximal edge. It never reaches d. Moreover, the reader may check that reverse face routing from this edge does not return to s either.

The Problem of Diversion. An additional problem in the presence of non-trivial border cycles is that face routing doesn't necessarily follow γ even within $\mathcal{R}(G)$ - it may jump from one part of γ to another part with higher t-values. We will refer to this as *diversion*. This is a phenomenon which occurs even on the sphere, but it doesn't complicate face routing there since there are no non-trivial border cycles. Diversion is illustrated in Fig. 3. This example also shows how reverse face routing does not in general undo (forward) face routing - from the maximal crossing of γ with a border cycle one cannot deduce where one originally entered the border cycle.

These phenomena make navigation in higher genus surfaces difficult. But it is possible to avoid the problem of diversion by two tricks. One is routing along a known subsegment of a connecting curve and the other is a variant of face routing which we call *slow face routing* (defined in the next section). The problem of meeting non-trivial border cycles inopportunely will be handled by using various connecting curves in different homotopy classes and a technique we call *backtracking*. Taken all together, these ideas are the basis for the generalized face routing which we define in the next section.

4   Generalized Face Routing

First we define *slow face routing*.

4.1   Slow Face Routing

Instead of leaving each border cycle β across the edge where γ crosses with maximal t value, we may leave β at the next (in terms of t) crossing of γ with

β (if this is an exit from the face or simply stop if not). This has the effect of slowing face routing down but makes it easier to handle. More precisely:

Algorithm 2 (Slow Face Routing). *Let γ be a generic connecting curve from a vertex s to a vertex d or a subsegment of this connecting curve and let $\mathcal{X} = \gamma(t_0)$ be an intersection of γ with G, which is not d. Let* entryVal, exitVal *be two variables recording t-values and* nextIn *a boolean. Slow face routing (along γ to d) starting from $\mathcal{X}$ is defined as follows:*

1. *Let β_0 be the border cycle determined by γ at $\mathcal{X}$. Set* entryVal $:= t_0$ *and $i := 0$.*
2. *Travel around β_i to locate the intersection of γ with β_i having next higher t-value after* entryVal, *if any - at each candidate intersection, notice its direction, setting* nextIn $:=$ true *if it is towards the inner side of β and setting* exitVal *to the t-value there. If d is encountered, stop there (exit with success). If no subsequent crossing (after $\gamma($*entryVal$)$*) exists or* nextIn $=$ true *then stop (exit with failure). Otherwise go to $\gamma($*exitVal$)$.*
3. *Increment i by one and set* entryVal *to the t-value at the current crossing. Let β_i be the border cycle determined by γ at this crossing. Go to step 2.*

Analogously we define reverse slow face routing *(along γ to s) starting from an intersection point $\mathcal{Y} = \gamma(t_0)$ of γ with G which is not s (replacing "next higher" with "next lower" and d with s in the algorithm description above).*

Definition 12 (Reaching of Slow Face Routing). *In either case - (forward) slow face routing or reverse slow face routing along γ - we will refer to the process as* reaching *a border cycle (or vertex or edge or intersection point of γ with G) $\mathcal{Z}$ if at some point during the respective algorithm there is a β_i equal to (or resp. containing) $\mathcal{Z}$.*

Lemma 7. *Given two intersections $\gamma(t_0)$ and $\gamma(t_*)$ of γ with G, if slow face routing does not reach $\gamma(t_*)$ from $\gamma(t_0)$, then there must be a non-trivial border cycle meeting γ at a t-value $t_0 \le t < t_*$.*

Also, it is easily verified that:

Lemma 8. *Given a crossing $\gamma(t_*)$ reached by slow face routing along γ starting from $\gamma(t_0)$, reverse slow face routing along γ starting from $\gamma(t_*)$ reaches $\gamma(t_0)$.*

In the absence of non-trivial border cycle chains, slow face routing gets to the same place eventually as (standard) face routing; it just may visit more intermediate faces, more often, on the way. It is however reversible and predictable and so provides a useful tool for defining a local routing algorithm on surfaces of higher genus.

4.2   Generalized Face Routing Algorithm (GFR)

We first define a generalized face routing algorithm GFR(γ) on a single connecting curve and then another algorithm GFR($\gamma_0, \gamma_1, \gamma_2$) which calls the first algorithm and is defined on a triple of connecting curves.

Algorithm 3 (GFR(γ) on one connecting curve). *We assume γ to be a generic connecting curve from one vertex, s, to another, d. Also assume for simplicity that $\gamma(0) = s$ and $\gamma(1) = d$.*

Four variables will be used for t parameter values. Firstly, entryVal *and* exitVal: *these will record respectively the t-value upon entering the current face and the best candidate t-value encountered so far for exiting that face. In addition, we use* maxVal, bktVal. *As well we use a boolean* failure *initialized to* false.

The algorithm GFR(γ) is as follows:

1. *slow face routing*
   Perform slow face routing along γ as far as possible. Besides the usual procedure using entryVal *and* exitVal, *we also set* bktVal := entryVal *each time we reset the latter. In addition we keep a record of the highest t-value encountered so far in each border cycle β using* maxVal *and we set* maxIn *to* true *or* false *depending on whether the crossing at this maximum t-value is towards the inner side of β or not. If slow face routing stops before reaching d then we have crossed to the inner side of some β and the next crossing with γ (along t), if any, is also a crossing to the inner side of β. Go to stage 2.*

2. *triage/backtracking*
   Here we handle the three possibilities:
   - *If there is no next crossing then go to the crossing where $t =$ entryVal occurs and set* failure := true *(we have failed). Perform reverse slow face routing starting at the current crossing, stopping upon reaching s, i.e. when* exitVal $= 0$ *(this must occur since we are just un-doing slow face routing).*
   - *If γ crosses to the inner side of β at the next crossing, and* maxIn $=$ false *then go to the crossing with* maxVal *and proceed to stage 3.*
   - *If γ crosses to the inner side of β at the next crossing (say c), and* maxIn $=$ true *then go to c. Start reverse face routing (not slow) restricted to the segment of γ with $t \in [$bktVal$, 1]$ (we are backtracking, see Fig. 4), until a t-value greater than* maxVal *occurs, in which case go to the crossing with maximal t value in the current border cycle and proceed to stage 3.*

3. *standard face routing*
   Here we finish up with (standard) face routing:
   - *If the current value of t is 1 then we have arrived at d so stop.*
   - *Otherwise start face routing from the edge where the current t value occurs.*

Algorithm 4 (GFR($\gamma_0, \gamma_1, \gamma_2$) on a triple of curves). *Assume $\gamma_0, \gamma_1, \gamma_2$ are connecting curves from s to d satisfying all the conditions for GFR on a single curve. Moreover suppose $\gamma_0\gamma_1^{-1}, \gamma_0\gamma_2^{-1}$ are respectively longitude and meridian of Σ_1 (see Fig. 5; this terminology is recalled in the next section).*

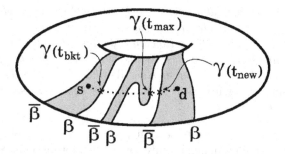

Fig. 4. Backtracking enables us to use the connection provided by γ between the two border cycles β and $\bar{\beta}$ (the rest of G is not shown, but $\mathcal{R}(G)$ is shaded). Here t_{max} and t_{bkt} represent `maxVal` and `bktVal` respectively. The first crossing of β with γ at a t value exceeding t_{max} occurs at $\gamma(t_{new})$; this is where we stop backtracking, having reached the other non-trivial border cycle.

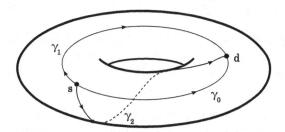

Fig. 5. Connecting curves required by Alg. 4 and Thm. 1

The algorithm GFR($\gamma_0, \gamma_1, \gamma_2$) is as follows:

0. Initialize `i` *:= 0.*
1. Call GFR with γ_i*.*
2. When GFR stops:
  - *If it returns* `failure` = `true`, *then we must be at s again, so increment* `i` *by 1, reset* `failure` *:=* `false` *and go to stage 1.*
  - *If it returns* `failure` = `false` *then we must be at d, so stop.*

5   Topological Background

We now recall some notions from topology before stating and proving our main result (Thm. 1) in Section 6. First we define a variant of *intersection number* (usually this is defined for closed curves; we generalize the notion to curve segments as well).

Definition 13 (Intersection Number). *Let β be a border cycle and γ a curve or curve segment such that $\beta \pitchfork \gamma$, i.e. β is transversal to γ (also at endpoints* *of γ if any).* To each side (along γ) of each intersection point of γ with β we assign the value $+.5$ if that side is pre-intersection and lies on the inner side of β or if that side is post-intersection and lies on the outer side, and $-.5$ if that side is pre-intersection and lies on the outer side of β or if that side is post-intersection and lies on the inner side. We then define the $Int\#(\gamma, \beta)$, the intersection number of γ with β, to be the (finite) sum of these values over all intersections of γ with β.

Since the number of sides of intersection points is always even, this number will be an integer. It also depends only on the homotopy classes of β and γ (when γ closed or the homotopy class of γ relative to the endpoints otherwise). Moreover it behaves well under concatenation and orientation reversal. Namely, $Int\#(\gamma\gamma', \beta) = Int\#(\gamma, \beta) + Int\#(\gamma', \beta)$ and $Int\#(\gamma^{-1}, \beta) = -Int\#(\gamma, \beta)$, where $\gamma\gamma'$ denotes the concatenation of these two curves (with the final endpoint of γ being the initial endpoint of γ'), and γ^{-1} is the same curve as γ but with opposite orientation. Note that the usual longitude λ and meridian μ of the torus have $Int\#(\lambda, \mu) = 1$. We also recall:

Lemma 9. *A closed curve homotopic to zero has zero intersection number with all other closed curves. And conversely, on a torus, any closed curve that is not homotopic to zero must meet either the longitude or meridan with non-zero intersection number.*

Lemma 10. *Let R be a region with one or more boundary components, and σ a closed curve. Then the sum of $Int\#(\sigma, \beta)$ over all boundary components β of R (each with its canonical orientation as a boundary of R) is zero.*

6   Main Results

Theorem 1. *Let s and d be vertices in G on the torus Σ_1. Let $\mathcal{S} = \{\gamma_0, \gamma_1, \gamma_2\}$ be three connecting curves from s to d such that $\lambda = \gamma_0\gamma_1^{-1}, \mu = \gamma_0\gamma_2^{-1}$ are respectively the longitude and meridian of the torus (see Fig. 5 and Sect. 5). Then $GFR(\gamma)$ succeeds for at least one $\gamma \in \mathcal{S}$ and for those γ where it fails, it returns to s with* `failure = true`.

An immediate consequence is:

Corollary 3. *Let $\gamma_0, \gamma_1, \gamma_2$ be connecting curves as required by the GFR alorithm on a triple of curves. Then $GFR(\gamma_0, \gamma_1, \gamma_2)$ succeeds.*

Proof. (of Thm. 1) We sketch the proof. Assume for simplicity that we have already perturbed $\gamma_0, \gamma_1, \gamma_2$ so these are all generic for G. If there are no non-trivial border cycles in Σ_1 then we are done since slow face routing will succeed along any connecting curve and thus so will GFR (in stage 1.). Otherwise, let β be a non-trivial border cycle. If it is not homotopic to zero, let $\bar{\beta}$ be its mate.

For each $\gamma \in \mathcal{S}$, let $first(\gamma)$ be either β or $\bar{\beta}$ if exists - whichever of these cycles is the first one crossed by γ or the one on whose inner side γ lies near s if

s belongs to both. Similarly, let $last(\gamma)$ be the last of these cycles crossed by γ or the one on whose inner side γ lies near d if d belongs to both. Note that by Lem. 7, slow face routing reaches $first(\gamma)$ from s and by Lem. 6 standard face routing reaches d from $last(\gamma)$.

It can be shown (by checking the various cases) that generalized face routing can only fail if case 1 in stage 2 occurs at $first(\gamma)$. This case occurs if and only if γ crosses to the inner side of $first(\gamma)$ never to return after possibly having crossed back *and* forth across $first(\gamma)$ various times. This in turn implies that $Int\#(\gamma, first(\gamma)) = -1$.

We remark that Lem. 10 implies each connecting curve $\gamma \in \mathcal{S}$ must have total intersection number κ with the boundary component(s) of $\mathcal{R}(G)$, where $\kappa \in \{0, -.5\}$ since all γ have same endpoints in $\mathcal{R}(G)$ (may be boundary or interior). Now suppose $Int\#(\gamma, first(\gamma)) = -1\ \forall \gamma \in \mathcal{S}$. Then $\mathcal{R}(G)$ must have two boundary components and $Int\#(\gamma, \beta) = -Int\#(\gamma, \bar{\beta}) + \kappa,\ \forall \gamma \in \mathcal{S}$.

Since there are two non-trivial border cycles $(\beta, \bar{\beta})$ and three elements of $\mathcal{S}$, there must be at least two elements of $\mathcal{S}$ with the same $first$ border cycle. Suppose $first(\gamma_0) = first(\gamma_1) = \beta$ (the other cases are similar). Then we compute $Int\#(\lambda, \beta) = Int\#(\gamma_0, \beta) - Int\#(\gamma_1, \beta) = Int\#(\gamma_0, first(\gamma_0)) - Int\#(\gamma_1, first(\gamma_1)) = 0$. But we know β is not homotopic to zero (since $\mathcal{R}(G)$ has two boundary components) so this implies that $[\beta] = \pm[\lambda]$ and thus $Int\#(\mu, \beta) = \pm 1$. However, computing as before, we have $Int\#(\mu, \beta) = -1 - Int\#(\gamma_2, \beta)$. So $Int\#(\gamma_2, \beta) = 0$ or -2. Thus in particular we cannot have $\beta = first(\gamma_2)$. But $\bar{\beta} = first(\gamma_2)$ (which implies $Int\#(\gamma_2, \bar{\beta}) = -1$) is also a contradiction since $Int\#(\gamma_2, \bar{\beta}) = -Int\#(\gamma_2, \beta) + \kappa \in \{0, 2, -.5, 1.5\}$. Thus in fact there must have been some $\gamma \in \mathcal{S}$ such that $Int\#(\gamma, first(\gamma)) \neq -1$.

This shows GFR(γ) succeeds for at least one $\gamma \in \mathcal{S}$. That it will return to s with `failure` set to `true` when it fails is trivial: as we have seen, failure can only occur if we encounter case 1 in stage 2 and in this case the algorithm indeed sets `failure` to `true`; we will return to s by Lem. 8 since we are performing reverse slow face routing from a point reached by (forward) slow face routing from s. □

7 Outlook for Higher Genus Surfaces and 3-D Local Routing

There are several limitations to using generalized face routing on surfaces Σ_g of genus $g > 1$. For convenience, let us view such a surface as a plane with various handles attached to it. In the case of the torus there is only one handle and we may view a suitable set of three connecting curves as follows: one curve from s to d not passing through the handle and two connecting curves passing through the handle (one winding around it as well). To cross the handle we use the trick of backtracking and the fact that we have two non-homotopic connecting curves through the handle. In the case of g handles in the plane, the simple correspondence of a non-trivial border cycle with its mate does not hold, so it is not immediately clear that a trick like backtracking works to cross a handle. But in addition, there is another problem - the *order* in which we cross the handles matters and in order to try various sequences, it appears one needs an amount of memory factorial in g.

Such considerations illustrate the dependence of face routing type algorithms on the topology of the underlying surface. Such a dependence is fine if we have geometric information for the arbitrary surface analagous to that assumed for the plane in standard face routing, namely if we have a global coordinate system for the surface (expressed for example as a region in the plane with various sides identified). However it makes using such algorithms as a tool for obtaining a local 3-D routing algorithm problematic, since the method of constructing such surfaces (containing a 3-D graph's vertices) would have to remain local.

References

1. Kranakis, E., Singh, H., Urrutia, J.: Compass Routing on Geometric Networks. In: Proceedings of 11th Canadian Conference on Computational Geometry, CCCG-99, August 15-18, 1999, Vancouver pp. 51–54 (1999)
2. Boone, P., Chavez, E., Gleitzky, L., Kranakis, E., Opatrny, J., Salazar, G., Urrutia, J.: Morelia Test: Improving the Efficiency of the Gabriel Test and Face Routing in Ad-hoc Networks. In: Kralovic, R., Sýkora, O. (eds.) SIROCCO 2004. LNCS, vol. 3104, Springer, Heidelberg (2004)
3. In: the 5th Workshop: Routing in Guanajuato comments of Jaroslav Opatrny (August 2006)
4. Kranakis, E., et al.: Routing in Mexico: The First Five Years and Beyond. In: Summary of problems discussed in the series of workshops. Routing in Morelia 2002, 2003, 2004 and Routing in Guanajuato 2005, 2006 with outlook to Routing in Oaxaca 2007 (August 2006)
5. Massey, W.S.: Algebraic Topology: An Introduction. Springer, New York (1967)
6. Fuchs, D.B., Fomenko, A.T., Gutenmacher, V.L.: A Course in Homotopic Topology, H Stillman Pub. (June 1986)

Topology Control and Geographic Routing in Realistic Wireless Networks

Kevin M. Lillis[1,2], Sriram V. Pemmaraju[1], and Imran A. Pirwani[1]

[1] Department of Computer Science,
University of Iowa, Iowa City, IA 52242-1419, USA
{lillis,sriram,pirwani}@cs.uiowa.edu
[2] Computer and Information Science,
St. Ambrose University, Davenport, IA 52803, USA
LillisKevinM@sau.edu

Abstract. We present a distributed topology control protocol that runs on a d-QUDG for $d \geq 1/\sqrt{2}$, and computes a sparse, constant-spanner, both in Euclidean distance and in hop distance. QUDGs (short for Quasi Unit Disk Graphs) generalize Unit Disk Graphs and permit more realistic modeling of wireless networks, allowing for imperfect and non-uniform transmission ranges as well as uncertain node location information. Our protocol is local and runs in $O(1)$ rounds. The output topology permits memoryless (geographic) routing with guaranteed delivery. In fact, when our topology control protocol is used as preprocessing step for the geographic routing protocol GOAFR$^+$, we get the routing time guarantee of $O(\ell^2)$ for any source-destination pair that are ℓ units away from each other in the input d-QUDG. The key idea is simple: to obtain planarity, we replace each edge intersection with a *virtual node* and have a real node serve as a proxy for the virtual node. This idea is supported by other parts of our protocol that (i) use clustering to keep the density of edge crossings bounded and (ii) guarantee that an edge between a virtual node and a neighbor is realized by a constant-hop path in the real network. The virtual node idea is simple enough to be useful in many contexts. For example, it can be combined with a scheme recently suggested by Funke and Milosavljević (INFOCOM 2007) to guarantee delivery under uncertain node locations. Similarly, the virtual nodes idea can also be used as a cheap alternative to edge-crossing removal schemes suggested by Kim et al. (DIALM-POMC 2005, SENSYS 2006).

1 Introduction

A wireless ad-hoc network typically consists of battery-powered individual nodes that are able to communicate with nodes in transmission range via radio broadcast and perform local computations. Because there is no centralized control and because each node has only a small amount of memory, *memoryless routing* protocols are highly desirable for wireless ad-hoc networks. In memoryless routing protocols, each node decides whom a message should be forwarded to based solely on the source and destination of the message and on information gathered

E. Kranakis and J. Opatrny (Eds.): ADHOC-NOW 2007, LNCS 4686, pp. 15–31, 2007.

from nearby nodes. As a result, the per node memory used by a memoryless routing protocol depends on the local density of nodes rather than on the total number of nodes. *Geographic routing protocols* are memoryless routing protocols that use geographic information such as coordinates of source and destination, and coordinates of nodes that are a small number of hops away. Geographic routing protocols take advantage of the fact that nodes in a wireless sensor network typically reside in low dimensional Euclidean space and may know their coordinates, at least approximately, in some agreed-upon global coordinate system. For the rest of the paper, we assume that nodes of the given wireless network reside in $\mathbb{R}^2$, the Euclidean plane.

Well-known methods in geographic routing are *greedy routing*, *face routing*, and *dual-mode routing*. The last method combines greedy and face routing. In greedy routing each node forwards a packet by choosing a neighbor that is closest to the destination. Greedy routing guarantees message delivery only in special circumstances, for example, when the communication graph of the wireless network happens to be a Delaunay triangulation [4]. In general, this scheme cannot guarantee delivery because a routed message may be trapped in a cycle [4]. In face routing it is assumed that the network is embedded in the plane with no edge crossings and using the *right-hand rule*, messages are forwarded along the boundaries of adjacent faces that intersect the line segment between the source node and the destination node. There exists a trade-off between greedy and face routing. Greedy routing is fast but cannot guarantee delivery, whereas face routing guarantees delivery, but is less efficient than the greedy method [18]. These two techniques have been combined in an effort to garner the benefits offered by each. Such dual-mode routing schemes [14,16] oscillate between face routing and greedy routing. A message is greedily forwarded until a *local minimum* is reached; this is a node that is closer to the destination than any of its neighbors. Then face routing is employed to forward the message until greedy routing can resume. Well known dual-mode routing protocols include GPSR [14] and GOAFR+ [16].

A starting point of much of the algorithmic research on wireless sensor networks is a graph-theoretic model of the pairwise communication between nodes in the network. The *unit disk graph* (UDG) may be the simplest of these models. Though popular because of its simplicity, the UDG model makes unrealistic assumptions such as radio transmission ranges being perfect disks, nodes having uniform transmission ranges, etc. The *Quasi Unit Disk Graph* model [2,17] alleviates some of the shortcomings of the UDG model. A Quasi Unit Disk Graph is parameterized by d, $0 \leq d \leq 1$ and is usually denoted d-QUDG. This graph consists of vertices in $\mathbb{R}^2$ and an edge set E satisfying the rules: (i) $\{u, v\} \in E$ if $\|u - v\|_2 \leq d$ and (ii) $\{u, v\} \notin E$ if $\|u - v\|_2 > 1$. Note that edges between pairs of vertices u, v with $d < \|u - v\|_2 \leq 1$ are left unspecified and are assumed to be provided by an adversary. By adjusting the parameter d, a d-QUDG can model physical obstructions, imprecise location information, and irregularity in transmission ranges. In this paper, we use the d-QUDG model with $d \geq 1/\sqrt{2}$.

It is easy to see that d-QUDGs and UDGs are locally dense graphs and are therefore far from being planar. On the other hand, all known guaranteed-delivery geographic routing schemes seem to require a planar embedding of the communication network. This has motivated a large body of literature on *topology control protocols*. Informally speaking, topology control usually refers to the process of dropping edges in the communication network so as to construct a sparse spanning subgraph that has desirable properties such as planarity. Using a sparse spanning subgraph for communication also has other benefits such as allowing nodes to transmit at lower power and reducing interference between radio signals. The disadvantage of topology control is that nodes u and v that might have been close to each other in the original communication network may end up being far away from each other following topology control. One precise statement of the topology control problem is this: given a network $G = (V, E)$ (modeled as a d-QUDG or as a UDG) embedded in $\mathbb{R}^2$, find a spanning subgraph $H = (V, E_H)$ such that H has the following properties:

Planarity: No two edges in E_H cross in the embedding of G.

Sparsity: The maximum degree of H is bounded, i.e., $\Delta(H) = O(1)$.

Spanner property: H is a t-spanner of G for $t = O(1)$. Recall that H is a t-spanner of G if for all $u, v \in V$, $d_H(u, v) \leq t \cdot d_G(u, v)$. Here, $d_H(\cdot, \cdot)$ and $d_G(\cdot, \cdot)$ refer to distances in H and G, respectively. These distances may be Euclidean distances or hop distances and when the difference is important we will call H a *Euclidean* t-spanner or a *hop* t-spanner respectively.

Given the many goals of topology control, the above stated problem is just one of many possible variants of the topology control problem. For example, in some topology control research, minimizing the total Euclidean weight of the spanner is a goal [19,7]; in other work, an important goal is to minimize an explicitly defined measure of signal interference [20].

Topology control protocols have been quite successful when the input graph is a UDG. For example, Wang and Li [22] present a distributed local protocol that takes a UDG as input and, in $O(1)$ communication rounds, computes a bounded degree, planar, Euclidean t-spanner, for constant t. This paper has been preceded by a number of papers on topology control that use versions of geometric proximity structures such as Delaunay triangulations, Gabriel graphs, Relative neighborhood graphs, Yao graphs, etc. The result that is critical to some of this work is a classical result from computational geometry due to Dobkin et al. [8] showing that "Delaunay graphs are almost as good as complete graphs" in the sense that a Delaunay triangulation of any planar point set is a Euclidean t-spanner of the complete graph on that point set. Topology control protocols for d-QUDG have been less common and less successful due to the somewhat fundamental reason that for d-QUDGs it is not always possible to obtain a spanning subgraph that is both planar and connected. See Figure 1(a) for a simple example. The graph shown in this figure represents a family of d-QUDGs for $d < 1$. This figure emphasizes the fact that even though UDGs (which are d-QUDGs with $d = 1$) always have a planar spanner, d-QUDGs may not, even for values of d arbitrarily close to 1. Thus, the standard approach to topology control in the plane - delete edges until a sparse, planar spanner is obtained, does not work. Barriére et al. [2] add *virtual edges* to the given d-QUDG to first obtain a supergraph and then run a standard topology control algorithm on the supergraph. The resulting output is a mix of real edges and virtual edges, that are realized by real paths. We propose adding *virtual nodes*, an approach that seems to be a simpler, more flexible alternative and one for which provable performance guarantees are relatively easy to obtain[1]. Next we describe our results in some detail.

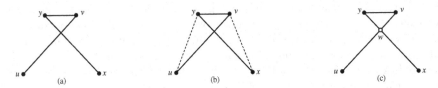

Fig. 1. (a) A d-QUDG G for which there is no connected, planar spanning subgraph. Here, $0 < d < 1$, $\|u-v\|_2 = \|x-y\|_2 = \|u-x\|_2 = 1$ and $\|y-v\|_2 = 1-d$. By the triangle inequality $\|u-y\|_2$ (and similarly $\|x-v\|_2) > d$. In order to make G planar, edge $\{u,v\}$ or $\{x,y\}$ must be removed, thus disconnecting G. (b) The algorithm of Barriére et al. [2] first adds virtual edges $\{u,y\}$ and $\{x,v\}$, the Gabriel Graph construction then drops edges $\{u,v\}$ and $\{x,y\}$. (c) Our algorithm adds virtual node w at the intersection of edges $\{u,v\}$ and $\{x,y\}$. The virtual node w is *controlled* by a real node in $\{u,v,x,y\}$.

Our Results. We present a distributed topology control protocol that runs on a d-QUDG for $d \geq 1/\sqrt{2}$, and computes a sparse, hop t-spanner, for constant t. As in other papers [2,11,17], the constraint $d \geq 1/\sqrt{2}$ is quite fundamental; otherwise two edges $e_1 = \{u_1, v_1\}$ and $e_2 = \{u_2, v_2\}$ may cross even though the hop-distance between an endpoint of e_1 and an endpoint of e_2 may be arbitrarily large. Our protocol is local and runs in $O(1)$ rounds. The output topology permits memoryless (geographic) routing with guaranteed delivery. In fact, when our topology control protocol is used as preprocessing step for the geographic routing protocol GOAFR$^+$ [16], we get a guarantee of $O(\ell^2)$ hops traveled by any message between a source-destination pair that are ℓ hops away from each other in the input d-QUDG. The key idea is simple: to obtain planarity, we replace each edge crossing by a *virtual node* and have a real node serve as a proxy for the virtual node (see Figure 1(c) for an example). This idea is supported by other parts of our protocol that (i) use clustering to keep the density of edge crossings bounded and (ii) guarantee that an edge between a virtual node and a neighbor is realized by a constant-hop path in the real network. A more complicated version of our protocol yields, in $O(1)$ rounds, a sparse *Euclidean t-spanner*, for constant t. When used in combination with GOAFR$^+$ [16], we get a routing guarantee of $O(\ell^2)$ Euclidean distance traveled by any message between a source-destination pair that are a Euclidean distance ℓ apart

---

[1] An anonymous referee has pointed out to us that this idea has been briefly mentioned by Kuhn et al. [17].

in the input d-QUDG. In a recent paper, Funke and Milosavljević [11] present a scheme, called *Macroscopic Geographic Greedy Routing* (MGGR), that they claim guarantees geographic routing under uncertain node locations. We show that MGGR can yield graphs with many connected components and therefore no message delivery guarantee can be provided; the virtual nodes idea provides a simple fix to this problem. However, a key idea from MGGR when combined with our scheme, adds load balancing properties to our routing protocol. Since most known guaranteed-delivery geographic routing schemes rely on planarity, some recent papers have focused on the problem of removing edge crossings, for example the CLDP protocol [15] and the LCR protocol [13]. The virtual nodes idea provides a simple, cheap alternative to these schemes. To simplify our presentation, we assume the $\mathcal{LOCAL}$ model of distributed computation [21] which assumes that nodes have unique IDs, nodes run synchronously, and there is no bound on message sizes. However, we do not misuse the unboundedness of message sizes; all messages exchanged in our protocols contain a constant number of node IDs, a constant number of node locations, and a constant amount of control bits. We further assume that the amount of memory available to each node is a constant times the amount of memory required to store a node's ID and its Euclidean coordinates.

1.1 Related Work

Most papers on topology control and geographic routing assume the UDG model for wireless networks. Here we review three papers that have considered topology control and memoryless routing on d-QUDGs and are therefore most relevant to our work.

Barriére et al. [2] solve the same problem as we do. The wireless network is modeled as a d-QUDG G with $d \geq 1/\sqrt{2}$ and the authors note that computing a Gabriel graph of G (which is a standard technique for topology control on UDGs, first introduced in [5]) may produce a disconnected spanning subgraph. To overcome this problem *virtual edges* are added to G and a Gabriel graph of the resulting supergraph H is computed. The output is a mixture of real and virtual edges and is connected and planar (see Figure 1(b)). The shortcomings of this approach are (i) virtual edges may be realized by arbitrarily long real paths, (ii) the Gabriel graph is not a constant-spanner and in fact there are families of planar point sets for which the Gabriel graph is an $\Omega(\sqrt{n})$ spanner [3,9], and (iii) the protocol is not local in the sense that information may have to travel between nodes that are more than constant hops from each other.

Kuhn et al. [17] model the wireless network as an arbitrary d-QUDG G, with $0 \leq d \leq 1$. Using a standard technique (as described by Alzoubi et al. [1]), a *connected dominating set (CDS)* is first extracted from G. Clustering is then used to reduce the number of edges of the CDS, resulting in a subgraph H of G. The authors show that using the *Echo Flooding Algorithm* [6,21] on H, yields asymptotically optimal routing on G. The authors then go on to consider d-QUDGs with $d \geq 1/\sqrt{2}$. For such d-QUDGs, it is shown that if the algorithm of Barriére et al. [2] is executed on H, rather than on the original graph G, then the resulting topology allows for geographic routing with good worst case guarantees. This result (at least as described by Kuhn et al. [17]) is specific to the hop metric and does not seem to immediately extend to the Euclidean metric. The paper also briefly mentions the idea of using virtual nodes for topology control on d-QUDGs.

Funke and Milosavljević [11] also attempt to solve the problem of topology control and memoryless routing on a d-QUDG input graph G, but do so via an interesting idea called *macroscopic routing*. Their starting point is the selection of *landmark nodes*. For an integer parameter $k > 0$, the landmarks form a *k-independent set* of $V(G)$. Each node in G is then "associated" with the landmark that is closest to it, in hops, effectively partitioning $V(G)$ into *Voronoi tiles*. From this combinatorial Voronoi tiling, a *Combinatorial Delaunay Map (CDM)* is constructed. The vertex set of the CDM is the set of landmarks, $\{L_i\}$. Edge $\{L_a, L_b\}$ is then added if the following local rule is satisfied: (i) there is a path in G from L_a to L_b consisting of a sequence of nodes associated with L_a followed by a sequence of nodes associated with L_b, and (ii) the one-hop neighborhood of this path contains only nodes associated with L_a and L_b. As can be seen in Figure 2, even though this local rule yields a planar graph there are instances of G for which the CDM has multiple connected components.

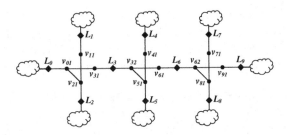

Fig. 2. Illustration of how the CDM constructed by Funke and Milosavljević [11] may have connected components. Consider the d-QUDG shown above; let parameter $k = 1$ and let the set of landmarks contain $L_i, 0 \le i \le 9$. The Voronoi tile of each landmark L_i contains nodes v_{ij}. None of the edges $\{L_i, L_j\}$, $0 \le i \ne j \le 9$ are added to the CDM because part (ii) of the local rule is violated and we get a CDM with at least 8 connected components. This construction can be easily extended to produce a CDM with arbitrarily many connected components. Similar examples can be constructed on UDGs and with any value of $k \ge 0$.

Once the CDM is constructed, a routing algorithm such as $GPSR$ [14] can be used to determine in which direction to send a message. The Funke-Milosavljević routing protocol only uses the CDM as a *macroscopic* guide for routing; a node u uses locations of nearby landmarks to determine which neighboring Voronoi tile to forward the message to and then the *microscopic* routing takes over and the message is forwarded to that neighboring tile using the gradient descent method similar to that used in the GLIDER protocol [10].

2 The Concept of Virtual Nodes

Rather than attempt to eliminate edge crossings to obtain planarity, we simply treat each edge crossing as a node — a *virtual node*. For the rest of the paper, we assume that $G = (V, E)$ is the input d-QUDG with $d \geq 1/\sqrt{2}$. Furthermore, we assume that G comes with an embedding in $\mathbb{R}^2$; each vertex $v \in V$ is at a point $p_v \in \mathbb{R}^2$ and each edge $\{u, v\}$ is a straight line segment connecting points p_u and p_v. We will think of vertices and edges as combinatorial objects as well as geometric objects depending on the context. Now consider a pair of edges $\{u, v\}$ and $\{x, y\}$ that cross. We deal with this edge crossing by modifying G as follows:

1. Add a virtual node w to G corresponding to the edge crossing. Associated with node w is a point $p_w \in \mathbb{R}^2$, namely the point of intersection of edges $\{u, v\}$ and $\{x, y\}$. Also associated with w is a *controller* node, denoted $w.controller$, belonging to the set $\{u, v, x, y\}$, which serves as the real proxy for w.
2. Delete edges $\{u, v\}$ and $\{x, y\}$ from G and add the four new edges $\{w, u\}$, $\{w, v\}$, $\{w, x\}$, and $\{w, y\}$. This transformation can be seen as going from the non-planar embedding of the graph in Figure 1(a) to the planar embedding in Figure 1(c).

We can eliminate all edge crossings by repeatedly applying the above two steps. The resulting *plane graph* (i.e., a planar graph along with a planar embedding) is denoted $VN(G)$. It is this plane graph on which we attempt to perform memoryless routing. First, we describe 3 properties that we seek for $VN(G)$. If $VN(G)$ were to satisfy all 3 properties, then we could perform fast memoryless routing on it.

Property 1: For every pair of adjacent nodes u and v in $VN(G)$, there is a constant-length path in G from $u.controller$ to $v.controller$. A routing step on $VN(G)$ that forwards a message from node u to its neighbor v is realized on the actual network G by a message going from $u.controller$ to $v.controller$. We want to show that the hop-distance in G between $u.controller$ and $v.controller$ is bounded by a constant. Otherwise, even though a pair of nodes u and v may be close together in $VN(G)$, there may be no cheap way of routing between u and v.

Property 2: Every real node in $VN(G)$ is the controller for a constant number of virtual nodes. For each node v in $VN(G)$ a routing table containing information on the neighbors of v in $VN(G)$ must be maintained at $v.controller$. This means the a real node u must store its own routing table as well as the routing table of each virtual node for which it is the controller. Hence we require that each real node in $VN(G)$ be the controller for only a constant number of virtual nodes.

Property 3: The size of the routing table for each node in $VN(G)$ is constant. For our routing to be memoryless, we must guarantee not only that each real node maintain a small number of routing tables, but also that the size of each routing table is small. One way to ensure this is to bound the degrees of nodes in $VN(G)$; a constant bound on the degrees would be ideal. This would result in each node's routing table consisting of the IDs and Euclidean coordinates of a constant number of nodes.

Now we prove the first of the three desired properties mentioned above for d-QUDGs with $d \geq 1/\sqrt{2}$.

Lemma 1. *Let $G = (V, E)$ be a d-QUDG with $d \geq 1/\sqrt{2}$. Then $VN(G)$ is a plane graph such that for adjacent nodes u and v in $VN(G)$ the hop distance between u.controller and v.controller in G is at most 5.*

Proof. We start with an observation that due to Kuhn et al. (see Lemma 8.1 in [17]): for intersecting edges $\{u, v\}, \{x, y\} \in E(G)$ and any $s, t \in \{u, v, x, y\}$ there is an st-path, that has hop-length at most 3. The lemma follows easily because (i) if both u and v are real nodes, then $u = u.controller$ and $v = v.controller$ are one hop apart, (ii) if one of u and v is a real node then there is a path from $u.controller$ to $v.controller$ whose length is at most 3 hops, and (iii) if both u and v are virtual nodes, then $u.controller$ and $v.controller$ are at most 5 hops apart. $\qquad\square$

It is easily checked that the upper bound of Lemma 1 is tight.

Corollary 1. *Let $G = (V, E)$ be a d-QUDG with $d \geq 1/\sqrt{2}$ and let H be a hop t-spanner of G. Then $VN(H)$ is a plane graph such that for any pair of adjacent nodes u and v in $VN(H)$, the hop distance between u.controller to v.controller in G is at most $5 \cdot t$.*

Property (2) above can be satisfied by ensuring that each edge in crossed at most a constant number of times, whereas Property (3) can be satisfied by guaranteeing a constant upper bound on the maximum vertex degree. This motivates the question of whether every d-QUDG has an $O(1)$-spanner with (i) constant degree and (ii) constant number of edges crossing every edge. An $O(1)$-*hop* spanner with constant degree is impossible, even for cliques since with a constant degree bound it takes $\Omega(\log n)$ hops to reach every vertex from any given vertex. Unfortunately, as shown below, the constant edge crossing requirement is incompatible with having an $O(1)$-*Euclidean* spanner. This example should be viewed as a generalization of the example in Figure 1(a). There, 0 crossings were allowed and we could not guarantee connectedness; here a constant number of crossings are allowed, but this still does not allow for the Euclidean t-spanner property, for any constant t.

Lemma 2. *Let G be a d-QUDG as shown in Figure 3 with $d < 1$, integer $\alpha > 0$, and $t \geq 1$. If H is an arbitrary spanning subgraph of G such that every edge in H is intersected by at most α other edges, then H is not a Euclidean t-spanner of G.*

Proof. Let $\ell = \|x_i - x_{i+1}\|_2 = \|y_i - y_{i+1}\|_2 = \frac{1-d}{\alpha}$, for $1 \leq i \leq \alpha$. Since G is connected and $N_G(u) = \{v\}$, edge $\{u, v\} \in E(H)$. Since $\{u, v\}$ is crossed by at most α other edges in H, there exists a vertex x_j that does not have an incident edge in H that intersects $\{u, v\}$. Specifically, edge $\{x_j, y_j\} \notin E(H)$. Hence $d_H(x_j, y_j) > \ell$. Combined with the fact that $d_G(x_j, y_j) = \varepsilon$, we get the ratio $\frac{d_H(x_j, y_j)}{d_G(x_j, y_j)} > \frac{\ell}{\varepsilon} = \frac{(1-d)/\alpha}{\varepsilon} > \frac{(1-d)/\alpha}{(1-d)/(t \cdot \alpha)} = t$. Therefore H is not a Euclidean t-spanner of G. $\qquad\square$

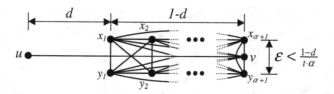

Fig. 3. This figure illustrates the proof of Lemma 2. In this graph G, $N_G(u) = \{v\}$ and the set of vertices $V(G) \setminus \{u\}$ form a clique. Note that each edge intersecting $\{u, v\}$ has the form $\{x_i, y_j\}$ and that the length of each edge $\{x_i, y_i\} = \varepsilon$.

3   A Hop-Spanner with Virtual Nodes

Going from G to $VN(G)$ gives us planarity and Lemma 1 tells us that adjacent nodes in $VN(G)$ are nearby in G (Property 1 holds for $VN(G)$). However, $VN(G)$ does not support *memoryless* routing because it does not guarantee Properties 2 or 3.

To get around this problem, we start by clustering G, sparsifying it and only then introducing virtual nodes (see Figure 4). Network clustering is a standard technique in routing and our approach is similar to that of Gao et al. [12]. We start by constructing a backbone graph G_B whose vertex set is a small subset of $V(G)$. A routing graph G_R is constructed by adding virtual nodes to G_B and then attaching non-backbone nodes to G_B. Algorithm `BuildBackbone` describes the backbone construction and Algorithm `Route1` describes how routing is performed on G_R.

Algorithm `BuildBackbone`
Input: $G = d$-QUDG with $d \geq 1/\sqrt{2}$
Output: Backbone Graph G_B

1. Place an infinite grid of size $d/\sqrt{2} \times d/\sqrt{2}$ on the plane. Since nodes know their locations in a global coordinate system, each node knows the identity of the grid cell to which it belongs. The nodes in each grid cell form a *cluster* and the grid induces a partition of $V(G)$ into clusters. Let $\widetilde{G}$ be the *cluster graph* of G in which the vertex set is the set of clusters and there is an edge from vertex C to C' iff there is an edge in G with one end node in cluster C and the other in cluster C'. Note that since the diagonal of each grid cell has length d, the nodes within any given grid cell form a clique.

2. For each cluster C_i the nodes belonging to the cluster select a representative *cluster head* h_i. Let $\mathcal{L}$ be the set of all cluster heads.

3. Each cluster head h_i (the head of cluster C_i) selects exactly one edge $\{u, v\}$ in G corresponding to each neighbor C_j of C_i in $\widetilde{G}$ such that u is in cluster C_i and v is in cluster C_j. The number of edges selected by h_i is therefore equal to the degree of C_i in $\widetilde{G}$. Let $\mathcal{S}$ be the set of end nodes of all such selected edges in G.

4. G_B is then defined as the subgraph of G induced by $\mathcal{L} \cup \mathcal{S}$.

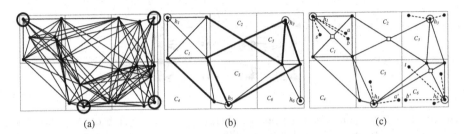

Fig. 4. (a) A d-QUDG G with $d = 1/\sqrt{2}$ along with a $1/2 \times 1/2$ grid and cluster heads for nonempty grid cells. (b) Backbone graph G_B. Edges to neighboring clusters selected by the cluster heads are bold. Other induced edges are non-bold. (c) Routing graph G_R, comprised of (i) $VN(G_B)$, i.e. G_B with virtual nodes (squares) at edge intersections and (ii) remaining nodes in G attached to cluster heads (dashed edges). To route a message M from s in cluster C_1 to t in cluster C_6, s forwards M to cluster head h_1, M is then routed from h_1 to h_6 using graph $VN(G_B)$ and a routing protocol such as GPSR, and h_6 broadcasts M to all of its neighbors, one of which is t.

Lemma 3. G_B has the following properties: (i) $\Delta(G_B)$ is $O(1)$, (ii) Each edge in G_B is crossed by $O(1)$ other edges, (iii) $V(G_B)$ is a connected dominating set of G.

Proof. Let us begin with a definition and two facts. **Definition:** For each grid cell C let the *surrounding* cells of C be the set of cells $\{C' \mid \exists x, y \in \mathbb{R}^2 : x \in C', y \in C, \|x-y\|_2 \leq 1\}$. **Fact 1:** For any grid cell C the number of surrounding cells of C is bounded above by a constant. **Fact 2:** A constant number of nodes from each grid cell C_i are included in G_B. These include cluster head h_i along with at most two nodes associated with each surrounding cell of C_i.

(i) Consider a node u in G_B that lies in grid cell C. Since G_B is an induced subgraph of G, the degree of u in G_B is bounded by the number of nodes in $V(G_B)$ that are at most distance 1 from u. Since there are $O(1)$ cells surrounding C (Fact 1) and since each such cell contributes $O(1)$ nodes to $V(G_B)$ (Fact 2), there are $O(1)$ nodes whose distance from u is at most 1. Hence degree of u in G_B is $O(1)$.

(ii) Consider an edge $\{u, v\}$ in G_B. Since the grid size is $d/\sqrt{2} \times d/\sqrt{2}$ and the maximum edge length is 1, edge $\{u, v\}$ passes through at most five grid cells; call them C_1 through C_5. Hence for any edge $\{x, y\}$ that intersects $\{u, v\}$, x and y both must lie in the surrounding cells of C_1 through C_5. From Facts 1 and 2 there are a constant number of such surrounding cells, each with a constant number of nodes in G_B. Therefore there are $O(1)$ edges that intersect $\{u, v\}$.

(iii) Since $V(G)$ is partitioned into clusters, each of which forms a clique, and since $V(G_B)$ includes at least one node from each cluster, $V(G_B)$ is a dominating set of G. We now show that $V(G_B)$ is connected. Consider nodes s and t in G_B. Since G is connected there is a path in G from s to t. Let this path be $s = x_0, x_1, \ldots, x_p = t$. This st-path corresponds to a sequence of adjacent clusters $C_0, C_1, \ldots, C_q, q \leq p$, with s being in cluster C_0 and t being in cluster C_q. To show s and t are connected in G_B we show the following pairs of nodes are connected in G_B: (i) s and h_0, (ii) t and h_q, (iii) h_i and $h_{i+1}, 0 \leq i \leq q - 1$. Since s and h_0 are in the same cluster there is an edge between them in G which means there is also an edge between them in G_B. Likewise, there is an edge between t and h_q in G_B. Since C_i and C_{i+1} are adjacent in $\widetilde{G}$ there is an edge $\{u, v\}$ in G_B where u is in cluster C_i and v is in cluster C_{i+1}. Also since u and h_i are in the same cluster there is an edge between them in G_B. Likewise there is an edge between v and h_{i+1}. Therefore h_i and h_{i+1} are connected in G_B. $\qquad\square$

Once G_B is constructed we add virtual nodes at each edge intersection and obtain $VN(G_B)$. The routing graph G_R is then defined as $VN(G_B)$ along with all nodes $V(G) \setminus V(G_B)$ with edges to their respective cluster heads. The following algorithm can then be used to route messages using G_R (see Figure 4(c)).

Algorithm Route1
Input: $s, t \in V(G)$, message M, geographic routing protocol $\mathcal{A}$

1. Node s forwards the message to its cluster head h.
2. The message is routed from cluster head h to cluster head h' of t using geographic protocol $\mathcal{A}$.
3. Cluster head h' broadcasts the message to its neighbors, one of which is t.

Route1 is memoryless because each real backbone node maintains constant-sized routing tables for itself and at most constant number of virtual nodes; each non-backbone node just maintains the ID and position of its cluster head. Now we claim (without proof to conserve space) a worst case upper bound on the number of hops it takes for a message to be delivered by Algorithm Route1. The lemma not only claims a relatively short st-path, but also that such a path uses only the backbone graph, with the possible exception of the two end nodes, s and t.

Lemma 4. *For any vertices s and t in G there is a path $L = (s = z_0, z_1, \ldots, z_\ell = t)$ in G_R such that $\ell \leq \alpha \cdot d_G(s,t)$, for some constant α. Furthermore, $z_1, z_2, \ldots, z_{\ell-1}$ are nodes in $VN(G_B)$.*

To obtain worst case guarantees on the number of hops it takes to route using Algorithm Route1, we use GOAFR$^+$ [16] as the geographic routing protocol $\mathcal{A}$ in Step 2 of Algorithm Route1. Like GPSR, GOAFR$^+$ is a dual-mode routing protocol, but it uses a clever way of controlling the face routing mode by using a set of bounding circles with geometrically varying area. Kuhn et al. [16] prove the following theorem about the performance of GOAFR$^+$.

Theorem 1. (Kuhn et al. [16]) *Let p^* be an optimal path from s to t. On a bounded degree UDG, GOAFR$^+$ reaches t with cost $O(c^2(p^*))$.*

Here $c : (0,1] \rightarrow \mathbb{R}^+$ is an arbitrary *cost function* that associates a cost $c(\|u-v\|_2)$ to every edge $\{u,v\}$ in the graph. The only restriction on c is that it is non-decreasing, i.e., $c(d') \geq c(d)$ if $d' > d$. For any path P in the graph, $c(P)$ is simply the sum of the costs of the edges in P. Note that $c(d) = 1$ for all $d \in (0,1]$ models hop distances, whereas $c(d) = d$ for all $d \in (0,1]$ models Euclidean distances. Theorem 1 is proved for bounded degree UDGs, however it easily extends to arbitrary subgraphs (not necessarily induced) of bounded degree UDGs. Since $VN(G_B)$ is clearly such a graph, we can use GOAFR$^+$ in Step 2 of Algorithm Route1 on $VN(G_B)$, even though $VN(()\,G_B)$ may not be a UDG. As a result, we get the following theorem providing worst case guarantees on the performance of Algorithm Route1.

Theorem 2. *Let s and t be nodes in G and let ℓ be the hop-length of a shortest st-path in G. If GOAFR$^+$ is used as the geographic routing algorithm $\mathcal{A}$ in Route1, then the hop-distance traveled by a message routed from s to t using Route1 is $O(\ell^2)$.*

4   A Euclidean Spanner with Virtual Nodes

The routing graph G_R constructed in Section 3 may not be a Euclidean t-spanner of G (see nodes a and b in cell C_1 of Figure 4(c) for an example). One possible solution is to compute a bounded degree, planar, Euclidean t-spanner of the clique induced by the vertices in each cell C rather than connecting all non-backbone nodes in C directly to C's cluster head. In order to account for nodes which may be arbitrarily close together yet lie in different grid cells (see nodes a' and b' of Figure 4(c)) we utilize three vertex partitions of G induced by three different grids obtained by "shifting" the grid of Section 3. This ensures every pair of vertices that are close to each other will be contained completely in a grid cell of at least one of the three grids. This construction is described in Algorithm BuildRoutingGraph and illustrated in Figure 5.

A key ingredient to this solution is the construction of a bounded degree, planar, Euclidean t-spanner of the cliques induced by the grid cells of the three grids. Wang and Li [22] have proposed a local algorithm for computing just such a spanner. For concreteness, we fix the Wang-Li algorithm and for any UDG H, let the *Wang-Li spanner* of H, denoted $WL(H)$, be the bounded degree, planar, Euclidean t-spanner of H computed by the algorithm of Wang and Li [22]. Let $G[C]$ be the subgraph of G induced by the vertices in cell C.

Algorithm `BuildRoutingGraph`
Input: $G = d$-QUDG with $d \geq 1/\sqrt{2}$
Output: Routing Graph G_R

1. Place a blue grid of $\frac{d}{\sqrt{2}} \times \frac{d}{\sqrt{2}}$ cells passing through $(0,0)$, a red grid of $\frac{d}{\sqrt{2}} \times \frac{d}{\sqrt{2}}$ cells passing through $(\frac{d}{3\sqrt{2}}, \frac{d}{3\sqrt{2}})$, and a green grid of $\frac{d}{\sqrt{2}} \times \frac{d}{\sqrt{2}}$ cells passing through $(\frac{2d}{3\sqrt{2}}, \frac{2d}{3\sqrt{2}})$.
2. For each edge e in G initialize $color(e)$ to the empty set.
3. For each non-empty grid cell C in each grid, construct the Wang-Li spanner $WL(G[C])$. If C belongs to a grid of color $x \in \{\text{blue}, \text{red}, \text{green}\}$ then add x to $color(e)$ for each edge e in $WL(G[C])$.
4. Construct G_B using the blue grid and algorithm `BuildBackbone` from Section 3.

5. G_R is the union of $VN(G_B)$ and each $WL(G[C])$ from step 3.

To route a message from s to t, node s first checks to see if it shares a cell with t in any of the three grids. If so, the message is routed from s to t using only edges of $WL(G[C])$. If not, then the backbone graph is used for routing. The details of this protocol are given below in Algorithm `Route2`.

Algorithm `Route2`
Input: $s, t \in G$, message M, geographic routing algorithm $\mathcal{A}$.

1. If s and t share a cell C in a grid of color x then route M on edges colored x, using algorithm $\mathcal{A}$. STOP.
2. Otherwise, let C be the blue cell containing s. Route M from s to the cluster head s' of C on edges colored blue using $\mathcal{A}$.
3. Let t' be the cluster head in the blue cell containing t. Route from s' to t' using $\mathcal{A}$ on $VN(G_B)$ constructed in Step 4 of `BuildRoutingGraph`.
4. Route from t' to t on edges colored blue using algorithm $\mathcal{A}$. STOP.

As in the case of `Route1`, it is easy to see that `Route2` is memoryless. The following lemma states that if s and t are nearby then one of the Wang-Li spanners will provide a short st-path.

Lemma 5. *For any pair of vertices u and v in G with $\|u - v\|_2 \leq \frac{d}{3\sqrt{2}}$, there is a cell of some color that contains both u and v.*

In the following, let $d'_H(s,t)$ denote the length of an st-path in H with smallest Euclidean length.

Lemma 6. *For any pair of vertices s and t in G, there exists a path $L = (s = z_0, z_1, z_2, \ldots, z_p = t)$ whose Euclidean length is at most $\beta \cdot d'_G(s,t)$ for some constant β. Furthermore, if $\|s - t\|_2 > \frac{d}{3\sqrt{2}}$ then there are vertices z_i and z_j in L, $i < j$, such that (i) z_i and z_j are cluster heads of the blue cells containing s and t respectively, (ii) the subpath $L_1 = (s = z_0, z_1, \ldots, z_i)$ lies in the blue cell that s belongs to, (ii) the subpath $L_2 = (z_i, z_{i+1}, \ldots, z_{j-1}, z_j)$ belongs entirely to $VN(G_B)$, and (iii) the subpath $L_3 = (z_j, z_{j+1}, \ldots, z_p = t)$ lies entirely in the blue cell containing t.*

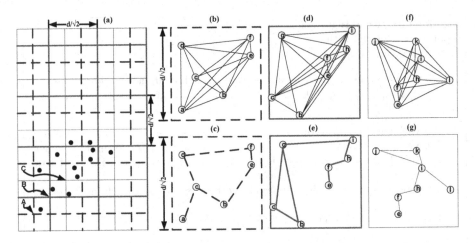

Fig. 5. (a) The three shifted grids with $\frac{d}{\sqrt{2}} \times \frac{d}{\sqrt{2}}$ dimensional cells. The grids are colored blue, red, and green and are shown with dashed, heavy, and light lines, respectively. (b) A clique induced by the blue cell C_b with lower left corner A. (c) A planar, constant-degree Euclidean t-spanner of $G[C_b]$. (d) A clique induced by the red cell C_r with lower left corner B. Note that vertices $b, c, e,$ and f are common to C_b and C_r. (e) A planar, constant-degree Euclidean t-spanner of $G[C_r]$. (f) A clique induced by the green cell C_g with lower left corner C. Note that vertices $e, f, h,$ and i are common to C_r and C_g. (g) A planar, constant-degree Euclidean t-spanner of $G[C_g]$. Note that vertices e and f are common to C_b, C_r, and C_g. Also note that edge $\{e, f\}$ happens to be common to the spanners in (c), (e), and (g) and hence has three colors in G_R.

Proof. Consider an st pair. If s and t share a grid cell of color x, then according to "Step 1" of Route2, there is a path of length at most $c_1 \cdot d_G(s,t)$, for some constant c_1, using only edges of color x.

So, let s and t not share any grid cell. By Lemma 5, $d'_G(s,t) > \frac{d}{3\sqrt{2}}$. Now, consider the blue grid and the partition induced by it. Let G_B be the backbone constructed in "Step 4" of BuildRoutingGraph. Let P be a shortest Euclidean length path in G having fewest vertices. Partition P into contiguous subsequences of vertices that belong to the same cell. Write these sequences as $B_0, B_1, \ldots, B_q$, each B_i belongs to C_i, and B_i and B_{i+1} belong to distinct cells. Now note that any vertex in B_i and any vertex in B_{i+2} are more than d units apart. Hence, $\frac{q \cdot d}{2} \le d_G(s,t)$. Let h_i denote the clusterhead of C_i. Consider $Q = (s \rightsquigarrow h_0, h_1, \ldots, h_q \rightsquigarrow t)$, where $s \rightsquigarrow h_0$ and $h_q \rightsquigarrow t$ are paths in the blue Wang-Li spanner between s and its clusterhead h_0, and t and its clusterhead h_q, respectively. Now, consider the backbone graph along with all the blue Wang-Li spanners; call it H. Note that H has no virtual nodes. Q can be realized as a path $R = (s \rightsquigarrow h_0, x_1, x_2 \ldots, x_{r-1}, h_r \rightsquigarrow t)$ in H with $r \le 3q$. This is because G_B contains a path of at most 3 hops between h_i and h_{i+1}, $0 \le i < q$. Furthermore, note that all the x_i's belong to G_B. G_R is obtained from H by the introduction of virtual nodes into G_B. The introduction of virtual nodes expands each edge in G_B by at most $(c+1)$ hops where c is the constant upper bound on the number of edges of G_B that can cross any edge in G_B. Thus the total Euclidean distance of the "expanded" path is $2c_1 + 3(c+1)q$, where c_1 is the upper bound on the Euclidean stretch guaranteed by the Wang-Li spanner. Using the upper bound of $q \leq 2d'_G(s,t)/d$, we get the result.   □

We can combine the spanning property of the routing graph, proved in the above lemma, with worst case performance guarantees for GOAFR$^+$ to obtain the following result.

Theorem 3. *Let s and t be nodes in G and let ℓ be the length of a shortest Euclidean st-path in G. If GOAFR$^+$ is used as the geographic routing algorithm $\mathcal{A}$ in* Route2, *then the Euclidean distance traveled by a message routed from s to t using* Route2 *is $O(\ell^2)$.*

Proof. If s and t are such that $\|s - t\|_2 \leq \frac{d}{3\sqrt{2}}$ then Lemma 5 tells us that s and t are in a cell C. In this case, Algorithm Route2 uses $WL(G[C])$ to route the message from s to t. By construction, $WL(G[C])$ is guaranteed to be planar and contain a path of Euclidean length at most $c \cdot \ell$ for some constant c. Therefore, if GOAFR$^+$ is used to route the message from s to t in $WL(G[C])$, the message travels $O((c\ell)^2) = O(\ell^2)$ Euclidean distance.

If $\|s - t\|_2 > \frac{d}{3\sqrt{2}}$, then Lemma 6 tells us that there exists a path $L = (s = z_0, z_1, z_2, \ldots, z_p = t)$ whose Euclidean length is at most $\beta \cdot \ell$ for some constant β. Now consider the subpaths L_1, L_2, and L_3 of L, and let ℓ_1, ℓ_2, and ℓ_3 respectively be the Euclidean lengths of these subpaths. Let C be the blue cell containing s. Since L_1 is a path from $s = z_0$ to z_i in $WL(G[C])$ and since Algorithm Route2 routes from $s = z_0$ to z_i using GOAFR$^+$ on the graph $WL(G[C])$, the message travels a distance of $O(\ell_1^2)$ from s to z_i. Similarly, the message travels a distance of $O(\ell_2^2)$ from z_i to z_j and $O(\ell_3^2)$ from z_j to t. Thus the message travels a total distance of $O(\ell_1^2 + \ell_2^2 + \ell_3^2)$ in G_R. Using the fact that $\ell_1 + \ell_2 + \ell_3 = \leq \beta \cdot \ell$ and the fact that $\ell_1^2 + \ell_2^2 + \ell_3^2 \leq (\ell_1 + \ell_2 + \ell_3)^2$, we get that $O(\ell^2)$ is the total distance traveled by the message.   □

5   Suggestions for Load Balancing

The topology control and routing schemes proposed in the previous two sections tend to overload the backbone nodes with the responsibility of forwarding most messages, with only a tiny fraction of the network nodes bearing most of the responsibility for forwarding messages. These schemes can be modified so the routing responsibility is more evenly distributed. Such modifications could include the standard technique of rotating the cluster head designation among all nodes in a cluster. Here we mention two other approaches. In the construction in the previous section, we used 3 shifted grids to ensure that nodes that are close enough share a grid cell in at least one of the three grids. This idea can also be used to construct 3 backbone graphs instead of one and routing could "rotate" among the different backbone graphs. An obvious extension would allow $k \geq 3$ distinct grids, leading to k different backbone graphs; this comes at the price of larger routing tables whose size would grow linearly in k. Another approach to load balancing may be achieved by combining the virtual nodes idea with a key idea from the MGGR protocol [11] mentioned in Section 1. An important idea in MGGR is to use the backbone graph simply as a "guide" rather than for actual routing. Suppose a node s has a message to send to node t. Rather than forwarding the message to its cluster head h, s determines *for itself* to which backbone node h', node h would have forwarded the message. Node s then attempts to send the message to some node in the cluster of h'. Thus the message moves from cluster to cluster without necessarily using the backbone, thereby reducing the load on backbone nodes.

Acknowledgment. We thank an anonymous referee for useful comments and pointers to the literature, especially for pointing out that the "virtual nodes idea" has been mentioned before.

References

1. Alzoubi, K.M., Wan, P.-J., Frieder, O.: Message-optimal connected dominating sets in mobile ad hoc networks. In: Proceedings of the 3rd ACM international symposium on Mobile ad hoc networking & computing (MobiHoc'02), pp. 157–164. ACM Press, New York (2002)
2. Barriére, L., Fraigniaud, P., Narayanan, L.: Robust position-based routing in wireless ad hoc networks with unstable transmission ranges. In: Proceedings of the 5th international workshop on Discrete algorithms and methods for mobile computing and communications (DIALM'01), pp. 19–27 (2001)
3. Bose, P., Devroye, L., Evans, W.S., Kirkpatrick, D.G.: On the spanning ratio of gabriel graphs and beta-skeletons. In: Rajsbaum, S. (ed.) LATIN 2002. LNCS, vol. 2286, pp. 479–493. Springer, Heidelberg (2002)
4. Bose, P., Morin, P.: Online routing in triangulations. In: Aggarwal, A.K., Pandu Rangan, C. (eds.) ISAAC 1999. LNCS, vol. 1741, pp. 113–122. Springer, Heidelberg (1999)
5. Bose, P., Morin, P., Stojmenović, I., Urrutia, J.: Routing with guaranteed delivery in ad hoc wireless networks. Wireless Networks 7(6), 609–616 (2001)
6. Chang, E.: Echo algorithms: Depth parallel operations on general graphs. IEEE Transactions on Software Engineering 8(4), 391–401 (1982)
7. Damian, M., Pandit, S., Pemmaraju, S.: Local approximation schemes for topology control. In: Proceedings of the twenty-fifth annual ACM symposium on Principles of distributed computing (PODC'06), pp. 208–217. ACM Press, New York (2006)
8. Dobkin, D.P., Friedman, S.J., Supowit, K.J.: Delaunay graphs are almost as good as complete graphs. Discrete and Computational Geometry 5(4), 399–407 (1990)
9. Eppstein, D.: Spanning trees and spanners. In: Sack, J.-R., Urrutia, J. (eds.) Handbook of Computational Geometry, ch. 9, pp. 425–461. Elsevier, Amsterdam (2000)
10. Fang, Q., Gao, J., Guibas, L., Silva, V., Zhang, L.: GLIDER: Gradient landmark-based distributed routing for sensor networks. In: Proceedings of the 24th Conference of the IEEE Communication Society (INFOCOM'05), vol. 1, pp. 339–350 (2005)
11. Funke, S., Milosavljević, N.: Guaranteed-delivery geographic routing under uncertain node locations. In: Proceedings of the 26th Annual Joint Conference of the IEEE Computer and Communications Societies (INFOCOM'07) (2007)

12. Gao, J., Guibas, L.J., Hershberger, J., Zhang, L., Zhu, A.: Geometric spanner for routing in mobile networks. In: Proceedings of the 2nd ACM international symposium on Mobile ad hoc networking & computing (MobiHoc'01), pp. 45–55 (2001)
13. Kim, Y.-J., Govindan, R., Karp, B., Shenker, S.: Lazy cross-link removal for geographic routing. In: Proceedings of the 4th international conference on Embedded networked sensor systems (SenSys'06), pp. 112–124. ACM Press, New York (2006)
14. Karp, B., Kung, H.T.: GPSR: Greedy perimeter stateless routing for wireless networks. In: Proceedings of the Sixth Annual ACM/IEEE International Conference on Mobile Computing and Networking (MobiCom'00), pp. 243–254 (2000)
15. Kim, Y.-J., Govindan, R., Karp, B., Shenker, S.: On the pitfalls of geographic face routing. In: Proceedings of the 2005 joint workshop on Foundations of mobile computing (DIALM-POMC'05), pp. 34–43. ACM Press, New York (2005)
16. Kuhn, F., Wattenhofer, R., Zhang, Y., Zollinger, A.: Geometric ad-hoc routing: Of theory and practice. In: Proceedings of the twenty-second annual symposium on Principles of distributed computing (PODC'03), pp. 63–72. ACM Press, New York (2003)
17. Kuhn, F., Wattenhofer, R., Zollinger, A.: Ad-hoc networks beyond unit disk graphs. In: Proceedings of the 2003 joint workshop on Foundations of mobile computing (DIALM-POMC'03), pp. 69–78. ACM Press, New York (2003)
18. Kuhn, F., Wattenhofer, R., Zollinger, A.: Worst-case optimal and average-case efficient geometric ad-hoc routing. In: Proceedings of the 4th ACM international symposium on Mobile ad hoc networking & computing (MobiHoc'03), pp. 267–278. ACM Press, New York (2003)
19. Li, X.-Y., Song, W.-Z., Wang, W.: A unified energy-efficient topology for unicast and broadcast. In: Proceedings of the 11th annual international conference on Mobile computing and networking (MobiCom'05), pp. 1–15. ACM Press, New York (2005)
20. Moscibroda, T., Wattenhofer, R.: Minimizing interference in ad hoc and sensor networks. In: Proceedings of the 2005 joint workshop on Foundations of mobile computing (DIALM-POMC'05), pp. 24–33. ACM Press, New York (2005)
21. Peleg, D.: Distributed computing: a locality-sensitive approach. In: Peleg, D. (ed.) Society for Industrial and Applied Mathematics, Philadelphia, PA, USA (2000)
22. Wang, Y., Li, X.-Y.: Localized construction of bounded degree and planar spanner for wireless ad hoc networks. In: Proceedings of the Joint Workshop on Foundations of Mobile Computing (DIALM-POMC'03), pp. 59–68 (2003)

Routing in Wireless Networks with Position Trees

Edgar Chávez[1], Nathalie Mitton[2], and Héctor Tejeda[1]

[1] Escuela de Ciencias Físico-Matemáticas
Universidad Michoacana de San Nicolás de Hidalgo
Av.Francisco J. Mujica
Morelia - Michoacán - México
[2] IRCICA/LIFL, Univ. Lille 1, CNRS UMR 8022, INRIA Futurs
Parc scientifique de la haute borne - 50, avenue Halley
59650 VILLENEUVE D'ASCQ - France

Abstract. Sensor networks are wireless adhoc networks where all the nodes cooperate for routing messages in the absence of a fixed infrastructure. Non-flooding, guaranteed delivery routing protocols are preferred because sensor networks have limited battery life. Location aware routing protocols are good candidates for sensor network applications, nevertheless they need either an external location service like GPS or Galileo (which are bulky, energy consuming devices) or internal location services providing non-unique virtual coordinates leading to low delivery rates. In this paper we introduce *Position Trees* a collision free, distributed labeling algorithm based on hop counting, which embed a spanning tree of the underlying network . The Routing with Position Trees (RTP) is a guaranteed delivery, non-flooding, efficient implicit routing protocol based on Position Trees. We study experimentally the statistical properties of memory requirements and the routing efficiency of the RPT.

Keywords: Location aware routing, virtual coordinates, wireless sensor networks.

1 Introduction

Sensor networks are wireless networks where all the nodes cooperate for routing messages in the absence of a fixed infrastructure. Here the nodes are low cost with limited computational resources and limited battery life. A simple distributed routing algorithm with small memory overhead, and a small CPU demand is thus mandatory for such networks [14]. Typical applications of sensor networks are environment sampling, monitoring disaster areas, security and inventory management.

A route is a sequence of nodes forwarding messages from the source node to the target node. The deployment of new applications in sensor networks heavily relies on the efficiency of route discovery mechanisms. Some (actually deployed) sensor networks use query distribution and data collection based on a model known as

E. Kranakis and J. Opatrny (Eds.): ADHOC-NOW 2007, LNCS 4686, pp. 32–45, 2007.
© Springer-Verlag Berlin Heidelberg 2007

data diffusion [15]. In this model, a sink node has a permanent connection with an outside network (*e.g.* the Internet or some wired network) and performs most of the data analysis, while the other nodes are only used for data acquisition or for simpler data processing. Several protocols have been proposed to express queries over the data sensed by the nodes and to aggregate them [20]. These concepts require support of efficient and robust routing protocols, more powerful than those used to support data diffusion.

Position awareness in sensor networks has improved the efficiency of route discovery and broadcasting algorithms in both power saving and latency measures. The fundamental idea behind position awareness (referred also as *geographic* or *geometric* information) is to provide each node in the network with a label having global information. This information is obtained through devices such GPS or Galileo. Routing protocols based on geographic coordinates of the sensors have revealed to be a very competitive alternative to the classical routing protocols for wireless ad hoc networks, reactive [16,25] or proactive [8].

Our contribution is a collision-free labeling algorithm based on the uniqueness of a path in a tree. This leads to a non flooding, guaranteed delivery routing protocol (as long as the nodes in the path does not vanish while the packet is in transit). Our routing algorithm presents the same good properties as classical geographical routing, yet it outperforms them, as shown with thorough experimentation, and they don't rely on external location services.

The rest of this paper is organized as follows. Section 2 reports geographic routing over virtual coordinates in literature. Section 3 describes our contribution: the *Routing with Position trees* (RPT). Simulation results are reported in 4 and Section 5 sketches some conclusions and future work.

2   Related Work

When nodes are aware of their geographic coordinates (for instance by the use of a GPS), a geographic routing based on these coordinates is feasible. The greedy approach is called *Most Forward Routing (MFR)* [29]. In MFR, the source node forwards the message to the node that is closest to the destination. This is a simple localized algorithm that does not guarantee delivery. A package can be trapped in a local minimum and the algorithm fails to find a path to the destination leading to low delivery rates. In dense networks the algorithm performs well. Other geographic approaches have then been proposed in order to minimize the energy consumption [17]. The local optima trap problem of the greedy approach *MFR* can be avoided by a guaranteed delivery routing schema with a hybrid approach [12,19] using greedy routing whenever possible and a mechanism to escape from local optima with face routing. Face routing is performed by extracting a planar subgraph of the network and forwarding the package through the faces. A different approach is used in [18] where a set of trees is used to escape from local minima instead of traversing the faces of a planar graph. Nevertheless, the geographic approaches assume that all nodes are equipped with a

GPS, which can become energy-costly and expensive. Moreover, GPS work only in out-door environments.

In order to reduce the cost, one can also equip only a subset of the nodes with a GPS and use these *special nodes* like in [9,5] to infer the position of the remaining nodes. In such case it suffices to know the distance relative to the *special nodes* using techniques such as time difference of arrival [26], angle of arrival [24], or signal strength [23]. Once the position of every node in the network in estimated one can use, in theory, any geographic routing algorithm. These approaches introduce, however, drawbacks to the geographic routing. It is possible, for example, that two nodes obtain the same coordinates leading to delivery failures.

Other approaches rely exclusively on the relative distances (or hop counting) to a set of nodes in the network, without the intervention of external location services. The general idea is to define a virtual coordinate system and use it to induce a routing protocol based on the virtual coordinates. We survey some of them below.

The authors of [6] introduce VCap (*Virtual Coordinate assignment protocol*) to support geographic routing. A system of virtual coordinates based on three *landmarks* is proposed. Nodes are assigned a triplet of coordinates given as the number of hops the node is distant from each *landmark*. A more accurate coordinate system can be established as the number of *landmarks* increases [2]. Then, nodes use a greedy routing, like in *MFR*, based on the Hamming distance computed on these coordinates (instead of the Euclidean distance in the original MFR). The storage overhead for each sensor is limited to the storage of its coordinates and the coordinates of its neighbors. If the routing reaches a node v with a local minimum, they give a *local detour* rule which consist in locally flooding within a finite neighborhood. A similar idea is presented in [11] with different tie break and recovery mechanisms. Another approach also based in hop counting is presented in [27] where loops are avoided recording the moving history, a time to live for dropping packages and elaborate tie break and recovery mechanisms. A different approach is used in [10] where landmarks are selected more carefully after partitioning the nodes into tiles, and elaborate gradient descent procedures are used to route packets, and high communication and storage overhead is required to increase the delivery rate.

All these geographic routings relying on virtual coordinates use landmarks and the number of hops each node is distant from each *landmark* to compute node coordinates. None of them guarantees delivery, unless significantly increasing the resource consumption (e.g. by flooding the network). The core problem of such routing is the amount of reference points needed to produce a unique reference framework. This unique reference framework can be attained by using $O(n)$ reference points, either in a implicit or explicit way. In particular if a underlying tree can be found, it is possible to obliviously route the packets by finding the unique path joining any two nodes in the tree, thus guaranteeing delivery. This would imply implicitly using $O(n)$ reference landmarks to fix n nodes under rigid transformations. The spanning tree must be found on-line, distributed and without central coordination, this approach has been addressed by several authors. We survey the most significant works in that direction.

In [22], the authors propose a spanning tree as a concrete example of a general framework to solve the routing problem. The spanning tree is obtained by an initial flooding from an arbitrary node, and recursively assigning as children of the current root the reachable unlabeled nodes; a second coordinate is given by assigning an *angle*, related to the number of nodes below the current root. The tree is then aligned to the topology of the network by resorting to either global coordinates (GPS-like) or measuring hops to a number of anchors. The routing is done by discovering an ancestor to the target node traveling up-tree or sideways using rings joining siblings or cousin nodes. The experimental setup in the paper support the claim of having close-to-shortest routes in the tree and good scalability.

A different approach is found in [1] where the canonical path between nodes is encoded by building a rooted tree where the *node coordinates* encode a route from the node to the root. Two nodes can exchange messages if they belong to the same tree by following the canonical path finding the least common ancestor up-tree and then following down-tree to the target node. The trees may be joined, and the labels reorganized to reflect changes in the root and the routes, or the nodes may disappear. Experiments are reported where the rooted trees approach outperforms the well known AODV (Ad Hoc on Demand Distance Vector Routing). In this paper we investigate the properties of label trees built top down, as opposed to bottom-up construction as described in [1], and systematically exploit the unique path joining any pair of nodes to route packets in the network. Top down construction have advantages such as a smaller number of messages to build the spanning tree, and a faster maintenance than using the fusion process.

In this paper, we propose a geographic routing algorithm based on virtual coordinates. We use the uniqueness of a path in a tree to assign labels to nodes and assure the labels uniqueness. As far as we know, this approach using this tree feature has only been used in [21] but it can be used only over a clustering structure.

3   Routing with Position Trees (RPT)

3.1   Position Tree Routing Algorithm

Basic Idea. In this paper, we propose a geographic routing algorithm based on virtual coordinates. The objective of a logical coordinate system is to fix all the nodes in a network making it invariant under rigid transformations. This is a theoretical requirement motivated by the observation that a dynamic routing algorithm must handle the packet to the next hop using only local information. If two nodes share the same description, then a local and deterministic decision can lead to only one route (unless the packet is divided, i.e. flooding the network), hence one of the nodes will not receive the packet. In a geographic routing using an external positioning service for each node, the network is fixed in the sense above. Our goal is then to fix the network in the sense above by using an **internal** positioning service while requesting the smallest amount of memory at each node.

Firstly we select and arbitrary root node and label hierarchically from root to child nodes (child nodes are all the unlabeled neighbors of the current root). Child nodes inherit their parent's label plus an extra number. Hence, the path from the root to each node is encoded in the label. A node's label is built from left to right, and it describes the lineage of the node. The size of the label is proportional to the length of the path from the nodes to the root, and is bounded by the height of the tree ($O(\log n)$ on the average if the tree is balanced).

The labeling is hierarchical. The root node arbitrarily enumerates its neighbors, it can even *skip a neighbor* (following an arbitrary enumeration heuristic optimizing an external goal). As a result of this observation, we can build a virtual topology that can be unrelated to the physical topology of the network, optimizing external goals (e.g. the reliability of the nodes), and the correctness of the algorithm will not be compromised.

Labels are thus used to perform an interval routing over the tree. Interval Routing was introduced in wired networks by Santoro and Khatib in [28] to reduce the size of the routing tables. It is based on representing the routing table stored at each node in a compact manner, by grouping the set of destination addresses that use the same output port into intervals of consecutive addresses. The main advantage of this scheme is the low memory requirements to store the routing on each node u: $O(\delta(u))$. The routing is computed in a distributed fashion with the following algorithm: at each intermediate node x, the routing process ends if the destination y corresponds to x, otherwise, it is forwarded with the message through an edge labeled by a set I such that $y \in I$. The algorithm is shown to be optimal for acyclic graphs, and it exhibits a worst-case complexity which is a factor of two from the optimal solution for an arbitrary topology. For naming nodes, authors of [28] construct a minimum-distance spanning tree and traverse in depth-first style assigning a distinct integer to each node. In our algorithm, this step is not necessary. For assigning labels to the neighbors, they use the spanning tree, in our case we use all the neighbors that are enabled, and this step is done locally. They gave some limitations when nodes are added o deleted, and with permanent disconnections a new tree must be constructed.

Finally, since we use only the neighborhood relationship both when labeling the nodes and routing packages, without additional hypothesis, it is possible to route in any arbitrary network, in particular in three dimensions.

Assumptions. We consider a sensor network composed of nodes uniformly scattered. The nodes are assumed static, or with a very low mobility with respect to the signal propagation speed. We assume that every node has a unique ID.

Once each node is labeled, the routing task can be run. To send a message to a destination node u, a node v needs to know the label of node u. It thus need a locating protocol such that [3,21] which retrieves a node label from its identity. We do not consider this locating part in this paper and assume that there exists such a locating service available in the network. Note that this service is required for any routing protocol.

Labeling the Nodes. The labeling process is very simple. First, a node of the network is chosen to be the *"root"*. It is labeled as the root R. The root node can be selected randomly, or with some heuristic. The selection of the root is not central to the correctness of our algorithm, although different selections will produce different dilations in a given path.

The root R advertises all its neighbors that they are the root's children. For it, it broadcasts a "Discovery" request. Each neighbor of R answers the root node by sending a "Tag Request" message containing its ID. Since this is the root which instances the labeling process, none of its children has already been labeled. The root sorts the ID of its children and is then able to assign a label to each of them. Labels are of the form Rm where m is a positive integer chosen by R, according to the children sort of R.

Once a node u has received its label, it is ready to name its own children in a similar fashion. Node u broadcasts a "Discovery" request to its neighbors. Only the ones of them which have not been labeled by another node answer with a "Tag Request" message. Parent node u sorts the ID of its non-tagged neighbors which become its children in the labeled tree. Node u assigns them a label. If the parent's label is a string s, its children's labels will be obtained by appending a positive integer to s. Since in wireless environments, a transmission by each node reaches all nodes within radius distance from it, node u can send the labels of all its children in only one message instead of sending one message per neighbor node.

This process is iterated until all the nodes in the network have been labeled. Once this happens, we must tell the root the labeling process is done. When a node becomes "childless", that is, all its neighbors are either been labeled or tagged by other nodes, it sends its parent a message informing it that its descendants have been completely determined. When a parent node has received similar messages from all its children, (*e.g* all descendants have been determined), it advertises its own parent, until the root gets the message from all its children. At that time, all the nodes of the network have been labeled.

Figure 1 illustrates the labeling process over an arbitrary topology. Dark lines represent the links in the tree. Dashed lines are the wireless links which have not been selected to be part of the tree during the labeling process.

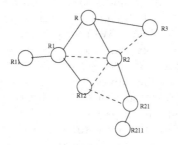

Fig. 1. Labeling process. Dark links are links of the tree, dashed links are the wireless underling links.

Storage Requirements. Once every node is tagged, the routes are computed and the network is ready for routing. The nodes need to store just their own label. As we will see below, nodes do not even need to store their neighbor's labels.

Sending a Packet. Once the labels are in place, sending a packet through the network is almost trivial, because the labels define a canonical a path joining any pair of nodes.

Let's suppose node A wants to send a packet to node B. The packet is forwarded up the tree to find the least common ancestor of A and B and then down the tree until finding node B. All the forwarding decisions are made locally, without flooding. To fix ideas, suppose A is labeled $R167895$ and B is labeled $R16774232$, the largest common prefix is $R167$ which corresponds to the least common ancestor of A and B. By broadcasting the source and destination labels there is only one node, labeled $R16789$, which will forward the packet. This process is repeated with the sequence $A = R167895, R167789, \cdots, R167$ (being the last one the common ancestor) and from there the packet is forwarded to the nodes with unique labels $R1677, R16774, \cdots, R16774232 = B$ to reach node B. The up tree forwarding could be an empty step if the source/destination is the least common ancestor of the destination/source respectively.

The above sketch of the routing protocol is as simple as the actual routing protocol. The actual implementation may rewrite the source in each step. In this case the current node will broadcast the new source, and only the new source node will forward the message rewriting the packet each time.

To illustrate this routing scheme, let's take the tree plotted on Figure 1. Let's suppose node $R211$ wants to send a message to node $R11$. The message will follow the path $R211 - R21 - R2 - R - R1 - R11$. But when node $R21$ forwards the message, nodes $R211, R12$ and $R2$ receive it whereas only node $R2$ needs to forward it. In this case R is the least common ancestor between the destination $R211$ and $R11$.

3.2   Analysis

Message Complexity. The total number of messages needed to label the nodes is linear in both the number of nodes and the total number of connections. To label its children, each node broadcasts a "Discovery" message, contacting at the same time all its neighbors. We thus generate N messages where N is the number of nodes in the network. After all unlabeled children reply to their parent, the parent node sends a unique message for the totality of its children giving them their corresponding labels. We thus generate N_1 messages where N_1 is the number of internal nodes in the labeled tree. To finish the labeling process, each node must inform its parents when all its children are tagged or if it is a leaf in the labeled tree. Hence the total number of messages is $N - 1$ (the root node does not need to send this message). Hence the total number of these messages is thus $2N + N_1 - 1$

Memory Complexity. Let t be the length of the longest path which starts at the root. The longest possible label has as many entries as t, so the largest possible amount of memory to store a label is t. The length of the path is $t = O(n)$ in worst case and $t = O(\log(n))$ if the tree is balanced. If a balanced tree can be found by e.g. selecting a node in the *center*, then the size of each label will be $O(\log(n))$.

3.3 Enhancements

Improving the Routing Paths. Since there is a single path in the tree from the source to the destination, and since the length of this path can be computed before the actual packet forwarding begins, multiple trees may help to discover more efficient routes. It also provides a simple way to recover from vanishing nodes or congestion problems. A way to improve our routing algorithm is to multiple roots to produce orthogonal labels (a vector of labels) and thus get a robust algorithm.

When running the routing algorithm from node A to B, every nodes' labels are compared and the couple of labels minimizing the path length (*e.g.* the ones which has the greatest common prefix) is chosen.

Nevertheless, maintaining several trees imply more memory requirements and more computations at each node. Yet, there is a trade-off to study between the number of roots and the node resource requirements.

Shrinking the Paths. Below the logical path defined by the label tree, there could be many physical connections (dashed links in Figure 1), defining a path shorter than the shortest path in the label tree. For example, in Figure 1, if the source node has label $R11$ and the target node has label $R21$, the shortest path in the tree would be $R11 - R1 - R - R2 - R21$. Since nodes $R1$ and $R2$ are neighbors, then we can avoid the upper part of the path.

A simple way to discover this shorter route is by examining the labels. Indeed, by examining the label, a node can compute the length of the path in the tree between itself and the destination. If R is the least common ancestor for nodes $R1$ and $R211$, the length of the path in the tree from $R1$ to $R211$ is equal to the length of the path from R to $R211$ more the length of the path from R to $R21$. Yet, the length of a path from node R and one of its descendant ($R1$ or $R211$) is equal to the difference between the size of the labels. R has size 1. $R1$ has size 2 (R and one integer), $R211$ has size 4 (R and 3 integers), thus hop distance in the tree from $R1$ and $R211$ is 4 $((2-1) + (4-1)$. Hence , a node is able to determine whether the path going through it is smaller than the one following the tree and thus decide whether directly forward the message. This will imply a slight modification of the routing algorithm because the shortcut node need to inform that he will be forwarding the message.

4   Experimental Results

We used our own C simulator assuming an ideal MAC layer, *i.e.* no interferences and no packet collisions. The nodes were randomly deployed in a 1×1 square using a Poisson Point Process (node positions are independent) with different intensity λ (λ represents the mean number of nodes per surface unit). These nodes have the same transmission range, R, therefore, two nodes are connected by an edge if and only if their Euclidean distance is at most R (assuming a Unit Disk Graph [7]). If there exists a link between nodes u and v, we say that nodes u and v are neighbors. We note $\delta(u)$ the degree of node u, *i.e.* the number of neighbors of node u. With such a Poisson node distribution, we have $\lambda\pi R^2 = \tilde{\delta}$. All results obtained are within a 95% - confidence interval.

4.1   Tree Features

First, we are interested in the intrinsic characteristics of the tree the routing is based on. We regard the mean tree depth. Indeed, the mean tree depth gives the latency needed to build the labeled tree. The labels are computed from the tree building and the deeper the tree, the bigger the maximum label size. Since each node only need to store its own label for routing, the maximum label size gives the memory requirements at each node. Tree depth gives a measure for the latency and for the memory requirements. We also computed the average number of messages sent per node for building the tree. This is proportional to the energy cost for building the tree.

First, we run simulations with fixing R to 0.10 with increasing λ. In this way, we can observe the behavior of the tree when the node degree increases while the network diameter remains constant. Figure 2 clearly shows that the parameters are independent of the node degree. This is a consequence of the building process, where each parent node sends only one message for its whole neighborhood, whatever its size. Furthermore, since a node only stores its own label, the memory requirements neither increase. We can thus deduce that when the number of nodes increases in a bounded environment, our labeling protocol scale well since it neither send more messages nor increase the storage needs.

We then consider 2 densities: low (mean node degree 8), and dense (degree 15). For each density, we make λ increase, and consequently the number of nodes increase, as well as the network diameter. Figure 3 shows the results. As

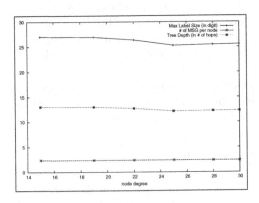

Fig. 2. RPT characteristics for $R = 0$

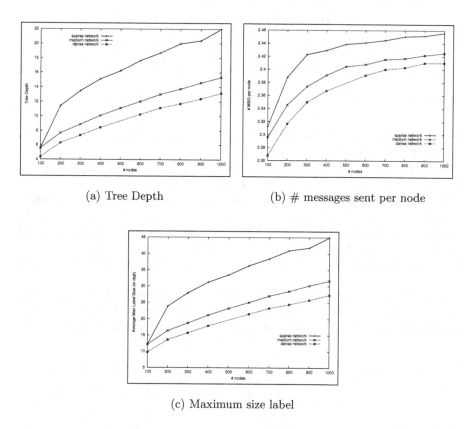

(a) Tree Depth

(b) # messages sent per node

(c) Maximum size label

Fig. 3. Characteristics of the tree for $\tilde{\delta} = 8$ and $\tilde{\delta} = 15$

expected when the network diameter increases, the tree also growths and thus, more memory is required at each node for storing labels. The number of messages sent remains very low and tends toward a constant asymptote. Simulation results match with the analysis of Section 3.2. The tree depth and the label size are in $O(log(n))$. Please note that we are using a random root node.

4.2   Routing Evaluation

We compared our algorithm RPT to other geographic routing protocols. Firstly we compared RPT with MFR [29] and VCap [6]. MFR uses geographic coordinates provided by a satellite receiver like GPS. MFR is thus more expensive than RPT in terms of equipments. Nevertheless, it provides routes with small number of hops and thus gives us a reference point. VCap uses virtual coordinates computed from landmarks nodes. Up to the best of our knowledge, this protocol gives the better results in terms of coordinates computing complexity and success rate among existing routing protocols based on virtual coordinates, Table 1 sums up the complexity in term of messages and storage overhead for these algorithms.

Table 1. Comparative complexity

	MFR	VCap	RPT
Cost	GPS	$N \times L$ msg	$2N + N_1$ msg
Node memory	own coordinates	own coordinates	own coordinates
	neighbors' coordinates	neighbors' coordinates	

N is the number of nodes in the network. L refers to the number of landmarks used in the VCap protocol.

Note that RPT requires the smaller amount of memory. On can also notice that RPT needs less message exchanges than VCap.

We run the simulation using the routing algorithms for the same samples of node distribution. We considered 2 densities: low (average node degree 8) and dense (average node degree 15). For each density, we make λ increasing, and hence increasing both, the number of nodes increased and the network diameter. To implement VCap, we use 5 *landmarks* randomly selected from the network nodes.

Figure 4(a) shows the success rate of the protocols studied. As expected, our protocol achieves a 100% of success, and far outperforms the other protocol based on labels (VCap) but also the geographic routing protocol MFR.

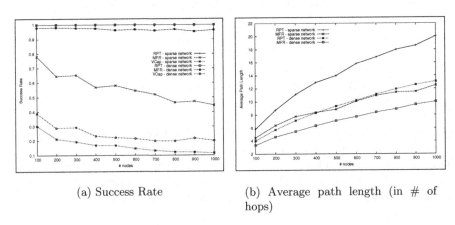

(a) Success Rate   (b) Average path length (in # of hops)

Fig. 4. Routing Characteristics of the tree for $\tilde{\delta} = 8$ and $\tilde{\delta} = 15$

Since we use a spanning tree of the network, routes are not optimal. We are thus interested in how far the routes provided by RTP are from the optimal. Figure 4(b) plots the average path length of RTP and MFR, MFR providing optimal paths. Since VCAP succeeds only for small paths, its average path length is not significant and we do not consider it in this figure. As expected, the average path length increases as the nodes have less neighbors. When comparing our algorithm vs. shortest paths (MFR), we can observe that the stretch factor is constant, which allows our algorithm to scale with an increasing network diameter.

We guess that the average path length provided by RTP could be greatly improved, particularly in dense networks, with the use of shortcuts as explained in Section 3.3. We keep this experimental analysis for future work.

We also compared RPT with other geographic routing algorithms that guaranteed delivery like Greedy Face Greedy (GFG) [12,19]. Since the path length depends on the planar subgraph extracted, we compared RPT with RNG [32], Gabriel Graph [13], Morelia Graph [4] and the virtual spanner [31,30]. Please note that GFG relies on an external location service like GPS. We tested two different network densities for all the algorithms and we show the results in figure 5.

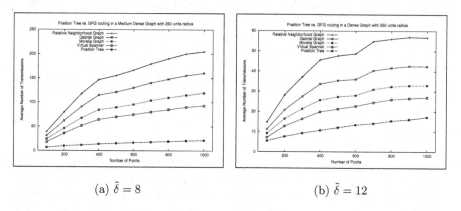

(a) $\tilde{\delta} = 8$ (b) $\tilde{\delta} = 12$

Fig. 5. Average path length (in # of hops) for $\tilde{\delta} = 8$ and $\tilde{\delta} = 12$ for RTP and GFG using several planarization methods, for $R = 0.25$

As Figure 5 shows, our RTP protocol outperform the GFG when regarding the length of the providing routes for any planarization method independently of the density.

5   Conclusions and Future Work

The label trees defined here are very similar to the coordinates reported in [1], the key difference is the top-down construction of a single tree in our approach. Our contribution also includes the investigation of statistical properties of the label trees. We showed our routing algorithm is very competitive when compared with other routing algorithms. It offers a very good trade-off between equipment cost, number of messages sent, stretch factor and memory requirement. In the extended version of this paper we will report a thorough comparison with other tree-based routing protocols. Below we enumerate some of the trends we may follow for future work:

– Assuming the nodes can adjust the transmission range we may try to build a tree that is least expensive in terms of the energy used for transmitting a package. As it is known that an optimal transmission radius can be found locally, we may choose edges that better approximate this radius.

- Since the underlying structure for routing is a tree, it is natural to explore broadcasting, optimizing e.g. the energy used or the latency in reaching the entire network. In this case a similar technique can be used as in the above case.
- For a fixed tree different root node selections may lead to different label sizes with impact in the amount of memory required for each node. The selection of a core node in a distributed manner is an interesting open problem.
- It will be interesting to study the behavior of RTP in a non-ideal MAC layer.
- The tree maintenance (i.e. taking care of vanishing nodes and the incorporation of new nodes to the network) is a challenging problem, we foresee some strategies based on the analysis of shortcuts in the tree, and the use of multiple roots where we need to balance the memory usage and the failure recovery (this also applies to the case of a non-ideal MAC layer).

We wish to thank the thorough review and suggestions of the anonymous referees who helped to improve the presentation.

References

1. Ben-Asher, Y., Feldman, M., Feldman, S.: Ad-hoc routing using virtual coordinates based on rooted trees. In: Sensor Networks, Ubiquitous, and Trustworthy Computing, 2006, pp. 6–13. IEEE Computer Society Press, Los Alamitos (2006)
2. Benbadis, F., Puig, J., de Amorim, M.D., Chaudet, C., Friedman, T., Simplot-Ryl, D.: JUMPS: Enhanced hop-count positioning in sensor networks using multiple coordinates. Elsevier, Amsterdam (2007)
3. Blazevic, L., Giordano, S., Le Boudec, J.: Anchored path discovery in terminode routing. In: Networking, Pisa, Italy (2002)
4. Boone, P., Chavez, E., Gleitzky, L., Kranakis, E., Opartny, J., Salazar, G., Urrutia, J.: Morelia test: Improving the efficiency of the gabriel test and face routing in ad-hoc networks. In: Kralovic, R., Sýkora, O. (eds.) SIROCCO 2004. LNCS, vol. 3104, pp. 23–24. Springer, Heidelberg (2004)
5. Capkun, S., Hamdi, M., Hubaux, J.: GPS-free positioning in mobile ad-hoc networks. In: HICSS (2001)
6. Caruso, A., Chessa, S., De, S., Urpi, A.: GPS free coordinate assignment and routing in wireless sensor networks. INFOCOM 1, 150–160 (2005)
7. Clark, B.N., Colbourn, C., Johnson, D.: Unit disk graphs. Discrete Math. 86(1-3), 165–177 (1990)
8. Clausen, T., Jacquet, P., Laouiti, A., Muhlethaler, P., Qayyum, A., Viennot, L.: Optimized Link State Routing Protocol (OLSR). RFC 3626 (October 2003)
9. Ermel, E., Fladenmuller, A., Pujolle, G., Cotton, A.: On selecting nodes to improve estimated positions. In: MWCM (2004)
10. Fang, Q., Gao, J., Guibas, L., Silva, V., Li, Z.: Glider: gradient landmark-based distributed routing for sensor networks. INFOCOM 1, 339–350 (2005)
11. Fonseca, R., Ratnasamy, S., Culler, D., Shenker, S., Stoica, I.: Beacon vector routing: Scalable point-to-point in wireless sensornets. Technical Report Tech. Rep. IRB-TR-04-012, Intel Research Berkeley (2004)

12. Frey, H., Stojmenovic, I.: On delivery guarantees of face and combined greedy-face routing in ad hoc and sensor networks. In: MOBICOM (2006)
13. Gabriel, K., Sokal, R.: A new statistical approach to geographic variation analysis. Systematic Zoology 18, 259–278 (1969)
14. Hill, J., Culler, D.: Mica: a wireless platform for deeply embedded networks. IEEE Micro 22(6), 12–24 (2002)
15. Intanagonwiwat, C., Govindan, R., Estrin, D.: Directed diffusion: a scalable and robust communication paradigm for sensor networks. In: MOBICOM (2000)
16. Johnson, D., Maltz, D., Broch, J.: Ad Hoc Networking, DSR The dynamic source routing protocol for multihop wireless networks, pp. 139–172 (2001)
17. Kuruvila, J., Nayak, A., Stojmenovic, I.: Progress and location based localized power aware routing for ah hoc sensor wireless networks. IJDSN 2, 147–159 (2006)
18. Leong, B., Liskov, B., Morris, R.: Gdstr: Geographic routing without planarization. In: Proceedings of the 3rd Symposium on Network Systems Design and Implementation (NSDI 2006), pp. 25–25 (2006)
19. Li, J., Gewali, L., Selvaraj, H., Muthukumar, V.: Hybrid greedy/face routing for ad-hoc sensor network. DSD 0, 574–578 (2004)
20. Madden, S., Franklin, M., Hellerstein, J., Hong, W.: TAG: a tiny aggregation service for ad-hoc sensor networks. SIGOPS Operating Systems 36, 131–146 (2002)
21. Mitton, N., Fleury, E.: Distributed node location in clustered multi-hop wireless networks. In: Cho, K., Jacquet, P. (eds.) AINTEC 2005. LNCS, vol. 3837, Springer, Heidelberg (2005)
22. Newsome, J., Song, D.: Gem: Graph embedding for routing and data-centric storage in sensor networks without geographic information. In: Proceedings of the 1st international conference on Embedded networked sensor systems, pp. 76–88 (2003)
23. Niculescu, D., Nath, B.: Ad hoc positioning system. In: GLOBECOM (2001)
24. Niculescu, D., Nath, B.: Ad hoc positioning system (APS) using AOA. In: INFO-COM (2003)
25. Perkins, C., Belding-Royer, E., Das, S.: Ad hoc On-demand Distance Vector Routing. RFC 3561 (July 2003)
26. Priyantha, N., Miu, A., Balakrishnan, H., Teller, S.: The cricket compass for context-aware mobile applications. In: MOBICOM, pp. 1–14 (2001)
27. Cao, T.A.Q.: A scalable logical coordinates framework for routing in wireless sensor networks. In: RTSS, pp. 349–358 (2004)
28. Santoro, N., Khatib, R.: Labelling and implicit routing in networks. The computer journal 28(1), 427–442 (1985)
29. Takagi, H., Kleinrock, L.: Optimal transmission ranges for randomly distributed packet radio terminals. IEEE transaction on communications com-22(3) (1984)
30. Tejeda, H., Chávez, E., Sánchez, J., Ruiz, P.: Energy-efficient face routing on the virtual spanner. In: Kunz, T., Ravi, S.S. (eds.) ADHOC-NOW 2006. LNCS, vol. 4104, pp. 101–113. Springer, Heidelberg (2006)
31. Tejeda, H., Chávez, E., Sánchez, J., Ruiz, P.: A virtual spanner for efficient face routing in multihop wireless networks. In: Cuenca, P., Orozco-Barbosa, L. (eds.) PWC 2006. LNCS, vol. 4217, pp. 459–470. Springer, Heidelberg (2006)
32. Toussaint, G.: The relative neighbourhood graph of a finite planar set. Pattern Recognition 12(4), 261–268 (1998)

Statistical Monitoring to Control a Proactive Routing Protocol

Kahkashan Shaukat and Violet R. Syrotiuk

Department of Computer Science and Engineering
Ira A. Fulton School of Engineering
Arizona State University
699 South Mill Avenue, Suite 553
Tempe, AZ 85281, USA
{kshaukat,syrotiuk}@asu.edu

Abstract. OLSR is a proactive routing protocol for mobile ad hoc networks. In OLSR, each node collects two-hop neighbourhood information and periodically sends topology control (TC) messages to update the link state. Rather than sending TC messages periodically, at each interval each node: (1) monitors *betweenness* of its two-hop neighbourhood graph; (2) if the measure is in-control no message is sent, otherwise a TC message is sent. Betweenness, a measure of centrality in a graph first used in social and biological network analysis, appears to correspond closely to the multi-point relay sets of OLSR. Hence a significant change in betweenness indicates a significant change in topology. Using this approach the control overhead in OLSR is reduced by 35-40%, with a corresponding savings in energy, and little impact on throughput or delay.

1 Introduction

Routing is a fundamental problem in *mobile ad hoc networks* (MANETs). As a result, a considerable number of routing protocols have been proposed [1,2,3] of which more than a dozen have been documented in the form of Internet-Drafts in the Internet Engineering Task Force (IETF) MANET working group [4]. As the IETF group moves forward with standardization, the field has been reduced to four candidate protocols, two proactive and two reactive.

One of the proactive candidates is *Optimized Link State Routing* (OLSR) [5]. Unlike a reactive protocol which responds to a link failure by repairing or rediscovering the affected path, a proactive protocol maintains its routes continuously through periodic transmission of state information. This can lead to unnecessary exchange of control overhead when the topology is changing slowly. Furthermore, the length of the period is difficult to determine since conditions in different parts of the network may necessitate a different length period for optimal performance.

In this paper, we directly address the problem of when to transmit the state information in a proactive routing protocol, selecting OLSR as our representative

E. Kranakis and J. Opatrny (Eds.): ADHOC-NOW 2007, LNCS 4686, pp. 46–58, 2007.

proactive protocol. This idea was inspired by the **Redback** network analysis tool [6] which uses passive monitoring techniques to transform the data collected into metrics that capture information about network behaviour. **Redback** takes as input a series of weighted graphs $G_1 G_2 \ldots G_k$ representing the network over time. The series is then processed using a suite of filtering tools, time series transforms, and other algorithms for network analysis.

Redback post-processes the graph series offline with each graph in the series including all nodes in the network. Our goal is to adapt the analysis for use in an online manner with each graph in the series corresponding to localized information, i.e., including only a subset of the nodes. In OLSR each node collects its two-hop neighbourhood periodically. Therefore, without any additional overhead, each node i can compute a graph series $G_1^i G_2^i \ldots G_k^i$ where each graph in its series corresponds to the topology of its two-hop neighbourhood. Each node i computes a *betweenness* measure on each graph $G_j^i, j = 1, \ldots, k$ in its series, indicating how often node i is on the shortest path between all pairs of nodes in G_j^i. This value seems to correspond well to the nodes in the *multi-point relay* (MPR) set that OLSR uses to propagate the state information through the network.

Each node monitors its value of betweenness, and if the observed value lies outside of the control region — indicating that the MPR set has likely changed — a topology control messsage is sent by the node. If the observed value is in-control, no action is taken. Using this approach in simulation, the control overhead in OLSR is reduced by 35-40%, with a corresponding saving in energy, and little impact on throughput or delay.

The remainder of this paper is organized as follows. Section 2 overviews related work on statistical process control. Section 3 overviews the OLSR routing protocol, defines the betweenness measure, and discusses how the measure can be integrated into OLSR using flip-flop filters. Section 4 describes the simulation set-up and presents the simulation results. Finally, we give conclusions and propose future work in Section 5.

2   Related Work

A collection of statistical techniques exist to monitor the performance of a system or process to determine if it has shifted away from some nominal operating state (the *in-control* state). The *Shewhart control chart* is a basic tool used for monitoring. It is a time-oriented representation of a characteristic that has been measured and is commonly used to provide *statistical process control* (SPC). Figure 1 shows an example Shewhart control chart. It plots the sample mean x of a controlled quantity against the desired population mean $\bar{x}$ over time. It consists of a center line that represents the average value of the characteristic $\bar{x}$ and two horizontal lines, the *upper* and *lower control limits* (UCL and LCL).

If the system is in control, nearly all of the samples fall between the UCL and the LCL. As long as the samples plot within the control limits, the system is assumed to be in control, and no action is necessary. However, a sample that plots outside of the control limits is interpreted as evidence that the system is

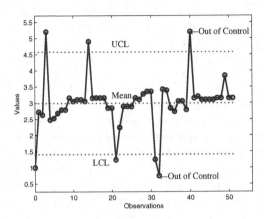

Fig. 1. An example Shewhart control chart

out of control, and adjustment or corrective action should be taken to return the system to desirable operating state.

SPC techniques have been developed to accommodate a wide variety of process types and behaviours. For example, *cumulative sum* control charts directly incorporate all the information in the sequence by plotting the cumulative sums of the deviations of the samples from a desired target value. EWMA control charts use an *exponentially weighted moving average* of the samples as the monitoring statistic. Cumulative sum and EWMA control charts are more effective than Shewhart control charts in quickly detecting small deviations from the in-control state [7].

Another promising approach useful for detecting process upsets is the *changepoint* model. This approach focusses on finding the point in time where the underlying model generating a series of observations has changed. Most of the work on changepoints has concentrated on detecting a sustained shift in the mean of the process. When all of the parameters of the underlying probability model are known the changepoint approach is equivalent to the cumulative sum control chart. An EWMA would also be appropriate, as it can be designed to be approximately equivalent in performance to the cumulative sum procedure. However, changepoints have not been widely used for other monitoring situations. Montgomery [7] gives a comprehensive treatment of methods of *statistical process control* (SPC).

Several MANET network management protocols use monitoring at the application layer of the protocol stack [8,9,10]. Badonnel et al. describe an information model and a probe-based architecture to integrate MANETs in a management framework to assess their operational state and behaviour. Kim and Noble [11] use EWMA control charts and techniques from SPC to estimate network capacity in MANETs. They present four filters designed to react quickly to persistent changes while tolerating transient noise. Such filters are agile when possible, but stable when necessary, adapting their behaviour to prevailing conditions.

Perhaps the work closest to ours is that of Barman et al. [12] where a new loss labelling technique using a flip-flop filter is introduced to estimate round trip time and to differentiate between congestion and loss in wireless networks.

3 Overview

This section provides a brief description of the OLSR protocol, a description of the betweenness measure, and then describes how we use betweenness to control when the OLSR protocol transmits topology control messages.

3.1 The Optimized Link State Routing Protocol

The *Optimized Link State Routing* (OLSR) protocol is a proactive routing protocol where each node periodically broadcasts its routing table so that its neighbouring nodes can achieve a complete view of the network state [5]. Important to the operation of OLSR are the *multi-point relay* (MPR) sets. An MPR is a subset of the one-hop neighbours of a node selected to forward its control packets. When each MPR forwards the message, a node is guaranteed communication with each of its two-hop neighbours. This provides an efficient implementation of network-wide broadcast.

Each node discovers and maintains topology information through the periodic exchange of HELLO and *topology control* (TC) messages. A HELLO message, exchanged between a node and its one-hop neighbours, contains a list of one-hop neighbours indicating those in the MPR set, a list of two-hop neighbours, and a list of neighbours that have selected this node as an MPR. The TC messages contain a list of all the nodes that have selected the sender as an MPR.

The MPRs periodically exchange network topology information. When exchanging link-state information, each node lists only the information of the nodes that have selected it as an MPR, that is, its *multi-point relay selector* (MS) set. OLSR reduces the flooding of control packets compared to other proactive routing protocols by selecting the MPRs [13], as only these MPRs can forward messages broadcast by its MSs.

3.2 The Betweenness Measure

In OLSR, each node collects its two-hop neighbourhood periodically via the HELLO messages. In this way, each node i obtains a graph series $G_1^i G_2^i \ldots G_k^i$ where each graph corresponds to its two-hop neighbourhood; each graph is just a subset of the network topology graph.

One measure of centrality in a graph is *betweenness*. For example, a node in a network is central to the extent that it falls on the shortest path between pairs of other nodes. This idea of was originally used in the study of social network analysis: when a person in a group is strategically located on the shortest communication paths connecting pairs of others, that person is in a central position; the others are assumed to be responsive to and can be influenced by that person. It is also used in biological network analysis contributing to protein-protein interaction, gene coexpression, and metabolic datasets [14,15]. In the networks we consider, betweenness can be viewed as a measure of the influence a node has over the flow of information between different points in the network.

The betweenness of a vertex t in a simple graph G was first defined by Freeman [16] as the number of shortest paths between pairs of other vertices that pass through t. This definition was generalized by Girvan and Newman [17] to edge betweenness, where the betweenness of an edge e is simply the number of shortest paths between pairs of vertices that run along e. In each definition, where there is more than one shortest path, the weight is equally distributed among them.

We define the *betweenness* of a vertex t for a graph $G = (V, E)$ as

$$C_B(G, t) = \sum_{u,v \in V} \frac{\sigma_{uv}(t)}{\sigma_{uv}}, \text{ where } u \neq t \neq v, \tag{1}$$

where σ_{uv} is the number of shortest paths from u to v in G, and $\sigma_{uv}(t)$ is the number of shortest paths from u to v in G that pass through vertex t. This may be normalized by dividing through by the number of pairs of vertices not including t, which is $(n-1)(n-2)$ where $n = |V|$.

Consider the graph $\mathcal{G}$ in Figure 2 with $n = 10$ vertices. The betweenness of vertices G and H is, respectively:

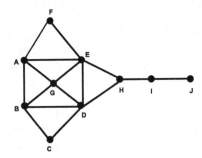

Fig. 2. An example graph $\mathcal{G}$ to illustrate the *betweenness* measure

$$C_B(\mathcal{G}, G) = \sum_{u,v \in V} \frac{\sigma_{uv}(G)}{\sigma_{uv}}, \text{ where } u \neq G \neq v$$

$$= \frac{2}{9 * 8} \left(\frac{\sigma_{AD}(G)}{\sigma_{AD}} + \frac{\sigma_{BE}(G)}{\sigma_{BE}} \right)$$

$$= \frac{1}{36} \left(\frac{1}{3} + \frac{1}{3} \right)$$

$$= 0.018$$

$$C_B(\mathcal{G}, H) = \sum_{u,v \in V} \frac{\sigma_{uv}(H)}{\sigma_{uv}}, \text{ where } u \neq H \neq v$$

$$= 0.389.$$

We see that node H has degree 3 which is less than the average node degree of 3.2 for the network. Yet from the view of betweenness, node H has a strong location in the network since it is a cut vertex; paths from the vertices in $\{A, B, C, D, E, F, G\}$ to vertices in $\{I, J\}$ must go through H. Hence H is essential for communication among vertices in these two sets.

3.3 Monitoring Betweenness to Control TC Messages in OLSR

The betweenness of a node i in a graph $G = (V, E)$ identifies the importance of the role that i plays in maintaining communication or connectivity in the graph. The higher the value of betweenness of a node i the more important i is to the graph. This can be associated with the MPRs of OLSR which play a role in communicating the control messages to the other nodes in the graph. Here, an MPR m for a node i is selected depending on its location in the graph such that m helps i to communicate to i's two-hop neighbours.

The two-hop neighbourhood is obtained with no extra control overhead directly from the HELLO messages. In our context, betweenness is a *local* measure not a global one, i.e., only the local topology of a node is used, not the topology of the entire network. Each node i monitors its value of betweenness in its two-hop neighbourhood graph. That is, for each graph $G_j^i, j = 1, \ldots, k$ in the graph sequence $G_1^i G_2^i \ldots G_k^i$, node i computes $C_B(G_j^i, i)$ and monitors this characteristic. Figure 3 shows an example control chart monitoring betweenness of a representative node of the network.

If the observed value falls within the region bounded by UCL and LCL (the in-control state) we need not take any action. This prevents the protocol from sending out unnecessary TC messages. If the value of the observed betweenness goes out-of-control it indicates that a significant change has occurred in the topology requiring a TC message to be sent. Thus, the importance of the node in terms of its location in the local network topology can be directly used to control

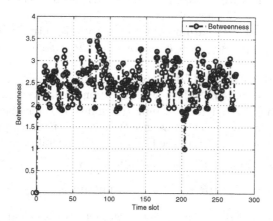

Fig. 3. A control chart monitoring *betweenness* of a representative node i from a graph series where each graph corresponds to the two-hop neighbourhood of i

the behaviour of the protocol by reducing the number of control messages when no significant topology changes occur.

We use the outcome of the betweenness measure being monitored to change how OLSR sends its TC messages. Instead of sending the TC messages periodically, we use an out-of-control indication to trigger their transmission. Specifically, at each node we keep track of the last ten values of betweenness calculated and use them to generate an estimated value for that node. We also observe the actual value of betweenness, and if the observed value lies outside the control region, a TC messsage is sent by the node. An observed value exceeding the control limits indicates that a significant change has occurred and that the node needs to send a TC message. We use flip-flop filters to calculate the deviation between the estimated value and the observed value and determine whether the observed value has exceeded its control limits. The stable filter of the flip-flop filter has a smoothing effect and filters out any erratic behaviour while the agile filter helps to adapt to the changes in the topology and make more accurate estimates. After filtering out the aberrant behaviour, if significant change is noticed, the protocol should adapt to the change.

Since the difference between the observed and estimated values of betweenness is small (because the number of nodes in the two-hop neighbourhood graph is small), we consider the upper and lower control limits to be one standard deviation away from the mean value rather than the more common three standard deviations. Thus, in our experiments, when the observed betweenness values fall within the one standard deviation control region, we use the stable flip flop that smooths out the effects implying that no TC message needs to be sent. Otherwise, we use the agile filter to quickly adapt to the situation and send out a TC message so that the neighbours can update their routes in the routing tables.

3.4   Flip-Flop Filters

Flip-flop filters [12] consist of two EWMA filters, one agile, with a gain of α, and the other stable with a gain β. An EWMA applies weight factors that decrease exponentially for each observation, giving more importance to recent observations while not disregarding older observations entirely. The degree of weight decrease is expressed as a constant *smoothing factor* or *gain*, a value between 0 and 1.

Given a new observation, an EWMA produces a new estimate as a linear combination of the old estimate plus the new observation, each given some weight. We use principles from SPC to detect a significant change in the neighbourhood of a node to trigger the sending of TC messages. We estimate the next predicted value $\bar{x}$ and the deviation $\bar{R}$ around $\bar{x}$, of the EWMA using the following equations:

$$\bar{x} = (1 - \alpha)\bar{x} + \alpha x_i, \text{ initially } \bar{x} = x_0$$
$$\bar{R} = (1 - \beta)\bar{R} + \beta|x_i \cdot x_{i-1}|, \text{ initially } \bar{R} = \frac{x_0}{2}.$$

Here $\bar{R}$ is used to estimate the deviation around $\bar{x}$ and is calculated only from samples x_i within the upper and lower control limits. Figure 4 shows the mechanism of a flip-flop filter.

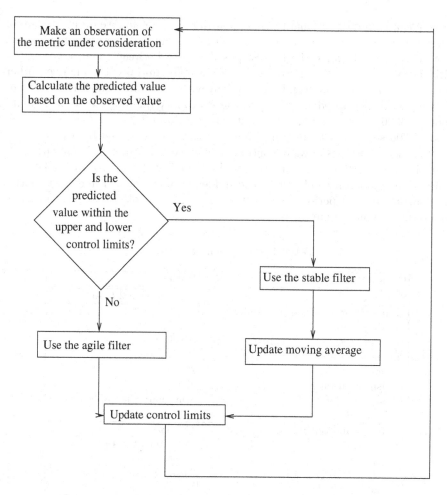

Fig. 4. Flow-chart explaining the operation of a flip-flop filter

Under normal operation, stable EWMA filters are employed to filter out short-term variations. A controller selects between the two filters; the underlying principle of this controller is to employ the stable filter when possible, but fall back to the agile filter when observations are unusually noisy. As long as the observed value x_i lies within the control limits, the topology is considered stable. Otherwise, x_i is considered an outlier and is used to indicate significant change in the topology; this triggers an immediate response in the system to take a corrective action.

4   Simulation Set-Up and Results

We use an extended version of **ns-2** [18] which provides the ability to simulate MANETs [19], as well as an implementation of the OLSR protocol. We create mobility scenarios which include 20 nodes in a $500 \times 500\,m^2$ area where every node has an omni-directional transmission radius of $250\,m$. We designate two of these nodes as the source and destination (s, t), and fix them diagonally apart from each other to force the longest possible path between them, i.e., s is positioned at $(0, 0)$ and t is positioned at $(500, 500)$. Initially, the eighteen other nodes are positioned uniformly at random within the area and the movement sequences are generated according to the stationary random waypoint mobility model [19,20] with a maximum velocity varying from $2\,m/s$ to $20\,m/s$ for a period of 700 seconds. The mobility model is also characterized by a pause time which is set to zero resulting in continuous movement. Our experiments use constant bit rate (CBR) data sources over the User Datagram Protocol (UDP) for data communication to eliminate the influence of congestion and flow control mechanisms on the performance of the routing protocol. Table 1 shows these and other simulation parameters.

Table 1. Simulation parameters

Parameter	Value
Simulator	**ns-2**, version 2.1b7a
Transport and routing protocol	UDP
Routing protocol	OLSR
Medium access control (MAC) protocol	IEEE 802.11
Simulation area	$500 \times 500\,m^2$
Number of nodes	20
Transmission range	$250\,m$
Mobility model	Stationary random waypoint
Node speed	$2, 10, 15, 20\,m/s$
Channel bandwidth	$11\,Mbps$
Traffic type	Constant bit rate (CBR)
Data packet size	$64\,bytes$
Packet arrival rate	$2\,packets/s$
Background traffic	None
Simulation duration	$700\,s$
Confidence interval	95%

For the source-destination pair (s, t), the data flow starts and ends at a random time such that the connection lasts for at least 650 seconds; this is the data transmission period. In order to achieve a high level of confidence in our results, we use ten randomly generated scenarios using ten different seeds for each of the four different speeds giving rise to 40 simulations for OLSR with and without monitoring.

We are interested in are observing the impact of using betweenness as a metric to trigger TC messages for OLSR. Thus, we look into how this mechanism affects the following common quantitative metrics [21]:

Average end-to-end delay: The end-to-end delay is the time taken for a packet to be transmitted from the source of the flow to the destination of the flow. The average end-end-delay (in seconds) is the end-to-end delay averaged over all the packets that are successfully received.

Throughput: The throughput is the total of all bits (or packets) successfully delivered to each flow destination over the total-time of the flow.

Energy consumption: The energy consumption is the total energy consumed during the data transmission period.

Control Overhead: The control overhead is the average number of control packets sent by the protocol during the data transmission period. In our experiments, the HELLO and TC packets are the control packets.

4.1 Simulation Results

Figure 5(a) plots the number of control packets transmitted during the simulation for the speeds $2, 10, 15$ and $20\,m/s$. We see that without monitoring the number of control packets transmitted is 8730 and 12517, while the number with monitoring is 5774 and 6760, for $2\,m/s$ and $20\,m/s$ respectively. This shows that when using betweenness as a measure to decide when to transmit TC messages, the control overhead decreases between 34% and 46% for $2\,m/s$ and $20\,m/s$, respectively.

Figure 5(b) shows the number of control packets transmitted per successful transmission of a data packet as a function of speed. The number of control packets per data packet is 1.34 and 2.41 for speeds $2\,m/s$ and $20\,m/s$, respectively, without monitoring; using monitoring reduces the number to 1.18 and 1.47 for speeds $2\,m/s$ and $20\,m/s$, respectively. The number of control packets that are transmitted to route a data packet successfully reduces by 12% to 39%. Thus, a dramatic reduction in the number of control packets required to route data packets is achieved.

However, what is important is whether the reduction of control overhead affects the packet delivery ratio. Figure 6(a) shows the packet delivery ratio. Here, we see that when using monitoring, the throughput decreases from 95.82% to 91.06% for $2\,m/s$ and from 76.32% to 75.23% for $20\,m/s$, respectively. The decrease in throughput may be attributed the control limits; we did not optimize these. Currently, the control limits are conservative, sending much fewer TC

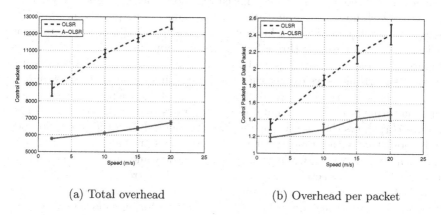

(a) Total overhead   (b) Overhead per packet

Fig. 5. Total control overhead and control overhead on a per packet basis

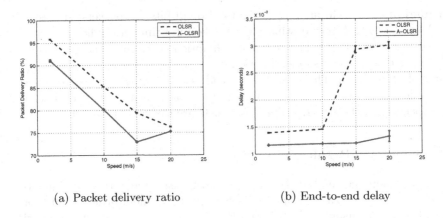

(a) Packet delivery ratio   (b) End-to-end delay

Fig. 6. Packet delivery ratio and average end-to-end delay

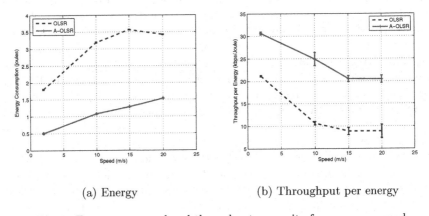

(a) Energy   (b) Throughput per energy

Fig. 7. Energy consumed and throughput per unit of energy consumed

messages than without monitoring. This causes more packets loss since the routing table at each node in the network is not updated as frequently as when TC messages are sent periodically. This conjecture is supported by the fact that, as speed increases, the drop in throughput is less pronounced.

Figure 6(b) plots the average end-to-end delay in seconds. The results are somewhat surprising as we expect that the delay using monitoring to be higher. However, this is not the case. While fewer data packets reach their destination when monitoring is used, those that do reach their destination do so more quickly.

Figure 7(a) shows the energy consumption in Joules during the simulation period. It shows that without monitoring, the energy consumed amounts to 1.80 and 3.42 for $2\,m/s$ and $20\,m/s$, while with monitoring the values are 0.50 and 1.53 for $2\,m/s$ and $20\,m/s$, respectively. Since the number of control packets decrease by 36% and 44% for $2\,m/s$ and $20\,m/s$, respectively, there is a corresponding savings in the energy usage of the nodes. Thus, OLSR with monitoring is much more energy efficient saving as much as 73% to 49% for $2\,m/s$ and $20\,m/s$, respectively.

While the throughput of OLSR is slightly reduced with monitoring, Figure 7(b) shows when energy is taken into account it achieves a much higher throughput per unit of energy consumed. This is because OLSR with monitoring transmits fewer control packets compared to OLSR where the TC messages are transmitted periodically.

5   Conclusions and Future Work

The main goal of this work was to change when a proactive routing protocol updates its state information. In OLSR, each node collects two-hop neighbourhood information and periodically sends topology control (TC) messages to update the link state. Rather than sending TC messages periodically, at each interval each node: (1) monitors *betweenness* of its two-hop neighbourhood graph; (2) if the measure is in-control no message is sent, otherwise a TC message is sent. Betweenness seems to corresponds closely to the multi-point relay sets of OLSR. Hence a significant change in betweenness indicates a significant change in topology. Using this approach, the control overhead in OLSR is reduced by 35-40% with a corresponding savings in energy and little impact on throughput or delay.

An important assumption made with control charts is that the observations are independent. It is well-known that conventional control charts do not work well if the monitored characteristic exhibits even low levels of correlation over time [7]. Specifically, these control charts signal too many false alarms if the data are positively correlated.

Our future work includes studying the correlation of the betweenness measure. Many techniques have been developed to adapt control charts to deal with moderate to high levels of autocorrelation in the data. See Montgomery [7, Chapter 9] for details. As well, a methodology to set the UCL and LCL are required.

Acknowledgments. This work was supported, in part, by NSF ITR-0220001 and DSTO contract 4500496906. The views and conclusions contained in this document are those of the authors and should not be interpreted as representing the official policies, either expressed or implied, of the National Science Foundation or Defence Sciences Technology Organisation (Australia).

References

1. Royer, E.M., Toh, C.-K.: Review of current routing protocols for ad hoc mobile wireless networks. IEEE Personal Communications, 46–55 (April 1999)
2. Broch, J., Maltz, D.A., Johnson, D.B., Hu, Y.C., Jetcheva, J.: A performance comparison of multi-hop wireless ad hoc network routing protocols. In: Proceedings of the 4th Annual International Conference on Mobile Computing and Networking (MobiCom'98), pp. 85–97 (October 1998)
3. Ramanathan, S., Steenstrup, M.: A survey of routing techniques for mobile communications networks. Baltzer/ACM Mobile Networks and Applications 1(2), 89–104 (1996)

4. Internet Engineering Task Force (IETF), Mobile Ad-Hoc Networking (MANET) Working Group, http://www.ietf.org/html.charters/manet-charter.html

5. Clausen, T., Jacquet, P.: Network working group RFC 3626, Optimized link state routing protocol. October (2003), http://hipercom.inria.fr/olsr/rfc3626.txt

6. Syrotiuk, V.R., Shaukat, K., Kwon, Y.J., Kraetzl, M., Arnold, J.: Application of a network dynamics analysis tool to mobile ad hoc networks. In: Proceedings of the 9th ACM/IEEE International Symposium on Modeling, Analysis and Simulation of Wireless and Mobile Systems (MSWiM'06), pp. 36–43 (October 2006)

7. Montgomery, D.C.: Introduction to Statistical Quality Control, 5th edn. JohnWiley & Sons, Inc, Chichester (2005)

8. Badonnel, R., State, R., Festor, O.: Management of mobile ad hoc networks: information model and probe-based architecture. International Journal of Network Management 15(5), 335–347 (2005)

9. Chen, W., Jain, N., Singh, S.: ANMP: Ad hoc network management protocol. IEEE Journal on Selected Areas in Communications 17, 1506–1531 (1999)

10. Shen, C.C., Srisathapornphat, C., Jaikaeo, C.: An adaptive management architecture for ad hoc networks. Proceedings of IEEE Communications Magazine 41(2), 108–115 (2003)

11. Kim, M., Noble, B.: Mobile network estimation. In: Proceedings of the 7th Annual International Conference on Mobile Computing and Networking (Mobicom'01'), pp. 298–309 (July 2001)

12. Barman, D., Matta, I.: Effectiveness of loss labeling in improving TCP performance in wired/wireless networks. In: Proceedings of the 10th IEEE International Conference on Network Protocols (ICNP'02), pp. 2–11 (November 2002)

13. Choi, J., Ko, Y.: A performance evaluation for ad hoc routing protocols in realistic military scenarios. In: Proceedings of the 3rd Workshop on Cooperative Internet Computing (CIC'04) (October 2004)

14. Pinney, J.W., McConkey, G.A., Westhead, D.R.: Decomposition of biological networks using betweenness centrality. In: Proceedings of 9th Annual International Conference on Research in Computational Molecular Biology (RECOMB/05) (May 2005)

15. Pinney, J.W., Westhead, D.R.: Betweenness-based decomposition methods for social and biological networks. In: Barber, S., Baxter, P.D., Mardia, K.V., Walls, R.E. (eds.) Interdisciplinary Statistics and Bioinformatics, pp. 87–90. Leeds University Press (2006)

16. Freeman, L.C.: A set of measures of centrality based on betweenness. Sociometry 40(1), 35–41 (1977)

17. Girvan, M., Newman, M.E.: Community structure in social and biological networks. Proceedings of National Academy of Science 99(12), 7821–7826 (2002)

18. The University of California, Berkeley, The network simulator – ns-2, http://www.isi.edu/nsname/ns/

19. C. M. University

20. Navidi, W., Camp, T.: Stationary distributions for the random waypoint model. IEEE Transactions on Mobile Computing 3, 99–108 (2004)

21. Corson, M.S., Macker, J.: Network working group RFC 2501, mobile ad hoc networking (MANET): Routing protocol performance issues and evaluation considerations (January 1999), http://www.ietf.org/rfc/rfc2501.txt

A Faster Distributed Approximation Scheme for the Connected Dominating Set Problem for Growth-Bounded Graphs

Beat Gfeller* and Elias Vicari**

Institute of Theoretical Computer Science,
ETH Zurich, Switzerland
{gfeller,vicariel}@inf.ethz.ch

Abstract. We present a distributed algorithm for finding a $(1 + \varepsilon)$-approximation of a *Minimum Connected Dominating Set* in the class of *Growth-Bounded* graphs, which includes *Unit Disk graphs*. In addition, the computed Connected Dominating Set guarantees a constant stretch factor on the length of a shortest path with respect to the original graph and induces a subgraph of constant degree. The nodes do not require any positioning or distance information.

The algorithm runs in $O\big(T_{\mathsf{MIS}} + 1/\varepsilon^{O(1)} \cdot \log^* n\big)$ synchronous rounds, where T_{MIS} is the time for computing a Maximal Independent Set (MIS) in the network graph. Using the fastest known deterministic algorithm for computing a MIS, the total running time is $O\big((\log \Delta + 1/\varepsilon^{O(1)}) \cdot \log^* n\big)$, where Δ is the maximum degree of the network graph. If one allows randomization, the running time reduces to $O\big((\log \log n + 1/\varepsilon^{O(1)}) \cdot \log^* n\big)$ rounds.

Keywords: Connected Dominating Set, Growth-Bounded Graphs, Distributed Approximation Scheme, Distributed Algorithms.

1 Introduction

Wireless sensor networks are used in an increasing number of current applied research projects, in a variety of applications ranging from agricultural management targeted at Indian farmers [18] to body area networks for human activity recognition [11]. To assist these developments, it is becoming ever more important to find solutions to the fundamental difficulties in employing such networks.

A *wireless ad hoc network* typically consists of a large number of nodes, each having wireless communication capability, modest computational power, and a battery with limited capacity. Two nodes can communicate directly if they are

* This author gratefully acknowledges the support of the Swiss SBF under contract no. C05.0047 within COST-295 (DYNAMO) of the European Union.
** This author is partially supported by the National Competence Center in Research on Mobile Information and Communication Systems NCCR-MICS, a center supported by the Swiss National Science Foundation under grant number $5005 - 67322$.

E. Kranakis and J. Opatrny (Eds.): ADHOC-NOW 2007, LNCS 4686, pp. 59–73, 2007.
© Springer-Verlag Berlin Heidelberg 2007

within mutual communication range, and forward messages via intermediate nodes to talk to distant nodes.

In contrast to wired networks, where usually a dedicated *backbone* infrastructure with high-throughput capabilities is available for long-distance routing, *ad hoc* networks do not have any *a priori* means to manage the routing and scheduling of messages. Another specialty of wireless ad hoc networks is mobility of the nodes, which may continuously cause changes in the topology.

In the absence of an organized routing scheme, simple flooding (i.e., the first time a message is received from any neighbor, it is forwarded to all other neighbors) could be used to transmit messages. However, this is very wasteful in terms of energy and causes interference problems if many nodes transmit a message at the same time. To organize routing more cleverly, a *virtual backbone* can be computed, i.e., a subset of the nodes which participate in multi-hop routing. In this fashion, even when messages are still forwarded by flooding inside the virtual backbone, the energy savings are significant.

When modelling the network as a graph, the most widely used concept for defining a backbone is the *Connected Dominating Set* (CDS). A CDS is a subset S of the nodes of a graph G that induces a connected subgraph of G, such that every node that is not in S has a neighbor in S.

Usually, the energy savings are higher if the number of nodes in the CDS is small. However, computing a *Minimum* CDS is NP-hard even on Unit Disk Graphs [4], and would require global information. Considering the highly dynamic nature of ad hoc networks, it is important that the CDS can be computed locally within a short time; a linear running time in the diameter of the network (as required to obtain global information) is clearly inappropriate. Moreover, a minimum CDS lacks some desirable properties which we demand from an efficient backbone. For instance, routing a message from a node v to node a w should not need many more intermediate hops than a shortest path in the original network. The maximum ratio over all vertex pairs between these two hop-distances is called the *stretch factor*. The stretch factor of a MCDS can be as bad as linear in the number of nodes (for instance, consider a ring network), and one would like to prevent this effect. Another desirable property is that nodes have constant degree in the CDS-induced graph, which might help to address interference issues. For these reasons, a natural trade-off is to find a CDS with only near-optimal size but fulfilling the aforementioned properties.

Related Work. The concept of a *virtual backbone* (in analogy to backbones in wired networks) was introduced in [6]. Since then, the construction of small connected dominating sets in Unit Disk Graphs (UDGs) has been intensively studied. A recent overview can be found in [2]. For the centralized setting with given coordinates of the nodes (which are embedded in the Euclidean plane), a *polynomial-time approximation scheme* (PTAS) was proposed in [3] (approximation schemes for related problems like the Minimum Dominating Set Problem in UDGs were given in [10]). The approach of [5] (as well as our own approach) yields a PTAS that does not require coordinate information about the nodes.

However, many of the first distributed algorithms either did not guarantee a good approximation ratio in the worst case, or had a linear running time (see [2]). The first approach achieving a constant approximation ratio in polylogarithmic time was [19]. Alzoubi et al. [1] were the first to provide an algorithm for computing a CDS with low stretch and low degree, which was coined *well-connected* CDS in [19]. Recently, Czygrinow et al. [5] presented a distributed algorithm with running time $O(1/\varepsilon^6 \cdot \log(1/\varepsilon) \cdot \log^3 n)$ achieving the approximation ratio $1 + \varepsilon$ for any $\varepsilon > 0$, but without proving any stretch or degree bounds.

In general graphs, the best known distributed algorithm for the MCDS problem has a polylogarithmic running time and achieves an approximation ratio of $O(\log \Delta)$, where Δ is the maximum degree of the network [7]. This approximation ratio is asymptotically optimal unless all problems in NP can be solved by deterministic algorithms with running time $n^{O(\log \log n)}$ [8].

Our Contribution. In this paper we present a distributed approximation scheme for the problem of finding a Minimum Connected Dominating Set in the class of growth-bounded graphs, which includes Unit Disk Graphs. An important feature of our algorithm is that the only information required by the nodes is the set of their direct neighbors. Distances between them or even coordinate information are not required.

The algorithm computes a *well-connected* $(1 + \varepsilon)$-approximation of a Connected Dominating Set, for any $\varepsilon > 0$. This takes $O(T_{\mathsf{MIS}} + 1/\varepsilon^{O(1)} \cdot \log^* n)$ rounds of synchronous computation, where T_{MIS} is the number of rounds needed to compute a Maximal Independent Set (MIS). Currently, the fastest deterministic distributed algorithm for computing a MIS in growth-bounded graphs is due to Kuhn et al. [12] and runs in $O(\log \Delta \cdot \log^* n)$ time. Using randomization, a MIS can even be computed in $O(\log \log n \cdot \log^* n)$ time in growth-bounded graphs [9]. Thus, we improve on the running time for computing a $(1 + \varepsilon)$-approximate CDS, while adding the guarantee that the computed CDS has constant stretch, constant degree, and therefore a linear number of edges.

The algorithm we propose builds substantially on the approach of Nieberg et al. [16] for computing a $(1 + \varepsilon)$-approximate Minimum Dominating Set (for growth-bounded graphs): In a nutshell, their solution partitions the graph into clusters of appropriate radius, computes an optimal DS on each of these, and takes the union of these sets to yield a DS for the graph. Employing the same idea, we cluster the graph and compute an optimal CDS for each cluster. However, we let the clusters overlap such that the union of the small CDS solutions forms a connected DS of the graph. We prove the approximation ratio of the CDS by adapting the proof from [16], which requires an additional non-trivial step, i.e., Lemma 7. In the proof of Lemma 7, ideas related to [5] are used. In addition, we prove the (well-)connectedness of the computed set, which has no equivalent in [16,5]. In order to turn our centralized approximation scheme for CDS into a distributed algorithm, we follow the lines of [13].

Intriguingly, such a relatively simple modification of known techniques yields a distributed approximation scheme which runs substantially faster than the previously known solution [5], and additionally guarantees well-connectedness.

2 Definitions

Network Model. We model a wireless network as a growth-bounded graph, a class which includes Unit Disk graphs. The precise definition follows in the next subsection. Each node has a unique identifier and knows which nodes are within its transmission range. When we say that a distributed algorithm computes a CDS, we mean that each node knows after executing its code whether it is part of the CDS or not. Assuming that a MAC layer for direct communication of neighbors has already been established, we do not consider collisions or other transmission failures, and we furthermore assume a synchronous network.

Terminology. Let $G = (V, E)$ be an undirected connected graph. For $V' \subseteq V$, we denote by $G[V']$ the subgraph of G *induced* by V': the vertex-set of $G[V']$ is V' and the edge-set consists of the edges of G with both endpoints in V'.

The *distance* $d(v, w)$ between two vertices $v, w \in V$ is the number of edges (or *hops*) that must be traversed to go from v to w on a shortest path. Similarly, we define the distance between two sets $A, B \subseteq V$ as the distance of two closest nodes $a \in A$ and $b \in B$, i.e., $d(A, B) = \min\{d(a,b) | a \in A, b \in B\}$. Two sets $S, T \subseteq V$ are called *2-separated*[1] if and only if $d(S, T) \geq 3$.

The *neighborhood* $N(v)$ of a vertex $v \in V$ is defined as the set all vertices $w \in V$ with $d(v, w) \leq 1$ (including v). The neighborhood of a set $S \subseteq V$ is defined as $N(S) := \bigcup_{s \in S} N(s)$. Likewise, the *i-neighborhood* of a set of vertices S is defined recursively as $N_0(S) := S$, and $N_i(S) := N_{i-1}(N(S))$ for $i \geq 1$. The *reduced* neighborhood $\Gamma_j(v, V')$ is defined for all $j \geq 0$ as $N_j(v)$ on the graph induced by the set $V' \subseteq V$.

A *maximal independent set* (MIS) M for a given graph $G = (V, E)$ is a subset $M \subseteq V$ such that for every $v, w \in M$ we have $v \notin N(w)$ (*independence*) and furthermore no superset $M' \supset M$ with the latter property exists (*maximality*). A *maximum independent set* (MaxIS) is a MIS of largest size. For a subset $S \subseteq V$, MaxIS(S) is a maximum independent set on the induced subgraph $G[S]$. A *dominating set* (DS) D for a given graph $G = (V, E)$ is a subset $D \subseteq V$ such that for every $v \in V$ there is a $w \in D$ such that $v \in N(w)$. Note that a MIS is always a DS. A *connected dominating set* (CDS) for a given graph $G = (V, E)$ is a dominating set M with the additional requirement that the nodes in M induce a connected subgraph of G. A *minimum* (or *optimal*) *connected dominating set* (MCDS) is a CDS of smallest possible cardinality, and the *minimum connected dominating set problem* is the problem of providing such a MCDS for a given graph. We define the function $\mathcal{C}(A)$, where $A \subseteq V$ is a subset of all nodes, as follows: $\mathcal{C}(A) \subseteq N(A)$ is a minimum connected set of nodes that dominates all nodes in A, under the condition that only nodes in $N(A)$ are used. It is crucial that this set may contain nodes from $V \setminus A$.

We are interested in a special class of graphs, called *growth-bounded graphs*.

---

[1] We use the term 2-separation because any shortest path between S and T contains two nodes outside $S \cup T$.

Definition 1. *A graph G is* growth-bounded *if there is a* polynomial *bounding function $f(r)$ such that for each node $v \in V$, the size of any MIS in the neighborhood $N_r(v)$ is at most $f(r)$, $\forall r \geq 0$.*

A subclass of growth-bounded graphs are *Unit Disk Graphs*, which are often used to model wireless communication networks. A graph $G = (V, E)$ is a Unit Disk Graph if it can be represented by placing a point p_v for each node $v \in V$ on the Euclidean plane $\mathbb{R}^2$, such that an edge $(u, v) \in E$ exists if and only if the distance $\|p_v - p_u\|_2$ is at most 1.

Unit Disk Graphs are growth-bounded with a bounding function $f(r) \in O(r^2)$, which is easily proved as follows: consider any node v and any independent set $I \subseteq N_r(v)$, for a $r \geq 0$. The corresponding points of two (independent) nodes in I must have distance greater than 1. Thus each point corresponding to a node in I exclusively occupies a disk of radius $1/2$ with an area of $(1/2)^2\pi$. As all these disks must lie inside a circle of radius $r + 1/2$, it follows that $|I| \leq \frac{(r+1/2)^2\pi}{(1/2)^2\pi} = 4r^2 + 4r + 1$. Furthermore, most other graph classes used to model wireless ad-hoc networks such as Quasi Unit Disk Graphs, Cover Area Graphs and other intersection graphs are growth-bounded [14,17].

To end this section, we introduce two properties of optimal connected dominating sets in growth-bounded graphs, which we use in our approach.

Lemma 1. *For any growth-bounded graph $G = (V, E)$ with bounding function f, there is a polynomial $p(r) \leq 3 \cdot f(r)$ such that for any $v \in V$, it holds $|\mathcal{C}(N_r(v))| \leq p(r)$, $\forall r > 0$.*

Proof. Let $f(r)$ be the polynomial bounding function of the growth-bounded graph G. Consider a maximal independent set I of $N_r(v)$ (for a fixed $r \geq 0$) and set $Q := I$. Let $k = |I|$ be the number of components of $G[I]$. We proceed by induction over k. If $k = 1$, $G[Q]$ is connected and the claim follows. Otherwise, since I is also a dominating set of $N_r(v)$, we can find two connected components $A, B \subseteq Q$ such that $d(A, B) \leq 3$. By adding to Q the vertices of a shortest path between them, we decrease the number of components by at least one and we increase $|Q|$ by at most two. We proceed inductively until Q induces a connected graph. Since $k = |I|$, we get $|\mathcal{C}(N_r(v))| \leq |Q| \leq |I| + 2|I| \leq 3f(r)$. As this holds for every $r > 0$, the claim is proved. $\qed$

Lemma 2. *Let $G = (V, E)$ be a growth-bounded graph with bounding function f, and choose a $S \subseteq V$ that induces a connected subgraph. Then for any MIS M of S, it holds: $|M| \leq f(1) \cdot |\mathcal{C}(S)|$.*

Proof. As $\mathcal{C}(S)$ dominates S, each node in M must have a neighbor in $\mathcal{C}(S)$ (or be in $\mathcal{C}(S)$ itself). But by definition, at most $f(1)$ nodes in M can have the same neighbor in $\mathcal{C}(S)$, so the claim follows. $\qed$

3   Finding a Small Connected Dominating Set

In the following, we describe a (sequential) procedure to construct for each $\varepsilon > 0$ a connected dominating set of size at most $(1 + \varepsilon)$ times the minimum. This procedure can be executed efficiently in a centralized way, and thus leads to a PTAS. Moreover, in Section 4, we show that the same procedure can be implemented efficiently in a distributed way, using the same technique as in [13].

The CDS is constructed by computing optimal connected dominating sets for small parts of the graph, and taking the union of these CDSs. We will construct the small CDSs such that their union leads to a connected set, as required. Each small CDS is an optimal solution of a small cluster specified as follows: choose any node $v \in V$, and consider the r-neighborhood of v for increasing values of $r = 0, 1, 2, \ldots$ until we find a large enough r^* such that

$$\left| \mathsf{MaxIS}\big(N(\Gamma_{r^*}(v))\backslash\Gamma_{r^*}(v)\big) \right| \leq \varepsilon \cdot \left| \mathsf{MaxIS}\big(\Gamma_{r^*}(v)\big) \right| \tag{1}$$

holds. We call this operation an *expansion* of v. As we show in Lemma 4 below, r^* is bounded by a function in $O(1/\varepsilon \cdot \log(1/\varepsilon))$ depending solely on ε for any class of growth-bounded graphs.

Our algorithm for finding a CDS for G proceeds as follows: starting with an empty set D, it chooses any node $v_1 \in V$ of G, finds a corresponding r_1^* such that Inequality 1 holds, and adds $\mathcal{C}(\Gamma_{r_1^*+4}(v_1, V))$ to the current solution D. After that, it removes all nodes in $\Gamma_{r_1^*+2}(v_1, V)$ from the graph G and we denote the set of remaining nodes by V'. Note that here we do not remove all nodes that are dominated by the current solution D from the graph. This is an important difference to the approach in [16]. As we will show, this modification guarantees that the final solution will be a *connected* dominating set.

In the reduced graph, the algorithm chooses another node $v_2 \in V'$, considers growing neighborhoods of v_2, until a r_2^* satisfying Inequality 1 is found. Note that the bounding function f of the original graph is still valid for the reduced graph, because any set which is independent in the reduced graph is also independent in the original graph. Furthermore, recall that $\mathcal{C}(\Gamma_{r_2^*}(v_2, V'))$ and $\mathcal{C}(\Gamma_{r_2^*+4}(v_2, V'))$ may contain some nodes from the original graph G as *dominators* which are outside V' because they are already dominated. Then, $\mathcal{C}(\Gamma_{r_2^*+4}(v_2, V'))$ is added to the current solution D, and all nodes in $\Gamma_{r_2^*+2}(v_2, V')$ are removed from the graph, just as before. Then, this procedure is repeated until all nodes have been removed from the graph.

The algorithm is described formally in Algorithm 1. Since the set of remaining nodes should always be clear from the context, we omit the second argument of $\Gamma(\cdot, \cdot)$ in the rest of the paper.

For proving Lemma 4, we need the following fact.

Lemma 3. *Consider any class of growth-bounded graphs with bounding function f, and a graph $G = (V, E)$ of this class. Then for any $r \geq 1$ and $v \in V$ it holds:*

$$\left| \mathsf{MaxIS}\big(N(\Gamma_r(v))\backslash\Gamma_r(v)\big) \right| \leq f(2) \cdot \left| \mathsf{MaxIS}\big(\Gamma_r(v)\backslash\Gamma_{r-1}(v)\big) \right|$$

Proof. Clearly, each node in $\mathsf{MaxIS}(N(\Gamma_r(v))\backslash\Gamma_r(v))$ has a neighbor in $\Gamma_r(v)$ in G. As $\mathsf{MaxIS}(\Gamma_r(v)\backslash\Gamma_{r-1}(v))$ is a dominating set of $\Gamma_r(v)\backslash\Gamma_{r-1}(v)$, each node in $\mathsf{MaxIS}(N(\Gamma_r(v))\backslash\Gamma_r(v))$ has a node in $\mathsf{MaxIS}(\Gamma_r(v)\backslash\Gamma_{r-1}(v))$ within distance at most two. However, the number of nodes in $\mathsf{MaxIS}(N(\Gamma_r(v)\backslash\Gamma_r(v))$ that lie

---

Algorithm 1. Computes a $(1 + O(\varepsilon))$-approximate MCDS

Input: A growth-bounded graph $G = (V, E)$, $\varepsilon > 0$
1 $D := \{\}$
2 { For analysis: $i := 1$ }
3 **while** $V \neq \{\}$ **do**
4   Choose any $v \in V$
5      $r := 0$
6   **while** $\big|MaxIS(N(\Gamma_r(v))\backslash\Gamma_r(v))\big| > \varepsilon \cdot \big|MaxIS(\Gamma_r(v))\big|$ **do**
7          $r := r + 1$
8      $D := D \cup \mathcal{C}(\Gamma_{r+4}(v))$ [2] (see footnote)
9      $V := V\backslash\Gamma_{r+2}(v)$
10   { For analysis: $S_i := \Gamma_r(v)$; $T_i := \Gamma_{r+4}(v)$; $i := i+1$ }
11 **return** D

---

within two hops of the same node in $MaxIS(\Gamma_r(v)\backslash\Gamma_{r-1}(v))$ can be at most $f(2)$, as otherwise these nodes could not be mutually independent.   □

Lemma 4. *Consider any class of growth-bounded graphs with bounding function f. Then, for any $\varepsilon > 0$, there is a $R_f^*(\varepsilon) = O\big(1/\varepsilon \cdot \log(1/\varepsilon)\big)$ such that for each graph G of this class, and each node v of G, it holds*

$$\big|MaxIS\big(N(\Gamma_{r^*}(v))\backslash\Gamma_{r^*}(v)\big)\big| \leq \varepsilon \cdot \big|MaxIS(\Gamma_{r^*}(v))\big|$$

for some $r^* \leq R_f^*$.

Proof. Fix an $\varepsilon > 0$ and assume in contradiction that no such R_f^* exists. This implies that for arbitrarily large values r', there is a graph in the class such that for some node v, $|MaxIS(N(\Gamma_{r'}(v))\backslash\Gamma_{r'}(v))| > \varepsilon \cdot |MaxIS(\Gamma_{r'}(v))|$ holds for all $0 \leq r \leq r'$. Consider such a value $r' \geq 2$. From

$$\big|MaxIS\big(N(\Gamma_{r'}(v))\backslash\Gamma_r'(v)\big)\big| > \varepsilon \cdot \big|MaxIS(\Gamma_{r'}(v))\big| \geq \varepsilon \cdot \big|MaxIS\big(\Gamma_{r'-2}(v)\big)\big|$$

and Lemma 3, we have

$$\big|MaxIS\big(\Gamma_{r'}(v)\backslash\Gamma_{r'-1}(v)\big)\big| > \bar{\varepsilon} \cdot \big|MaxIS\big(\Gamma_{r'-2}(v)\big)\big|, \quad \text{for } \bar{\varepsilon} = \varepsilon/f(2).$$

Hence, for all $r: 2 \leq r \leq r'$ we have

$$\big|MaxIS\big(\Gamma_r(v)\big)\big| \geq \big|MaxIS\big(\Gamma_r(v)\backslash\Gamma_{r-1}(v)\big)\big| + \big|MaxIS\big(\Gamma_{r-2}(v)\big)\big|$$
$$> (1 + \bar{\varepsilon}) \cdot \big|MaxIS\big(\Gamma_{r-2}(v)\big)\big|.$$

Assume for the moment that r' is an even number. Then we have

$$\big|MaxIS\big(\Gamma_{r'}(v)\big)\big| > (1 + \bar{\varepsilon}) \cdot \big|MaxIS\big(\Gamma_{r'-2}(v)\big)\big| > (1 + \bar{\varepsilon})^2 \cdot \big|MaxIS\big(\Gamma_{r'-4}(v)\big)\big|$$
$$> \ldots > (1 + \bar{\varepsilon})^{\frac{r'}{2}} \cdot \big|MaxIS\big(\Gamma_0(v)\big)\big| = (1 + \bar{\varepsilon})^{\frac{r'}{2}}.$$

---

[2] Recall that $\mathcal{C}(\Gamma_{r+4}(v))$ represents the optimal CDS of $\Gamma_{r+4}(v)$.

Since $|\mathsf{MaxIS}(\Gamma_{r'}(v))|$ grows only polynomially in r' (Lemma 1), but the term $(1+\bar{\varepsilon})^{\frac{r'}{2}}$ grows exponentially in r' (note that $(1+\bar{\varepsilon})^{1/2} > 1$), the above inequality will be violated for some large enough r', which is a contradiction. If r' is odd, the same reasoning can be applied.

The claimed bound on R_f^* follows easily from the inequality $(1+\bar{\varepsilon})^{1/\bar{\varepsilon}} > e - 1$, for $\bar{\varepsilon}$ small enough. $\qquad\square$

It is clear that Algorithm 1 terminates, and that D then contains a dominating set, because only dominated nodes are removed from the graph.

Let $S_1, S_2, \ldots, S_k$ and $T_1, T_2, \ldots, T_k$ be the sets $\Gamma_r(v_i)$ and $\Gamma_{r+4}(v_i)$ respectively as chosen in each iteration of the outer while-loop of Algorithm 1. We now show that the computed solution D, which consists of the union of the $\mathcal{C}(T_i)$, forms a *connected* subgraph of G.

Lemma 5. *The union* $D := \bigcup_{i=1}^k \mathcal{C}(T_i)$ *induces a connected subgraph of* G.

Proof. First, we show that any two nodes $a, b \in D$ with distance $d(a, b) = 2$ in G are part of the same connected component in (the subgraph induced by) D. To that end, consider any node $w \in V$ and its neighbors $N(w)$ (see the left part of Figure 1). Let s be the *first* node among $N(w)$ (including w) that is removed in line 9 of Algorithm 1. When s is removed in the i-th iteration, it holds that $s \in \Gamma_{r_i^*+2}(v_i)$ and so all nodes in $N(w)$ are in $T_i = \Gamma_{r_i^*+4}(v_i)$, whereby $v_1, v_2, \ldots$ represent the centers of the expansions. Hence they are dominated by $\mathcal{C}(\Gamma_{r_i^*+4}(v_i))$. Therefore, any pair of nodes in $D \cap N(w)$ is connected by a path of length at most $p(R^*) + 1$ hops consisting only of nodes in D (recall p from Lemma 1).

Second, consider any pair $u, v \notin D$ of nodes adjacent in G (see the right part of Figure 1). We show that there must exist nodes $u' \in N(u) \cap D$ and $v' \in N(v) \cap D$ such that u' and v' are connected by a path of length at most $p(R^*) - 1$ consisting only of nodes in D. To that end, assume w.l.o.g. that u is removed (line 9) before or at the same time as v. When u is removed in the j-th iteration (i.e., $u \in \Gamma_{r_j^*+2}(v_j)$), also v is dominated by $\mathcal{C}(\Gamma_{r_j^*+4}(v_j))$. The second claim follows. Combining this with the first claim, we have that any two nodes $a, b \in D$ with distance $d(a, b) = 3$ in G are connected by a path of length at most $3p(R^*) + 1$ consisting only of nodes in D.

These two facts together imply that D induces a connected subgraph. Indeed, if D were disconnected, then the shortest path in G between two closest components of D would consist solely of nodes in $V \backslash D$. However, such a path cannot be of length two or three by the above facts, and not longer either, because then D would not dominate all the nodes. $\qquad\square$

Now that we have shown that the set D computed by Algorithm 1 is a connected dominating set, we prove that its size is at most $1 + \varepsilon$ times larger than the optimum. To this end, we need two lemmas.

Lemma 6. *Let* $\varepsilon > 0$ *and* $r > 0$ *be such that the following inequality is fulfilled:*

$$\left| \mathsf{MaxIS}\big(N(\Gamma_r(v)) \backslash \Gamma_r(v)\big) \right| \leq \varepsilon \cdot \left| \mathsf{MaxIS}\big(\Gamma_r(v)\big) \right|.$$

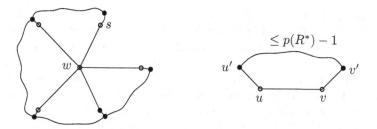

Fig. 1. Illustration of Lemma 5. The filled dots represent nodes that have already joined D. The other nodes are represented by empty dots.

Then, for $\varepsilon' := \varepsilon \cdot (3f(4) + 3) \cdot f(1)$, it follows:

$$|\mathcal{C}(\Gamma_{r+4}(v))| \leq (1 + \varepsilon') \cdot |\mathcal{C}(\Gamma_r(v))|.$$

Proof. We show how to extend $\mathcal{C}(\Gamma_r(v))$ to a connected dominating set of $\Gamma_{r+4}(v)$ by adding only relatively few nodes so that the claim follows. Let M be a maximal independent set of $\Gamma_{r+1}(v) \backslash \Gamma_r(v)$. Any node in $\Gamma_{r+4}(v)$ lies within 3 hops of some node in $\Gamma_{r+1}(v) \backslash \Gamma_r(v)$, and thus within 4 hops of some node in M. Thus, all nodes in $\Gamma_{r+4}(v)$ are dominated if we add to our solution the set $\mathcal{C}(\Gamma_4(w))$ for each $w \in M$ (note that $|\mathcal{C}(\Gamma_4(w))| \leq p(4) \leq 3f(4)$). In order to connect these sets to $\mathcal{C}(\Gamma_r(v))$, we need to add at most 3 additional nodes for each $w \in M$. Thus in total, we obtain a connected dominating set of $\Gamma_{r+4}(v)$ of size at most $|\mathcal{C}(\Gamma_r(v))| + (3f(4) + 3) \cdot |\mathsf{MaxIS}(\Gamma_{r+1}(v) \backslash \Gamma_r(v))| \leq |\mathcal{C}(\Gamma_r(v))| + (3f(4)+3) \cdot |\mathsf{MaxIS}(N(\Gamma_r(v)) \backslash \Gamma_r(v))| \leq |\mathcal{C}(\Gamma_r(v))| + \varepsilon \cdot (3f(4)+3) \cdot |\mathsf{MaxIS}(\Gamma_r(v))| \leq |\mathcal{C}(\Gamma_r(v))| + \varepsilon \cdot (3f(4) + 3) \cdot f(1) \cdot |\mathcal{C}(\Gamma_r(v))|$, using Lemma 2 for the last inequality. □

Let V^* be the set of nodes chosen as centers v for the growing neighborhoods in the algorithm. As for each $v \in V^*$, $\Gamma_{r+2}(v)$ is removed from the graph before choosing a new node, the collection $\{S_1, S_2, \ldots, S_k\}$ consists of 2-separated sets. We have the following lower bound for the size of an optimal CDS for G.

Lemma 7. *Let $S_1, S_2, \ldots, S_k$ be the collection of 2-separated sets in $G = (V, E)$ computed by Algorithm 1 and $\varepsilon'' := 2f(1)\varepsilon$. Then,*

$$(1 + \varepsilon'') \cdot |\mathcal{C}(V)| \geq \left| \bigcup_{i=1}^{k} \mathcal{C}(S_i) \right|.$$

Proof. Since the S_i are 2-separated, the sets $N(S_i)$ are disjoint, so the sets $\mathcal{C}(V) \cap N(S_i)$ are disjoint, too. Furthermore, as $\mathcal{C}(V)$ must dominate all nodes of G, including those in S_i, the set $\mathcal{C}(V) \cap N(S_i)$ must dominate all nodes in S_i. To complete the proof, we now show that $|\mathcal{C}(S_i)| \leq (1 + \varepsilon'') \cdot |\mathcal{C}(V) \cap N(S_i)|$ for all i, and thus, $|\mathcal{C}(V)| \geq \sum_{i=1}^{k} |\mathcal{C}(V) \cap N(S_i)| \geq \frac{1}{1+\varepsilon''} \cdot \sum_{i=1}^{k} |\mathcal{C}(S_i)|$.

To that end, we add some nodes to $\mathcal{C}(V) \cap N(S_i)$ in order to obtain a *connected* set which dominates S_i. Let x be the number of connected components in $\mathcal{C}(V) \cap$

$N(S_i)$ and suppose that $x \geq 2$, otherwise $\mathcal{C}(V) \cap N(S_i)$ is connected and hence $|\mathcal{C}(S_i)| \leq (1 + \varepsilon'') \cdot |\mathcal{C}(V) \cap N(S_i)|$ is trivial. Then the individual connected components of $\mathcal{C}(V) \cap N(S_i)$ can be connected by adding at most $2x$ nodes (see the proof of Lemma 1). Note that each connected component of $\mathcal{C}(V) \cap N(S_i)$ must contain one node from $N(\Gamma_r(v)) \backslash \Gamma_r(v)$ to ensure the global connectivity of the solution. Thus, by choosing one such node for each connected component of $\mathcal{C}(V) \cap N(S_i)$, we obtain an independent set of size x. Therefore, we have $x \leq |\mathsf{MaxIS}(N(S_i) \backslash S_i)|$. By construction of the S_i, $|\mathsf{MaxIS}(N(S_i) \backslash S_i)| \leq \varepsilon \cdot |\mathsf{MaxIS}(S_i)|$, and by Lemma 2, $|\mathcal{C}(S_i)| \geq |\mathsf{MaxIS}(S_i)| / f(1)$.

So we can obtain a connected set that dominates S_i by adding to $\mathcal{C}(V) \cap N(S_i)$ at most $2\varepsilon f(1) \cdot |\mathcal{C}(S_i)|$ nodes. The claim now follows by choosing $\varepsilon'' = 2f(1)\varepsilon$.   □

Theorem 1. *The set D computed by Algorithm 1 is a $\left(1 + O(\varepsilon)\right)$-approximation for the connected dominating set problem.*

Proof. Let $\{S_1, S_2, \ldots, S_k\}$ and $\{T_1, T_2, \ldots, T_k\}$ be as defined in Algorithm 1. By Lemma 6, it holds that $|\mathcal{C}(T_i)| \leq (1 + \varepsilon') \cdot |\mathcal{C}(S_i)|$ for all $i = 1, \ldots, k$, and $D = \bigcup_{i=1}^{k} \mathcal{C}(T_i)$. Hence we have

$$|D| = \left| \bigcup_{i=1}^{k} \mathcal{C}(T_i) \right| \leq \sum_{i=1}^{k} |\mathcal{C}(T_i)| \leq (1 + \varepsilon') \cdot \sum_{i=1}^{k} |\mathcal{C}(S_i)|$$

$$= (1 + \varepsilon') \cdot \left| \bigcup_{i=1}^{k} \mathcal{C}(S_i) \right| \leq (1 + \varepsilon')(1 + \varepsilon'') \cdot |\mathcal{C}(V)|,$$

where the last inequality follows from Lemma 7.   □

We now shortly discuss how Algorithm 1 can be implemented in a centralized fashion to obtain a PTAS. Most steps of the algorithm can be trivially computed efficiently. The crucial part is the computation of the *maximum* independent sets $\mathsf{MaxIS}(N(\Gamma_r(v)) \backslash \Gamma_r(v))$ and $\mathsf{MaxIS}(\Gamma_r(v))$, and of $\mathcal{C}(\Gamma_{r^*+4}(v))$. Note that r is bounded by the constant R^* for any fixed ε, so from the growth-bounded property we know that the size of the MaxIS is bounded by a constant. Thus, by enumerating all node subsets of cardinality at most $f(r)$, and selecting the largest of those which is both independent and maximal, a *maximum* independent set is found in polynomial time. Since the considered subsets have only constant size, independence and maximality of the subsets can be checked in constant time. The same arguments apply to the computation of $\mathcal{C}(\Gamma_{r^*+4}(v))$, due to Lemma 1. Hence, Algorithm 1 has polynomial time complexity for any fixed $\varepsilon > 0$, but exponential time complexity in $1/\varepsilon$.

4   A Fast Distributed Approximation Scheme

The main goal of this paper is to provide a *fast* distributed algorithm that computes a $(1 + \varepsilon)$-approximation for the MCDS problem in growth-bounded graphs. The algorithm we describe here is an adaptation from [13], adjusted to computing a CDS instead of a DS.

A naive distributed implementation would be that in each round all nodes which have the highest ID within their $2R^* + 9$-hop neighborhood are expanded concurrently. However, this approach requires a linear number of rounds in the worst case, because there can be a linear waiting chain of nodes. The observation that every expansion affects only neighbors within a small radius leads to a more efficient algorithm: expansions of nodes with sufficient mutual distances can be scheduled concurrently. Roughly, this can be achieved by computing a MIS of G, and then coloring this MIS with few colors such that two nodes with the same color are distant enough. The coloring is achieved using a *clustergraph* $\bar{G}$ of radius $c = c(\varepsilon)$ with centers $W \subseteq V$: $\bar{G} = (\bar{V}, \bar{E})$, where $\bar{V} := W$ and for every $u, v \in W$ $(u, v) \in \bar{E}$ if and only if $d(u, v) \leq c$. Note that if W is an independent set and G is growth-bounded, $\bar{G}$ has a maximum degree of $\Delta_{\bar{G}} = O(f(c))$. Hence using a MIS of G to construct a clustergraph $\bar{G}$ of radius c, $\bar{G}$ can be colored with $O(\Delta_{\bar{G}}^2)$ colors in $O(c \cdot \log^* n)$ time [15]. Note that a communication round in the cluster graph costs $O(c)$ rounds in the original graph. The coloring is then used to schedule the expansion of neighbors of MIS-nodes. We choose $c = 2R^* + 11$ for reasons becoming apparent in the proof of Lemma 9. A more detailed description is given in Algorithm 2.

Algorithm 2. Computes a $(1 + O(\varepsilon))$-approximate MCDS distributively

---

Input: A growth-bounded graph G, $\varepsilon > 0$, R^* (according to Lemma 4)
1 Compute a MIS I of G;
2 Construct the clustergraph $\bar{G}$ of I using radius $2R^* + 11$;
3 Color $\bar{G}$ with $\gamma = O(\Delta_{\bar{G}}^2)$ colors;
4 $D := \{\}$;
5 **for** $k = 1$ *to* γ **do**
6   **for every** $v \in I$ *with color* k **do** in parallel
7   **if** $N(v) \cap V \neq \{\}$ **then**
8   For some $u \in N(v) \cap V$, find the smallest r^* such that $|\mathsf{MaxIS}(N(\Gamma_{r^*}(v)) \setminus \Gamma_{r^*}(v))| \leq \varepsilon \cdot |\mathsf{MaxIS}(\Gamma_{r^*}(v))|$;
9   Compute $\mathcal{C}(\Gamma_{r^*+4}(u))$;
10   Inform $\Gamma_{r^*+4}(u)$ about r^* and $\mathcal{C}(\Gamma_{r^*+4}(u))$;
11             $D := D \cup \mathcal{C}(\Gamma_{r^*+4}(u))$;
12             $V := V \setminus \Gamma_{r^*+2}(u)$;
13 **return** D

Lemma 8. *Algorithm 2 terminates in* $O(T_{\mathsf{MIS}} + 1/\varepsilon^{O(1)} \cdot \log^* n)$ *time.*

Proof. Computing a MIS of G takes time T_{MIS}. Then, the clustergraph can be constructed in constant time, as its edges only span at most distance $2R^* + 11 = O(1/\varepsilon \cdot \log(1/\varepsilon))$. Furthermore, $\bar{G}$ can be colored with $O(\Delta_{\bar{G}}^2) = O(f^2(R^*))$ colors in $O(R^* \cdot \log^* n)$ time using the well-known algorithm of [15].

The outer for-loop is executed $O(\Delta_{\bar{G}}^2)$ times. Inside the for-loop, the number of different MaxIS that each node u (as in line 8) must compute is $2r^* = O(R^*) =$

$O(1/\varepsilon \cdot \log(1/\varepsilon))$. For computing each MaxIS and MCDS for a neighborhood of radius r, u collects all information about $\Gamma_r(u)$ and then computes the set locally. As $r \leq R^*$, all steps in lines 8 to 12 can be executed in $O(1/\varepsilon \cdot \log(1/\varepsilon))$ time. □

Lemma 9. *The set D computed by Algorithm 2 is a $(1 + O(\varepsilon))$-approximate minimum connected dominating set.*

Proof. By construction, any two nodes that are concurrently used for an expansion have distance at least $2R^* + 9$, because they are respective neighbors of two MIS nodes of distance at least $2R^* + 11$. The radius used by either expansion is at most R^*, and since each expansion only involves the nodes within a radius of at most $R^* + 5$, all concurrent expansions would have the same result if they were executed sequentially. Therefore, there exists an execution of the sequential Algorithm 1 which computes the same set D as Algorithm 2. It follows that Algorithm 2 achieves the same approximation ratio as Algorithm 1. □

These two lemmas lead to our main theorem.

Theorem 2. *For any $\varepsilon > 0$, Algorithm 2 computes a $(1 + O(\varepsilon))$-approximate minimum connected dominating set in $O(T_{MIS} + 1/\varepsilon^{O(1)} \cdot \log^* n)$ time.*

5   Well-Connectedness

The connected dominating set computed by our Algorithm 1 is not only a $(1 + \varepsilon)$-approximation of a minimum CDS, but has additional properties which are desirable for its usage as a backbone in a wireless network. Let $G'(V', E')$ be the graph induced by the CDS_A of the (growth-bounded) graph $G = (V, E)$ computed by Algorithm 1. Then, it holds for any $\varepsilon > 0$:

1. The backbone graph G' has maximum degree $O(1/\varepsilon^{O(1)})$, and therefore it has only $O(1/\varepsilon^{O(1)} \cdot |V'|)$ edges;
2. Using G' as a routing backbone guarantees a $O(1/\varepsilon^{O(1)})$ stretch.

We assume in the following that source s and destination d of a routing request are both members of the CDS. If this is not true for either or both of them, then we can easily choose a neighbor inside the CDS as a representative. This will add at most two hops to the routing path, so if the stretch is low for any pair s, d inside the CDS, the stretch of any pair s, d inside G is also low.

To make the second statement precise, define $\lambda := \max_{u,v \in V'} \frac{d_{G'}(u,v)}{d_G(u,v)}$ as the hop-stretch of G'. Furthermore, if G is a UDG and if $D_G(u,v)$ denotes the geometric length of a shortest path in G, then the *geometric stretch* of G' is $\mu := \max_{u,v \in V'} \frac{D_{G'}(u,v)}{D_G(u,v)}$.

Lemma 10. *Let CDS_A be the CDS computed by Algorithm 1. The subgraph G' of G induced by the nodes in CDS_A has maximum degree $O(1/\varepsilon^{O(1)})$.*

Proof. First, note that each partial CDS T_i computed by the algorithm covers a subgraph with diameter $O(R^*)$, so according to Lemma 1, $|T_i|$ and therefore its degree is at most $O(f(R^*))$. Second, each node u of G can only be contained in $O(f(R^*))$ many different T_i, because any expansion that leads to some T_j containing u must have as its center a MIS-node in distance at most $R^* + 2 = O(1/\varepsilon \cdot \log(1/\varepsilon))$ from u (and there are only $O(f(R^*))$ MIS nodes in distance $O(R^*)$ from v). □

Lemma 11. *For any $\varepsilon > 0$, the hop-stretch γ of G' is $O(f(R^*)) = O(1/\varepsilon^{O(1)})$. Further, if G is a UDG, then the geometric stretch λ of G' is also $O(1/\varepsilon^{O(1)})$.*

Proof. Consider any source s and destination d in V'. Let $\mathcal{P} = \langle p_1, p_2, \ldots, p_k \rangle$ be the sequence of nodes in a shortest path in G from $s = p_1$ to $d = p_k$. We define a new path $\mathcal{Q}$ going through nodes $\langle q_1, q_2, \ldots, q_k \rangle$ as follows: $q_i := p_i$ if $p_i \in D$. Otherwise, let q_i be any node in $D \cap N(p_i)$ (such a node exists because D is dominating). Note that $q_i \in D, \forall i : 1 \leq i \leq k$. From the proof of Lemma 5, we can conclude that between any pair $(q_i, q_{i+1}), \forall i : 1 \leq i \leq k - 1$, there is a path in D of length at most $3p(R^*) + 1 = O(f(R^*))$. Hence there is a path $\mathcal{Q}$ of length $\leq k(3p(R^*) + 1) = k \cdot O(f(R^*)) = k \cdot O(1/\varepsilon^{O(1)})$ from q_1 to q_k solely consisting of nodes in D.

For the geometric stretch in UDGs, note that in the path $\mathcal{R} := \langle t_1 = s, t_2, \ldots t_k = d \rangle$ of shortest geometric length, the outer two of any three consecutive nodes t_i, t_{i+1}, t_{i+2} must have distance at least 1: $\|t_i, t_{i+2}\|_2 \geq 1, \forall i \in \{1, \ldots, k - 2\}$. So $\mathcal{R}$ with k hops has length at least $(k - 1)/2$. On the other hand, the path with the fewest number of hops (at most k) has length at most k. Since we have shown just before that G' includes a path with hop-stretch $\gamma = O(1/\varepsilon^{O(1)})$ between any pair of nodes, it follows that the geometric stretch λ of G' is at most $2k \cdot \gamma/(k - 1) \leq 4\gamma = O(1/\varepsilon^{O(1)}), \forall k \geq 2$. For $k = 1$, the path with fewest hops has length at most $\gamma = O(1/\varepsilon^{O(1)})$, which completes the claim. □

Summarizing, we have the following.

Theorem 3. *The CDS computed by Algorithm 1/Algorithm 2 is well-connected.*

6 Conclusion

The distributed algorithm we presented computes a $(1 + \varepsilon)$-approximate *Minimum Connected Dominating Set* with $O(1/\varepsilon^{O(1)})$ degree and stretch in time $O(T_{MIS} + 1/\varepsilon^{O(1)} \cdot \log^* n)$ for growth-bounded graphs. Hence, a faster algorithm for computing a MIS in growth-bounded graphs would directly improve the running time of our solution for fixed $\varepsilon > 0$. Clearly, any approach that requires a MIS for constructing a CDS must have a running time of $\Omega(\log^* n)$ because of the lower bound by Linial [15]. It is an intriguing open question whether a constant approximation for the MCDS problem can be achieved in $o(\log^* n)$ or even constant time using a fundamentally different approach.

Acknowledgements. We thank an anonymous reviewer for pointing out an error in our original analysis of the algorithm's approximation ratio.

References

1. Alzoubi, K.M.: Connected dominating set and its induced position-less sparse spanner for mobile ad hoc networks. In: Proceedings of the 8th IEEE Intern. Symposium on Computers and Communications (ISCC), pp. 209–216. IEEE Computer Society Press, Los Alamitos (2003)
2. Blum, J., Ding, M., Thaeler, A., Cheng, X.: Connected dominating set in sensor networks and MANETs. In: Handbook of Combinatorial Optimization, pp. 329–369. Kluwer Academic Publishers, Dordrecht (2004)
3. Cheng, X., Huang, X., Li, D., Wu, W., Du, D.-Z.: A polynomial-time approximation scheme for the minimum-connected dominating set in ad hoc wireless networks. Networks 42(4), 202–208 (2003)
4. Clark, B.N., Colbourn, C.J., Johnson, D.S.: Unit disk graphs. Discrete Mathematics 86(1-3), 165–177 (1990)
5. Czygrinow, A., Hańćkowiak, M.: Distributed approximation algorithms in unit-disk graphs. In: Dolev, S. (ed.) DISC 2006. LNCS, vol. 4167, pp. 385–398. Springer, Heidelberg (2006)
6. Das, B., Sivakumar, R., Bharghavan, V.: Routing in ad-hoc networks using a virtual backbone. In: 6th Intern. Conference on Computer Communications and Networks (IC3N '97), pp. 1–20 (1997)
7. Dubhashi, D.P., Mei, A., Panconesi, A., Radhakrishnan, J., Srinivasan, A.: Fast distributed algorithms for (weakly) connected dominating sets and linear-size skeletons. Journal of Computer and System Sciences 71(4), 467–479 (2005)
8. Feige, U.: A threshold of ln n for approximating set cover. In: Proceedings of the 28th Annual ACM Symposium on Theory of Computing (STOC), pp. 314–318. ACM Press, New York (1996)
9. Gfeller, B., Vicari, E.: A Randomized Distributed Algorithm for the Maximal Independent Set Problem in Growth-Bounded Graphs. In: Proceedings of the 26th Annual ACM SIGACT-SIGOPS Symposium on Principles of Distributed Computing (PODC). ACM Press, New York (2007)
10. Hunt III, H.B., Marathe, M.V., Radhakrishnan, V., Ravi, S.S., Rosenkrantz, D.J., Stearns, R.E.: NC-approximation schemes for NP- and PSPACE-hard problems for geometric graphs. J. Algorithms 26(2), 238–274 (1998)
11. Junker, H., Stäger, M., Tröster, G., Blättler, D., Salama, O.: Wireless networks in context aware wearable systems. In: Proceedings of the 1st European Workshop on Wireless Sensor Networks (EWSN), pp. 37–40 (2004)
12. Kuhn, F., Moscibroda, T., Nieberg, T., Wattenhofer, R.: Fast deterministic distributed maximal independent set computation on growth-bounded graphs. In: Fraigniaud, P. (ed.) DISC 2005. LNCS, vol. 3724, pp. 273–287. Springer, Heidelberg (2005)
13. Kuhn, F., Moscibroda, T., Nieberg, T., Wattenhofer, R.: Local Approximation Schemes for Ad Hoc and Sensor Networks. In: 3rd ACM Joint Workshop on Foundations of Mobile Computing (DIALM-POMC). ACM Press, New York (2005)
14. Kuhn, F., Wattenhofer, R., Zollinger, A.: Ad-Hoc Networks Beyond Unit Disk Graphs. In: 1st ACM Joint Workshop on Foundations of Mobile Computing (DIALM-POMC). ACM Press, New York (2003)
15. Linial, N.: Locality in distributed graph algorithms. SIAM Journal on Computing 21(1), 193–201 (1992)
16. Nieberg, T., Hurink, J.L.: A PTAS for the minimum dominating set problem in unit disk graphs. Memorandum 1732, University of Twente, Enschede (2004)

17. Nieberg, T., Hurink, J.L.: Wireless communication graphs. In: Proceedings of DEST Intern. Workshop on Signal Processing for Sensor Networks, ISSNIP'04, pp. 367–372 (2004)
18. Panchard, J., Rao, S., Prabhakar, T., Jamadagni, H., Hubaux, J.-P.: COMMON-Sense Net: Improved Water Management for Resource-Poor Farmers via Sensor Networks. In: Intern. Conference on Communication and Information Technologies and Development ICTD (2006)
19. Parthasarathy, S., Gandhi, R.: Distributed algorithms for coloring and domination in wireless ad hoc networks. In: Foundations of Software Technology and Theoretical Computer Science (FSTTCS), pp. 447–459 (2004)

Coordinating Concurrent Transmissions: A Constant-Factor Approximation of Maximum-Weight Independent Set in Local Conflict Graphs

Petteri Kaski[1], Aleksi Penttinen[2], and Jukka Suomela[1]

[1] Helsinki Institute for Information Technology HIIT
Department of Computer Science, University of Helsinki
P.O. Box 68, FI-00014 University of Helsinki, Finland
`firstname.lastname@cs.helsinki.fi`
[2] Networking Laboratory, Helsinki University of Technology
P.O. Box 3000, FI-02015 TKK, Finland
`firstname.lastname@tkk.fi`

Abstract. We study the algorithmic problem of coordinating transmissions in a wireless network where radio interference constrains concurrent transmissions on wireless links. We focus on pairwise conflicts between the links; these can be described as a conflict graph. Associated with the conflict graph are two fundamental network coordination tasks: (a) finding a nonconflicting set of links with the maximum total weight, and (b) finding a link schedule with the minimum total length. Our work shows that two assumptions on the geometric structure of conflict graphs suffice to achieve polynomial-time constant-factor approximations: (i) bounded density of devices, and (ii) bounded range of interference. We also show that these assumptions are not sufficient to obtain a polynomial-time approximation scheme for either coordination task.

Keywords: Geometric graphs, maximum-weight independent set, radio interference.

1 Introduction

A fundamental challenge in wireless networking is the shared transmission medium, which in many cases prevents concurrent transmissions due to radio interference. This brings forth the algorithmic problem of coordinating the transmissions so that performance loss due to interference does not occur. In this work, we investigate the polynomial-time approximability of network coordination within the following framework.

Interference in Wireless Networks. A wireless network consists of devices which communicate with each other by radio transmissions. We study unicast networks where each radio transmission has one designated recipient, see Fig. 1 for an illustration.

E. Kranakis and J. Opatrny (Eds.): ADHOC-NOW 2007, LNCS 4686, pp. 74–86, 2007.

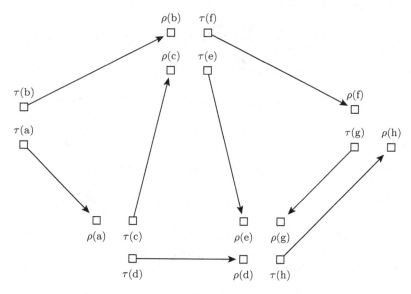

Fig. 1. A wireless network. Wireless communication links are marked with arrows. Devices are marked with boxes, τ denotes a transmitter and ρ denotes a receiver. For clarity, each device in this illustration takes part in only one transmission.

A radio transmission may interfere with other transmissions. We focus on systems where the radio interference is dominated by the near-far effect: radio reception from a distant transmitter may be blocked by other transmitters which are much closer to the receiver.

Figure 2 illustrates transmitter-receiver pairs where the near-far effect might occur in our example. In the illustration, the link from $\tau(a)$ to $\rho(a)$ and the link from $\tau(d)$ to $\rho(d)$ cannot be active simultaneously: the transmitter of the latter blocks the receiver of the former.

We focus on pairwise conflicts between the links. The pairwise conflicts can be described as a *conflict graph* [1]. A conflict graph $G = (V, E)$ is an undirected graph where each vertex $v \in V$ corresponds to a communication link and an edge $\{u, v\} \in E$ describes that the links u and v are mutually conflicting. Figure 3 illustrates a conflict graph; here the vertex a $\in V$ corresponds to the communication link from $\tau(a)$ to $\rho(a)$ and the vertex d $\in V$ corresponds to the communication link from $\tau(d)$ to $\rho(d)$. As these two links cannot transmit simultaneously, there is an edge $\{a, d\} \in E$ in the conflict graph.

Algorithmic Problems and Earlier Work. Associated with a conflict graph are the following network coordination tasks.

(1) Given some weights (such as priorities or utilities) on each vertex, find an *independent set* of the maximum total weight; in other words, find a nonconflicting set of links of the maximum total weight. See Fig. 3 for an example.
(2) Given some data transmission needs on each link, find a *link schedule* of the minimum length such that at each point in time, the set of active links is

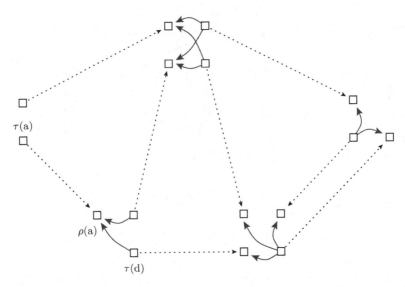

Fig. 2. The near-far effect in a wireless network. Solid arrows point from an interfering transmitter to the interfered receiver. For example, if the transmitter $\tau(d)$ is active, the device $\rho(a)$ cannot receive the transmission from $\tau(a)$. The signal power received from $\tau(a)$ is too low in comparison with the interfering power received from $\tau(d)$.

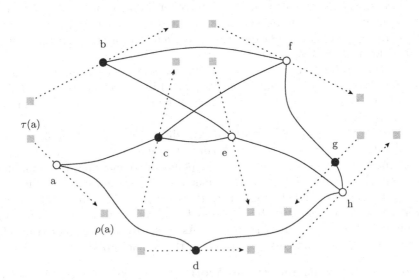

Fig. 3. The conflict graph for the example in Fig. 2. There is one vertex for each communication link and one edge for each pair of conflicting links; vertices are marked with circles and edges are marked with solid lines. Note that even though interference in Fig. 2 was highly localised, such locality is no longer immediately visible in the conflict graph. An independent set $\{b, c, d, g\}$ is highlighted; these communication links can be active simultaneously.

nonconflicting, and each link is active for a time that suffices to cover its data transmission needs. See Sect. 6 for a precise linear programming formulation.

An approximation algorithm for maximum-weight independent set also implies an approximation algorithm for the link scheduling problem in the same class of graphs [2,3]. Unfortunately, both problems are prohibitively hard to approximate in general graphs [4,5,6]. Jain et al. [1] have put forth the question of whether there is a family of conflict graphs which arises in realistic network deployments and which makes the problem of finding an independent set easier.

Contribution. Our work shows that two assumptions on the structure of conflict graphs suffice in order to achieve a polynomial-time constant-factor approximation of maximum-weight independent set and link scheduling:

(i) Bounded density of the devices. Radios are points in a low-dimensional space and they are not located in an arbitrarily dense manner.
(ii) Bounded range of interference. Conflicts are caused by the near-far effect; if there is a conflict, the interfering transmitter is close to the interfered receiver.

Note that we do not need to assume that there has to be interference in certain situations, say, between nodes close to each other; indeed, such assumptions are often not valid in practice [7,8]. We can make measurements in the deployed physical system to determine whether a pair of links is mutually conflicting; there is no need to use a simplifying model of radio propagation and interference.

We also show that the assumptions (i) and (ii) alone are not sufficient in order to achieve an arbitrarily small approximation ratio; further assumptions such as a bounded range of the wireless link are required.

2 Statement of Results

Let N be a constant which controls the relative density of the nodes.

Definition 1. *An N-local conflict graph is a tuple (G, τ, ρ) where $G = (V, E)$ is an undirected graph and τ, ρ are functions $V \to \mathbb{R}^2$ such that*

(i) *the function $v \mapsto (\tau(v), \rho(v))$ is an injection, and no unit disk in $\mathbb{R}^2$ contains more than N points in $\tau(V) \cup \rho(V)$*
(ii) *for all $\{v, u\} \in E$ it holds that $d(\tau(v), \rho(u)) < 1$ or $d(\tau(u), \rho(v)) < 1$ where $d(\cdot, \cdot)$ is the Euclidean distance.*

We call $\tau(v)$ the *transmitter* and $\rho(v)$ the *receiver*, the intuition being that a pair $(\tau(v), \rho(v))$ corresponds to a data transmission link. Note that $d(\tau(v), \rho(v))$ is unrestricted and that some receivers and transmitters may coincide; however, the pair $(\tau(v), \rho(v))$ must be unique for every vertex.

If we required $\tau(v) = \rho(v)$ for each $v \in V$, we would obtain what we call $(2, N)$-*local graphs* [9]; these are similar to *civilised graphs* [10, §8.5]. Thus, N-local conflict graphs can be interpreted as a natural generalisation of the families of local graphs and civilised graphs.

In Sect. 3, we derive some basic properties of N-local conflict graphs. We show that N-local conflict graphs are not contained in families such as bounded-degree, bounded-density, bipartite, planar, or disk graphs.

In Sect. 4, we prove our main result; *MWIS* refers to the problem of finding a maximum-weight independent set:

Theorem 1. *MWIS for N-local conflict graphs admits a polynomial-time $(5 + \epsilon)$-approximation algorithm for any constants $\epsilon > 0$ and N.*

While the time complexity of the algorithm depends on the parameter N, we emphasise that the approximation ratio does not depend on N. This is unlike families such as bounded-degree graphs where achievable approximation ratios typically depend on the parameters of the family [11]; for example, MWIS in graphs of maximum degree Δ can be approximated within a factor of $O(\Delta \log \log \Delta / \log \Delta)$ [12].

In Sect. 5, we show that approximating beyond a certain constant factor remains hard:

Theorem 2. *MWIS for N-local conflict graphs admits no polynomial-time approximation scheme (PTAS) for any N unless $P = NP$.*

This is unlike families such as disk graphs; for example, MWIS in disk graphs admits a PTAS [13].

In Sect. 6, we consider the problem of fractional covering by independent sets in local conflict graphs, obtaining analogous approximability and inapproximability results for the covering LP, which captures the link scheduling problem.

3   Representability

It is not immediate from Definition 1 which graphs admit representation as a local conflict graph. The purpose of this section is to shed some light on this question. We begin by showing that local conflict graphs are not contained in families of graphs such as planar graphs, bounded-degree graphs, or disk graphs.

A first observation is that the family of N-local conflict graphs is closed under deletion of edges and vertices. Furthermore, an N_1-local conflict graph is an N_2-local conflict graph for any $N_2 \geq N_1$.

Theorem 3. *Any bipartite graph can be represented as a 1-local conflict graph.*

Proof. Consider a bipartite graph $G = (V, E)$. The set V can be partitioned into $A = \{a_1, a_2, \ldots, a_m\}$ and $B = \{b_1, b_2, \ldots, b_n\}$ for some m, n such that all edges are between A and B. Let $\tau(a_i) = (3i, -3)$ and $\rho(a_i) = (0, 0)$ for all i; let $\tau(b_j) = (0, 0)$ and $\rho(b_j) = (3j, 3)$ for all j.

It follows that the maximum degree of a vertex in a 1-local conflict graph can be as high as $|V| - 1$ (consider the complete bipartite graph $K_{1,n}$), the average degree and the minimum degree can be as high as $|V|/2$ (consider $K_{n,n}$), and there are 1-local conflict graphs that are not planar and not disk graphs (consider $K_{3,3}$).

A local conflict graph need not be bipartite. To illustrate the rich substructure that can occur in a local conflict graph, we show that a local conflict graph may contain relatively large but not arbitrarily large cliques.

Theorem 4. *A complete graph on N^2 vertices is representable as an N-local conflict graph.*

Proof. Consider a complete graph with vertices $V = \{v_{i,j} \mid i, j \in \{1, 2, \ldots, N\}\}$. Let $\tau(v_{i,j}) = (0, i/N)$ and $\rho(v_{i,j}) = (0, j/N)$ for all i and j.

Lemma 1. *For every tournament (complete oriented graph) $G = (V, A)$ on n vertices, there are $s \in V$, $t \in V$ and $X \subseteq V$ with $\{s\} \times X \subseteq A$, $X \times \{t\} \subseteq A$, and $|X| \geq (n-2)/6$.*

Proof. Let $v, u \in V$, $v \neq u$. Let

$$Q = \{x \mid (v, x) \in A, (x, u) \in A\}, \qquad R = \{x \mid (u, x) \in A, (x, v) \in A\},$$
$$S = \{x \mid (x, v) \in A, (x, u) \in A\}, \qquad T = \{x \mid (v, x) \in A, (u, x) \in A\}.$$

If $|Q| \geq (n-2)/6$ or $|R| \geq (n-2)/6$, we are done. Otherwise, $|S| + |T| \geq 2(n+1)/3$. If $|S| \geq (n+1)/3$, the subgraph induced by S contains a vertex a with outdegree at least $(n-2)/6$; let $s = a$, $t = u$, and let $X \subseteq S$ consist of the successors of a. The case $|T| \geq (n+1)/3$ is analogous.

Theorem 5. *A complete graph on $6N^2 + 8$ vertices cannot be represented as an N-local conflict graph.*

Proof. To reach a contradiction, assume that there is a complete graph on $6N^2 + 8$ vertices that is an N-local conflict graph. We say that $P(v, u)$ holds if $d(\tau(v), \rho(u)) < 1$. Orient the graph as follows: if $P(v, u)$ and not $P(u, v)$, assign the direction (v, u) on $\{v, u\} \in E$; if $P(u, v)$ and not $P(v, u)$, assign the opposite direction; otherwise both $P(v, u)$ and $P(u, v)$ hold, in which case assign an arbitrary direction.

Choose s, t, and X as in Lemma 1; $|X| \geq N^2 + 1$. Now, $P(s, x)$ and $P(x, t)$ hold for all $x \in X$. A unit disk centred at $\tau(s)$ contains all points $\rho(X)$. There can be at most N distinct points; thus, there is a set $X' \subseteq X$ and a point $\rho' \in \mathbb{R}^2$ such that $\rho(X') = \{\rho'\}$ and $|X'| \geq N + 1$.

A unit disk centred at $\rho(t)$ contains all points $\tau(X')$. Again, there can be at most N distinct points; thus, there are two distinct vertices $v, u \in X'$ and a point $\tau' \in \mathbb{R}^2$ with $\tau(v) = \tau(u) = \tau'$. This is a contradiction because $v \mapsto (\tau(v), \rho(v))$ is an injection by Definition 1.

It follows immediately that the family of N-local conflict graphs is not closed under taking minors (form a bipartite graph by splitting each edge of a large complete graph in two).

4 Proof of Theorem 1

The input of the algorithm consists of a graph $G = (V, E)$, points $\tau(v)$ and $\rho(v)$ for each $v \in V$, and a weight $w(v)$ for each $v \in V$.

We present the algorithm in a somewhat more general setting than required by the MWIS problem. Write $C(v, I)$ for the *contribution* of the vertex $v \in V$

in the proposed solution $I \subseteq V$. Let $W(v, I) = w(v)C(v, I)$ for each $I \subseteq V$ and $v \in V$, and let $W(A, I) = \sum_{v \in A} W(v, I)$ for any $A \subseteq V$. The objective is to find a solution $I \subseteq V$ that maximises $W(V, I)$, that is, maximises weighted contributions.

In the case of MWIS, we define $C(v, I) = 1$ if $v \in I$ and there is no $u \in I$ with $\{v, u\} \in E$ and $d(\tau(u), \rho(v)) < 1$, otherwise $C(v, I) = 0$. The set I that maximises $W(V, I)$ is (after removing vertices with zero contribution) a maximum-weight independent set in G.

Define the set of possibly interfering vertices

$$U(v) = \{v\} \cup \{u \in V \mid d(\tau(u), \rho(v)) < 1\}.$$

The algorithm makes use of the following assumptions on C; these are immediate in the case of MWIS for local conflict graphs: $C(v, I)$ can be evaluated in polynomial time; $C(v, I) = 0$ for all $v \notin I$; $C(v, I_1) \geq C(v, I_2)$ for all $I_1 \subseteq I_2$ with $v \in I_1$ (contributions are nonincreasing); and $C(v, I) = C(v, I \cap U(v))$ for all I and $v \in V$ (locality).

In the algorithm, the full problem is divided into *subproblems*. Each subproblem is defined by a subset $A \subseteq V$, and the associated task is to find a set $I \subseteq A$ that maximises $W(A, I)$.

Let $\hat{W}(v) = W(v, \{v\})$ for each $v \in V$, and let $\hat{W}(A) = \sum_{v \in A} \hat{W}(v)$. We make use of the following properties. As $C(v, I) = 0$ for $v \notin I$, we have $W(A, I) = W(A \cap I, I)$ for all $A \subseteq V$. By nonincreasingness, $W(v, I_1) \geq W(v, I_2)$ for all $I_1 \subseteq I_2 \subseteq V$ and $v \in I_1$; in particular, $\hat{W}(v) \geq W(v, I)$ for all $v \in I$ and $I \subseteq V$. As $C(v, I) = 0$ for $v \notin I$, we have $\hat{W}(v) \geq W(v, I)$ for all $v \in V \setminus I$ and $I \subseteq V$. In summary, $\hat{W}(A) \geq W(A, I)$ for all $A, I \subseteq V$, and $\hat{W}(I) \geq W(I, I) = W(V, I)$ for all $I \subseteq V$.

To create the subproblems, we apply a shifting strategy [14,15]. We make the following initial assignments. Choose an integer $k \geq 3$ such that $5k^2/(k-2)^2 < 5 + \epsilon$. Let

$$
\begin{aligned}
A_i' &= \{(x, y) \in \mathbb{R}^2 \mid i \leq x < i + 1\}, & i &\in \mathbb{Z}, \\
B_j' &= \{(x, y) \in \mathbb{R}^2 \mid j \leq y < j + 1\}, & j &\in \mathbb{Z}, \\
A_i &= \bigcup\{A_t' \mid t \in \mathbb{Z}, t \equiv i \pmod{k}\}, & i &= 0, 1, \ldots, k-1, \\
B_j &= \bigcup\{B_t' \mid t \in \mathbb{Z}, t \equiv j \pmod{k}\}, & j &= 0, 1, \ldots, k-1, \\
D_{ij} &= \mathbb{R}^2 \setminus A_i \setminus B_j, & i, j &= 0, 1, \ldots, k-1.
\end{aligned}
$$

Each D_{ij} consists of squares $k - 1$ units wide and high. We write $D_{ij1}, D_{ij2}, \ldots$ for the nonempty squares of D_{ij}. Let $Z_{ij} \subseteq V$ be the set of vertices v with both $\tau(v)$ and $\rho(v)$ in D_{ij}, and let $X_{ij\beta} \subseteq Z_{ij}$ be the set of vertices v with both $\tau(v)$ and $\rho(v)$ in $D_{ij\beta}$. Form the set of "short links" $X_{ij} = \bigcup_\beta X_{ij\beta}$ and the set of "long links" $Y_{ij} = Z_{ij} \setminus X_{ij}$.

Now we can use the following procedure to find an approximately optimal solution. In the first part we solve small subproblems by exhaustive search; in the second part we apply the standard greedy algorithm for finding a large cut.

1. Initialise $\mathcal{N}$ to an empty set.
2. For all $i = 0, 1, \ldots, k-1$ and $j = 0, 1, \ldots, k-1$:
   (a) For each nonempty square $D_{ij\beta}$, find a subset $I \subseteq X_{ij\beta}$ which maximises $W(X_{ij\beta}, I)$. Call this set $S_{ij\beta}$.
   (b) Let $S_{ij} = \bigcup_\beta S_{ij\beta}$. Insert S_{ij} into $\mathcal{N}$.
3. For all $i = 0, 1, \ldots, k-1$ and $j = 0, 1, \ldots, k-1$:
   (a) Initialise Γ and Λ to empty sets.
   (b) For each nonempty square $D_{ij\beta}$, in an arbitrary order: Write $[\beta, \Gamma]$ for the set of vertices $v \in Y_{ij}$ such that one of the points $\tau(v), \rho(v)$ is located in $D_{ij\beta}$ and the other point is located in $D_{ij\Gamma} = \bigcup_{\gamma \in \Gamma} D_{ij\gamma}$; define the set $[\beta, \Lambda]$ similarly. Add β to Λ if $\hat{W}([\beta, \Gamma]) > \hat{W}([\beta, \Lambda])$; otherwise add β to Γ.
   (c) Let $T_1 = \{v \in Y_{ij} \mid \tau(v) \in D_{ij\Gamma}, \rho(v) \in D_{ij\Lambda}\}$ and let $T_2 = \{v \in Y_{ij} \mid \tau(v) \in D_{ij\Lambda}, \rho(v) \in D_{ij\Gamma}\}$. Let $T \in \{T_1, T_2\}$ be the set that maximises $\hat{W}(T)$. Call this set R_{ij}, and insert R_{ij} into $\mathcal{N}$.
4. Return $\tilde{I} = \arg\max_{I \in \mathcal{N}} W(V, I)$.

The time complexity of the algorithm is polynomial in the size of the input since the number of nonempty squares $D_{ij\beta}$ is bounded by $2|V|$, the number of distinct transmitters or receivers in each square is bounded by a constant, and a pair $(\tau(v), \rho(v))$ uniquely determines v for all $v \in V$. The following three lemmata establish the correctness of the algorithm. We denote by $I^*(A)$ an optimal solution to the subproblem A.

Lemma 2. *Each S_{ij} is an optimal solution of the subproblem X_{ij}.*

Proof. To reach a contradiction, assume that S_{ij} is not optimal, i.e., there exists an $I \subseteq X_{ij}$ with $W(X_{ij}, I) > W(X_{ij}, S_{ij})$. Then there is a β with $W(X_{ij\beta}, I) > W(X_{ij\beta}, S_{ij})$. As the squares $D_{ij\delta}$ are separated by stripes of width one, $U(v) \cap X_{ij} \subseteq X_{ij\beta}$ for all $v \in X_{ij\beta}$. Thus, $W(X_{ij\beta}, I) = W(X_{ij\beta}, I \cap X_{ij\beta})$ and $W(X_{ij\beta}, S_{ij}) = W(X_{ij\beta}, S_{ij} \cap X_{ij\beta}) = W(X_{ij\beta}, S_{ij\beta})$. Thus, $W(X_{ij\beta}, I \cap X_{ij\beta}) > W(X_{ij\beta}, S_{ij\beta})$, contradicting the choice of $S_{ij\beta}$.

Lemma 3. *Each R_{ij} is a 4-approximate solution of the subproblem Y_{ij}.*

Proof. The greedy algorithm in parts (3a) and (3b) finds a cut (in the directed graph with an arc of weight $\hat{W}(v)$ from $\tau(v)$ to $\rho(v)$ for each $v \in Y_{ij}$) of a total weight at least $\hat{W}(Y_{ij})/2$, implying by (3c) that $\hat{W}(Y_{ij})/4 \leq \hat{W}(R_{ij})$. All transmitters of R_{ij} are in $D_{ij\Gamma}$ and all receivers of R_{ij} are in $D_{ij\Lambda}$ or vice versa. The distance between any receiver and transmitter is larger than 1. Thus, $U(v) \cap R_{ij} = \{v\}$ for each $v \in R_{ij}$, implying that $W(v, R_{ij}) = W(v, \{v\}) = \hat{W}(v)$. Therefore, $\hat{W}(R_{ij}) = W(Y_{ij}, R_{ij})$. In summary, it holds that $W(Y_{ij}, I^*(Y_{ij})) \leq \hat{W}(Y_{ij}) \leq 4W(Y_{ij}, R_{ij})$.

Lemma 4. *The set $\tilde{I}$ is a $\left(5k^2/(k-2)^2\right)$-approximate solution.*

Proof. There is exactly one set A_i and exactly one set B_j that contains any given point in $\tau(V) \cup \rho(V)$. For each vertex $v \in V$, there are at most two sets A_i and at most two sets B_j that contain $\tau(v)$ or $\rho(v)$. Thus, there are at least $(k-2)^2$ pairs (i,j) such that D_{ij} contains both $\tau(v)$ and $\rho(v)$. In notation, $|\{(i,j) \mid v \in Z_{ij}\}| \geq (k-2)^2$ for all $v \in V$.

Let $I_{ij}^* = I^*(V) \cap Z_{ij}$. As the contributions are nonincreasing, we have $W(v, I^*(V)) \leq W(v, I_{ij}^*)$ for all $v \in I_{ij}^*$ and $W(Z_{ij}, I^*(V)) \leq W(Z_{ij}, I_{ij}^*)$ for all Z_{ij}, implying that $(k-2)^2 W(V, I^*(V)) \leq \sum_{i,j} W(Z_{ij}, I^*(V)) \leq \sum_{i,j} W(V, I_{ij}^*)$. Thus, there is a pair (i,j) satisfying $(k-2)^2 W(V, I^*(V)) \leq k^2 W(V, I_{ij}^*)$. As the sets X_{ij} and Y_{ij} partition Z_{ij}, we have $W(V, I_{ij}^*) = W(X_{ij}, I_{ij}^*) + W(Y_{ij}, I_{ij}^*) \leq W(X_{ij}, I^*(X_{ij})) + W(Y_{ij}, I^*(Y_{ij}))$. By Lemmata 2 and 3, we obtain

$$W(X_{ij}, I^*(X_{ij})) + W(Y_{ij}, I^*(Y_{ij})) \leq W(V, S_{ij}) + 4W(V, R_{ij})$$
$$\leq (1+4)\max\{W(V, S_{ij}), W(V, R_{ij})\} \leq 5W(V, \tilde{I}).$$

Remarks. As the above divide-and-conquer approach gives a constant-factor approximation, it is of interest whether a similar result can be obtained by considerably simpler greedy algorithms. However, this does not seem to be the case. First, maximal independent sets may be arbitrarily small: $K_{1,n}$ is a 1-local conflict graph with a maximal independent set of size 1 and an independent set of size n. Second, vertices which are part of a large independent set may also be relatively rare: take a complete graph on N^2 vertices and remove the edges of a clique, possibly with some extra edges.

The next section proves that MWIS in local conflict graphs does not admit a PTAS unless P = NP. Note that if we restricted to the subfamily of N-local conflict graphs where there is no vertex v with $d(\tau(v), \rho(v)) > 1$, a simplification of our algorithm would give a PTAS for MWIS.

5   Proof of Theorem 2

The maximum-weight directed cut problem is defined as follows: given a directed graph $G = (V, A)$ and a nonnegative weight $w(a)$ for each arc $a \in A$, find a subset $S \subseteq V$ such that the total weight of the resulting cut $\delta^+(S) = \{(u,v) \in A \mid u \in S, v \notin S\}$ is maximised. The problem is APX-complete already in the unweighted case [16,17].

We show that for any N, approximating MWIS in N-local conflict graphs within factor α implies approximating maximum-weight directed cut in an arbitrary directed graph within factor α. The reduction is similar to the one used by Chvátal and Ebenegger [18]; applied here, the reduction actually shows that the underlying undirected graph of the directed line graph of an arbitrary directed graph is a 1-local conflict graph.

Consider an instance of maximum-weight directed cut $G = (V, A)$ with the arc weights $w(a)$. Without loss of generality, we may arbitrarily relabel the vertices so that $V = \{(0, 3), (0, 6), \ldots, (0, 3|V|)\} \subseteq \mathbb{R}^2$. Construct an N-local conflict graph

$G' = (V', E')$ with weights w' as follows. Let $V' = A$ and $E' = \{\{(t, u), (u, v)\} \mid (t, u) \in A, (u, v) \in A\}$. For each $(u, v) \in A$, let $\tau((u, v)) = u$, $\rho((u, v)) = v$, and $w'((u, v)) = w((u, v))$. The construction is a valid N-local conflict graph for any $N \geq 1$.

Let $S \subseteq V$ be a directed cut of G. Now, $\delta^+(S)$ is an independent set of the same weight in G' because there is no pair of arcs $(t, u), (u, v)$ in $\delta^+(S)$. Conversely, let $I' \subseteq V'$ be an independent set in G'. Let S consist of all transmitters of the vertices in I'. Note that S contains no receiver of a vertex in I'. Thus, $I' \subseteq \delta^+(S)$, i.e., S defines a directed cut with weight at least that of I'.

It follows that if W^* is the maximum weight of a directed cut in G, there is an independent set with weight W^* in G'. An α-approximation algorithm for MWIS finds an independent set of weight at least W^*/α in G', which transforms to a directed cut with weight at least W^*/α in G.   □

6   Link Scheduling

We conclude this paper by considering fractional covering by independent sets in local conflict graphs. For an independent set I, we let $I(v) = 1$ if $v \in I$ and $I(v) = 0$ if $v \notin I$.

Definition 2. *In the* link scheduling problem, *the input consists of an N-local conflict graph (G, τ, ρ) and a* requirement *$r(v) \geq 0$ for each vertex v in G. The task is to*

$$\text{minimise} \quad \sum_I x(I)$$
$$\text{subject to} \quad \sum_I I(v)x(I) \geq r(v) \quad \text{for all } v,$$
$$x(I) \geq 0 \qquad \text{for all } I,$$

where v ranges over all vertices in G and I ranges over all independent sets in G. The value $L = \sum_I x(I)$ is called the length *of the schedule.*

If G is a conflict graph and $r(v)$ is the amount of data that has to be transmitted on the link from $\tau(v)$ to $\rho(v)$, the link scheduling problem corresponds to finding an optimal schedule of data transmissions in a wireless communication network. In this setting, the vector x is interpreted as a schedule that assigns to independent set I the time slice $x(I)$.

The special case $r(v) = 1$ for all v corresponds to fractional colouring; the minimum schedule length is the fractional chromatic number of G.

Example 1. Consider the example in Fig. 3, and let $r(v) = 1$ for each v. Let $I_1 = \{b, c, d, g\}$, $I_2 = \{b, c, h\}$, $I_3 = \{a, f, h\}$, $I_4 = \{a, e, f\}$, and $I_5 = \{d, e, g\}$; each of these sets is an independent set. Choose $x(I_1) = x(I_2) = x(I_3) = x(I_4) = x(I_5) = 1/2$. Now x is a solution to the link scheduling problem, and the length of the schedule equals $5/2$.

The number of variables in the above LP may be exponential in the size of the input. However, it is not necessary to construct the LP explicitly: the link scheduling problem can be solved $(\alpha + \epsilon)$-approximately in polynomial time for any $\epsilon > 0$ as long as there is a polynomial-time α-approximation algorithm for finding the maximum-weight independent set of G for arbitrary weights [2,3]. Thus, Theorem 1 has the following corollary.

Corollary 1. *The link scheduling problem for N-local conflict graphs admits a polynomial-time $(5 + \epsilon)$-approximation algorithm for any constants $\epsilon > 0$ and N.*

The main result of this section shows that approximating beyond a certain constant factor remains hard.

Theorem 6. *The link scheduling problem for N-local conflict graphs admits no PTAS for any N unless $P = NP$.*

We begin with the following lemma.

Lemma 5. *There are constants k, $c > 1$, and Δ such that the following problem is NP-hard: Given a graph with maximum vertex degree at most Δ, decide whether its fractional chromatic number is at most k or at least ck.*

Proof. Khot [5] established that it is NP-hard to colour a k-colourable graph with $k^{\log(k)/25}$ colours for all sufficiently large constants k, even for graphs of bounded degree. In fact, Khot's proof shows that distinguishing between the following two cases is NP-hard for graphs of bounded degree: (i) There is a k-colouring. Thus, also the *fractional* chromatic number is at most k. (ii) The ratio of the number of vertices to the maximum size of an independent set is at least $k^{\log(k)/25}$. Thus, the *fractional* chromatic number is at least $k^{\log(k)/25}$.

Proof of Theorem 6. We show that a PTAS for link scheduling in N-local conflict graphs can be used to solve the NP-hard problem in Lemma 5. Let $G = (V, E)$ be an arbitrary graph with maximum vertex degree at most Δ. Label the vertices so that $V \subseteq \mathbb{Z}$. We associate with each $v \in V$ three points, v_1, v_2, and v_3, by setting $v_j = (3v, 3j)$. No unit disk contains more than one point.

Construct an instance of the link scheduling problem: an N-local conflict graph $G' = (V', E')$ and the corresponding requirements $r(v')$ for each $v' \in V'$. We refer to the elements $v' \in V'$ as *links*, the intuition being that they correspond to a pair of transmitter and receiver; we reserve the word *vertex* for the elements $v \in V$.

For each vertex $v \in V$, introduce two links $(v_1, v_2) \in V'$ and $(v_2, v_3) \in V'$ with the requirements $r((v_1, v_2)) = k - 1$ and $r((v_2, v_3)) = 1$. For each edge $\{u, v\} \in E$, introduce two links $(u_2, v_2) \in V'$ and $(v_2, u_2) \in V'$ with the requirements $r((u_2, v_2)) = r((v_2, u_2)) = 1$. Finally, let $\tau((x, y)) = x$, $\rho((x, y)) = y$, and $E' = \{\{(x, y), (y, z)\} \mid (x, y) \in V', (y, z) \in V'\}$.

Select a positive $\epsilon' < \min\{(\sqrt{c} - 1)/2, (1 - 1/\sqrt{c})/(2k\Delta)\}$, and use the PTAS to solve the constructed link scheduling instance within factor $(1 + \epsilon')$. Let the length of the schedule be L.

Let the fractional chromatic number of G be χ_f. If $\chi_f \leq k$, we may use a fractional colouring of size k to construct a feasible schedule of length k for the link scheduling instance as follows. Interpret the fractional colouring as a schedule in which the vertices may be active or inactive. Without loss of generality we may assume that the schedule is exact; that is, each vertex is active for exactly 1 time unit. Whenever $v \in V$ is active, transmit data on links (v_2, v_3) and (v_2, u_2) for each vertex u adjacent to v in G; this is possible since the vertices adjacent to v cannot be active and thus are not transmitting at the same time. Whenever $v \in V$ is inactive, transmit data on (v_1, v_2). Note that each vertex is active for 1 time unit and inactive for $k-1$ time units, implying that the requirements $r(v')$ are met. Observe that if $\chi_f \leq k$, we have $L \leq (1 + \epsilon')k$ since the length of the schedule cannot be more than k and we solved the link scheduling instance within factor $(1 + \epsilon')$.

On the other hand, if $L \leq (1 + 2\epsilon')k$, we may use a schedule of length $(1 + 2\epsilon')k$ to construct a fractional colouring. Let $T(v')$ be the union of all time intervals when the link $v' \in V'$ is active in the schedule, and let $|T(v')|$ be the total length of these time intervals. Consider an arbitrary vertex $v \in V$. Let $V'(v)$ be the set of all links $(v_2, u_2) \in V'$ with $u \in V$. We have $T((v_1, v_2)) \cap T(v') = \emptyset$ for all $v' \in V'(v)$. Furthermore, $|T(v_1, v_2)| \geq k - 1$, and the total schedule length equals $k + 2\epsilon'k$. Thus, $|\bigcup_{v' \in V'(v)} T(v')| \leq 1 + 2\epsilon'k$. On the other hand, $|T(v')| \geq 1$ for each $v' \in V'(v)$. Observe that $|\bigcap_{v' \in V'(v)} T(v')| \geq 1 - 2\epsilon'k|V'(v)|$ since each new edge introduced to the intersection shortens the intersection by at most $2\epsilon'k$ units. Also observe that $|V'(v)| \leq \Delta$.

A non-isolated vertex v in the fractional colouring problem may be active in time intervals $\bigcap_{v' \in V'(v)} T(v')$ because none of its neighbours may be active at the same time; for an isolated vertex (the case of an empty $V'(v)$), colouring is trivial. Thus, we have a partial fractional covering of length $(1 + 2\epsilon')k$ that covers each vertex for at least $1 - 2\epsilon'k\Delta$ units of time. Multiply all time assignments by $1/(1 - 2\epsilon'k\Delta)$ to obtain a fractional colouring of size $(1 + 2\epsilon')k/(1 - 2\epsilon'k\Delta) < (1 + (\sqrt{c} - 1))k/(1 - (1 - 1/\sqrt{c})) = ck$ that covers each vertex for at least 1 unit of time. Thus, $\chi_f < ck$.

To summarise, $\chi_f \leq k$ implies $L \leq (1 + \epsilon')k$ and $\chi_f \geq ck$ implies $L > (1 + 2\epsilon')k$. This shows that we can use a PTAS to distinguish in polynomial time between the two cases in Lemma 5.

Remark. It is not known whether Theorem 6 could be obtained as a simple corollary of Theorem 2. For example, the conversion method by Erlebach and Jansen [19] cannot be applied directly as it requires that the family of graphs is closed not only under the deletion of vertices but also under the duplication of vertices.

Acknowledgements. We thank Harri Haanpää, Patrik Floréen, Pekka Orponen, and Marja Hassinen for useful discussions and comments.

This work was supported in part by the Academy of Finland, Grants 116547, 117499, 202203, 202204, and 202205, by Helsinki Graduate School in Computer Science and Engineering (Hecse), and by the IST Programme of the European Community, under the PASCAL Network of Excellence, IST-2002-506778.

References

1. Jain, K., Padhye, J., Padmanabhan, V.N., Qiu, L.: Impact of interference on multi-hop wireless network performance. Wireless Networks 11(4), 471–487 (2005)
2. Jansen, K.: Approximate strong separation with application in fractional graph coloring and preemptive scheduling. Theoretical Computer Science 302(1–3), 239–256 (2003)
3. Young, N.E.: Sequential and parallel algorithms for mixed packing and covering. In: Proc. 42nd Annual Symposium on Foundations of Computer Science FOCS, Las Vegas, NV, USA, October 2001, pp. 538–546. IEEE Computer Society Press, Los Alamitos (2001)
4. Håstad, J.: Clique is hard to approximate within $n^{1-\epsilon}$. Acta Mathematica 182, 105–142 (1999)
5. Khot, S.: Improved inapproximability results for MaxClique, chromatic number and approximate graph coloring. In: Proc. 42nd Annual Symposium on Foundations of Computer Science FOCS, Las Vegas, NV, USA, October 2001, pp. 600–609. IEEE Computer Society Press, Los Alamitos (2001)
6. Lund, C., Yannakakis, M.: On the hardness of approximating minimization problems. Journal of the ACM 41(5), 960–981 (1994)
7. Goldsmith, A.: Wireless Communications. Cambridge University Press, Cambridge, UK (2005)
8. Krishnamachari, B.: Networking Wireless Sensors. Cambridge University Press, Cambridge, UK (2005)
9. Suomela, J.: Approximability of identifying codes and locating-dominating codes. Information Processing Letters 103(1), 28–33 (2007)
10. Doyle, P.G., Snell, J.L.: Random Walks and Electric Networks. The Mathematical Association of America, Washington, DC, USA (1984)
11. Halldórsson, M.M.: Approximations of independent sets in graphs. In: Jansen, K., Rolim, J.D.P. (eds.) APPROX 1998. LNCS, vol. 1444, pp. 1–13. Springer, Heidelberg (1998)
12. Halldórsson, M.M.: Approximations of weighted independent set and hereditary subset problems. Journal of Graph Algorithms and Applications 4(1), 1–16 (2000)
13. Erlebach, T., Jansen, K., Seidel, E.: Polynomial-time approximation schemes for geometric intersection graphs. SIAM Journal on Computing 34(6), 1302–1323 (2005)
14. Hochbaum, D.S., Maass, W.: Approximation schemes for covering and packing problems in image processing and VLSI. Journal of the ACM 32(1), 130–136 (1985)
15. Hunt III, H.B., Marathe, M.V., Radhakrishnan, V., Ravi, S.S., Rosenkrantz, D.J., Stearns, R.E.: NC-approximation schemes for NP- and PSPACE-hard problems for geometric graphs. Journal of Algorithms 26(2), 238–274 (1998)
16. Khanna, S., Motwani, R., Sudan, M., Vazirani, U.: On syntactic versus computational views of approximability. SIAM Journal on Computing 28(1), 164–191 (1999)
17. Papadimitriou, C.H., Yannakakis, M.: Optimization, approximation, and complexity classes. Journal of Computer and System Sciences 43(3), 425–440 (1991)
18. Chvátal, V., Ebenegger, C.: A note on line digraphs and the directed max-cut problem. Discrete Applied Mathematics 29(2–3), 165–170 (1990)
19. Erlebach, T., Jansen, K.: Conversion of coloring algorithms into maximum weight independent set algorithms. Discrete Applied Mathematics 148(1), 107–125 (2005)

Information Brokerage Via Location-Free Double Rulings*

Stefan Funke and Imran Rauf

Max-Planck-Institut für Informatik,
Stuhlsatzenhausweg 85, 66123 Saarbrücken, Germany
{funke,irauf}@mpi-inf.mpg.de

Abstract. The in-network aggregation and processing of information is what sets a sensor network apart from a pure data acquisition device. One way to model the exchange of information between the network nodes is to distinguish between nodes that are *producers* of information, i.e., those that have collected data, detected events, etc., and nodes that are *consumers* of information, i.e., nodes that seek data or events of certain types. In this paper we aim to support that exchange of information via a so-called *information brokerage* scheme. Main features of our proposed scheme are that 1) it works in a location-free setting where nodes are unaware of their geographic locations 2) it is robust to non-regular network topologies and 3) it does not require the information producers and consumers to know of each other. Our proposed scheme employs boundary detection algorithms which only quite recently have been developed to extract geometry and topology information even in location-free network deployments.

1 Introduction

In their first generation, sensor networks were primarily considered a data acquisition device, where the data acquired at the sensor nodes was transferred to a central authority for evaluation and storage. Due to the rapid growth of the size of sensor network deployments, such a centralized model of operation becomes restricting as the amounts of data that can be transferred through the network is naturally limited. Recently, more emphasis is put on processing and interpreting the acquired data *within* the network. That is, sensor networks have made the step from a pure data acquisition device to a new form of computing device. For the *in-network processing* of sensor data, novel schemes for the information exchange between the sensor nodes are necessary which are more focused on the data itself rather than the identities of the individual network nodes. *Data-centric* processing, storage and retrieval of information has been the focus of several recent papers, see [10,13,8]. For example in GHT [13], each

---

* This work was supported by the Max Planck Center for Visual Computing and Communication (MPC-VCC) funded by the German Federal Ministry of Education and Research (FKZ 01IMC01).

E. Kranakis and J. Opatrny (Eds.): ADHOC-NOW 2007, LNCS 4686, pp. 87–100, 2007.
© Springer-Verlag Berlin Heidelberg 2007

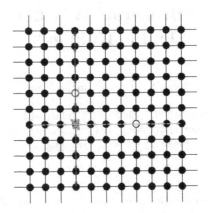

Fig. 1. Simplest double ruling scheme where informations producers and consumers are brought together at the node indicated by a green cross

type σ_A of information (like occurrence of some event A) is mapped to a location using a *geographic hash function* ϕ, which is known to all nodes within the network. Upon occurrence of event A, a nearby sensor node detecting this occurrence sends a message to a node close to the location determined by $\phi(\sigma_A)$, indicating that is has gathered data of type σ_A. Any other node interested in information of type σ_A can obtain that information by sending a query message to $\phi(\sigma_A)$. Note that in this scheme, the network nodes do not have to be aware of their identities, i.e., a directory service as usually required for point-to-point routing is not necessary; the routes of the messages are purely determined by the *type of the data* (and the associated geographic locations). The presence of location information at the network nodes for this and several related approaches is fundamental, though.

In this paper we take a slightly different view on the information exchange between the nodes within a sensor network. We consider the network nodes as *producers* and *consumers* of information. A *producer* of information v_p, upon detection of some event A, sends messages along two paths within the network, leaving a trail indicating that v_p has gathered information of type σ_A[1]. A *consumer* of information v_c that is interested in retrieving information of some type σ_A sends messages along two paths querying for data of type σ_A. If one of the messages on its way visits a node which has information about type σ_A stored by some producer, it sends back a message to the consumer following the trail left by the consumer's message. What a *double ruling scheme* guarantees is that the trajectories of the messages of producer and consumer always meet each other. In Figure 1 we have depicted a very simple double ruling scheme where producers of information distribute their data on vertical paths and consumers look for information on horizontal paths. Observe that this simple scheme enjoys a certain *distance-sensitivity*, that is, if a consumer is looking for an event that has happened nearby, the two trajectories of consumer and producer are

---

[1] The trail data expires after some time depending on the lifetime of the event itself.

guaranteed to meet quickly; distance-sensitivity is not provided by schemes like GHT ([13]): the location where data about a certain type of event is stored can be arbitrarily far away from both consumers and producers. This drawback is partly alleviated by the GLS [1] approach at a higher maintenance cost. All the aforementioned approaches require geographic location information at the network nodes, though.

The aim of this paper is to develop a information brokerage scheme that does not require location-information at the network nodes, yet it is based on the same simple intuition of the double ruling scheme in Figure 1. Due to the absence of location information, the directional information (vertical/horizontal) has to be replaced by topology-based gradient fields, and the property that the trajectories of information consumers and producers meet has to be ensured by a setup phase which extracts important topological properties of the network. The latter is achieved by the employment of boundary detection algorithms which only quite recently have been developed.

1.1   Related Work

The most prominent representatives of data-centric information storage and retrieval schemes are GHT (Geographic Hash Table, [13]) and GLS ([1]) as already sketched above. In very recent work, the concepts of geographic hashing and double rulings have been generalized in [11], where the authors propose a scheme in which both information dissemination as well as information retrieval is performed by sending messages along a closed replication curve. The respective curve is guaranteed to pass through a certain location as defined by a hash function (similar to GHT), still retrieval can successfully achieved in a distance-sensitive manner since retrieval and replication curve typically meet each other quite quickly. GHT, GLS, and the latter approach are crucially dependent on *geographic location information* at the network nodes. Fewer approaches have been proposed that also work in a location-free setting. Fang et al. in [2] combine the GLIDER routing scheme with a hashing approach on the top level similar to GHTs and local double rulings (within the topologically simple routable tiles). Since their approach is built on top of a routing scheme, though, it requires the permanent maintenance of rather complex global topology information throughout the network. Another approach, based on a hierarchical naming and routing scheme, was presented in [5]. It is essentially an extension of the GLS scheme ensuring distance-sensitivity in location-free scenarios if the underlying routing scheme produces close-to-optimal routes. Again, being based upon a point-to-point routing scheme, a considerable amount of global topology information has to be constantly maintained throughout the network. Unfortunately, the simple approach of vertical and horizontal message trajectories as sketched in Figure 1 does not easily extend to more complex network topologies (i.e. non-grid-like, potentially with holes).

1.2 Our Contribution

We develop an information brokerage scheme that is based on the same simple intuition of the double ruling scheme in Figure 1. Main features of our scheme are that it is *location-free*, i.e. does not rely on geographic location information at the network nodes (like [13,1,11] do). Our scheme is *distance-sensitive*, i.e. the cost for retrieving information about some event is proportional to the distance to where the event happened (unlike [13,1]), and it is not built on top of a point-to-point routing scheme (like [2,5]) and hence does not require the maintenance of extensive global topology information during operation. Instead, after a setup phase, the only information that needs to be maintained are two scalar values at each network node corresponding to the two directions of the double ruling scheme. Changes in the network topology can be easily dealt with by repeatedly checking and updating the scalar values based on the values stored with communication neighbors. Due to the absence of location information, the directional information (vertical/horizontal) as used in Figure 1 has to be replaced by topology-based gradient fields, and the property that the trajectories of information consumers and producers meet has to be ensured by a setup phase which extracts important topological properties of the network. We also evaluate our information brokerage scheme in simulations and show that it behaves favorably compared to for example GHT or simple location-based double ruling schemes. In all producer-consumer brokerage schemes there is a trade-off between the time and space cost of information diffusion when producer nodes record new data and have new detections, vs. the query time cost that consumer nodes have to pay to discover this information. In this work we are mainly interested in allowing consumers to quickly obtain the desired information while keeping the cost of information diffusion for the producer reasonable.

2 Location-Free Double Rulings

2.1 Intuition in the Continuous Case

To explain our approach we first consider a continuous (not necessarily simply connected) domain where all the sensors are deployed; see Figure 2, left, for an illustration (the holes in the domain represent communication voids as for example induced by obstacles like buildings). The idea of our double ruling scheme is to choose one of the boundary cycles of the domain and partition it into four pieces G_-, G_+, H_-, H_+ such that the G and H pieces alternate, see Figure 2, center. A producer of information upon detection of an interesting event, sends a message to both G_- and G_+, essentially simulating the vertical paths in the simple scheme from Figure 1. A node searching for information, on the other hand, sends a message two messages to H_- and H_+ (corresponding to the horizontal paths); see Figure 2, right.

Several interesting observations are worth noting: 1) It does not matter which boundary cycle (outer boundary cycle or boundary cycle of a hole) we choose, if we partition it in the above described way, the paths of producers and consumers

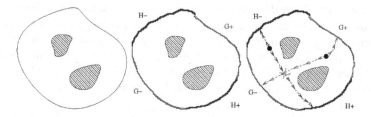

Fig. 2. Intuition of our approach in the continuous setting

are guaranteed to cross. For reasons of load-balancing, though, it is preferable to partition the largest boundary cycle which typically is the outer boundary cycle; partitioning of a small cycle leads to a lot of producers' or consumers' paths that have to go through a small part of the network. 2) It would also be possible – instead of partitioning a boundary cycle into 4 pieces – to simply choose 4 *points* on the boundary cycle, and defining the producers' or consumers' paths towards those single points. But again, this would lead to a high load imbalance within the network. 3) For correctness it is irrelevant on which paths the messages reach G_-/G_+ (H_-/H_+ respectively); to limit the overall traffic load it is generally advisable to prefer relatively short paths, though.

2.2 Translation to the Discrete Setting

Translating the idea from the continuous setting to real wireless sensor network deployments incurs two major difficulties. First of all, in a real network deployment is only a discrete sampling of the continuous domain. Secondly, the above description heavily relies on the *geometry* of the domain. But in many application scenarios the network nodes only know approximately about their geographic positions or not at all; equipping every single node with a GPS receiver is typically not feasible due to cost reasons, in particular for a large number of deployed nodes. To overcome these difficulties we will employ algorithms and techniques

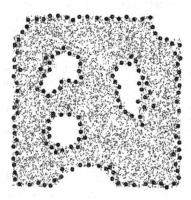

Fig. 3. Output of a boundary detection algorithm

that have recently been developed under the topic of *boundary detection*. Here one assumes that the communication graph of a wireless network resembles a *(quasi)-unit-disk graph* and the goal is to identify nodes close to network boundaries just by inspection of the communication graph without using any geometric location information. Fortunately the communication graph of a wireless network implicitly contains enough information such that this is possible as has been shown in a sequence of papers [4,3,6,15]. See Figure 3 for a sketch of the output of the boundary detection algorithm in [6]; note that the algorithm actually produces connected sets of nodes which form the boundaries, the subsampling as a set of few boundary nodes was only chosen for visualization purposes.

The intersection of producer's and consumer's paths in the discrete setting either happens at a common node on these paths, or an edge from one path crosses an edge of the other. Since one of the nodes in a pair of crossing edges must have edges to all three other nodes, the successful rendezvous of consumer's and producer's paths is guaranteed by replicating the message to the 1-hop neighbors of the nodes on these paths.

The main steps of our approach are as follows:

1. use a boundary detection algorithm to identify the outer boundary of the network (we use the algorithm proposed in [6])
2. partition the outer boundary into 4 well-behaved pieces (see later for potential caveats)
3. construct gradient-fields between opposing boundary pieces for the double ruling scheme

In the following we go into a bit more detail regarding each of these steps.

Identification of the Outer Boundary Cycle. As we noted earlier, any boundary cycle is sufficient to implement our double-ruling scheme but for better load-balancing we prefer an outer boundary cycle. Note that in the discrete setting, it is impossible to identify exactly the true geometric boundary without the location information of nodes at hand. Fortunately, boundary-detection algorithms can be employed to identify nodes which are 'near' the boundary. For example, the boundary detection algorithm by Funke et al. [6] guarantees to mark a node close to every point on the true geometric boundary.

To determine an outer cycle, we first use the algorithm in [6] to mark nodes near all boundaries. The marked nodes would also contain nodes near hole boundaries, however it is reasonable to assume that the largest connected component B of those marked nodes constitutes the outer boundary. Our algorithm then starts with a cycle in B and tries to 'grow' it until all other network nodes are connected in one component when removing B including its 1-hop neighborhood. To begin with, we arbitrarily pick some node u in B. Let $v \in B$ be the node at maximum distance from u, p_1 be the shortest path from u to v in B, and $w \in p_1$ be the node halfway between u and v on p_1. We mark w along with nodes that are up to 4-hop away from it. This essentially cut p_1 in the middle (see [6] for a more detailed description of this procedure), the alternative path p_2 from u to v combined with p_1 yields the initial cycle C. For the next phase,

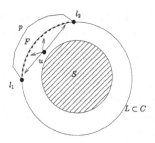

Fig. 4. Growing C by updating the current cycle with the one formed by L and p

let $\mathcal{S}$ be the set of connected components after removing C together with its 1-hop neighbors and $S \in \mathcal{S}$ be the largest of these connected components. We try to grow S as much as possible by introducing more nodes to it and updating the outer cycle as needed. Every node u which is in the 1-hop neighborhood of S (and C as well) tries to include itself into S. Let $F \subseteq C$ be the set of neighbors of u in C, L be the largest connected component of $C \setminus F$, and l_1 and l_2 be the extreme points of L. If there is l_1-l_2 path p which do not pass through nodes in $S \cup F$ then we replace the current cycle with the one formed by L and p. We keep updating C in this manner until no more node can be added to S.

Partitioning of the Outer Boundary Cycle. An important aspect that influences the load-balancing property of the producers' and consumers' paths is the partitioning of the outer boundary cycle. For a badly chosen partition these paths may concentrate in certain regions and thus cause congestion in the network. Figure 5, left shows the temperature gradient field (as described in the following) when the partition of outer-cycle of square-shaped network is taken to be the sides of the square, while Figure 5, right considers the other case when the partitions extend across the corners of the square. Note that the contours in the former case are uniform while in later case they are more concentrated at corners which will lead to higher load on those nodes. This observation suggests to select the partition that somehow respect the geometry of the network to have better load-balancing (without using the geographic location information, though).

Our heuristic tries to identify convex corners in the network boundary and cut the cycle at these corners. For a node x on the outer cycle C, let $u, v \in C$

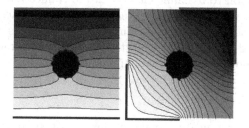

Fig. 5. Gradient field for square-shaped network and different partitions

be two nodes which are k distance apart from x on C in the opposite directions, where $k \leq |C|$. Consider the ratio $r(x) = \frac{2k}{\text{dist}(u,v)}$, where $\text{dist}(u, v)$ represents the hop count distance between u and v in the communication graph. Note that if x is near a convex corner of the sensor network, the distance $\text{dist}(u,v)$ is much smaller than $2k$. This observation suggest to select four corners (cut-points) which are sufficiently far apart and have highest $r(x)$ value. Using $k = |C|/8$, we pick them one by one and selecting the next which is at least $k + 1$ distance from previously selected corners and has the highest ratio.

Construction of Temperature Gradient-Fields for the Double Ruling. Given a partition of outer cycle G_-, G_+, H_-, H_+ such that the G and H pieces alternate, a straight forward way to define producers' and consumers' paths is to use the hop count towards these pieces. More precisely the producer announces its message on shortest paths towards G_- and G_+, whereas the consumer look for the desired information by sending a message on shortest paths towards H_- and H_+. The required hop counts (distances to the respective pieces) can be computed by 4 breadth-first searches each from a piece of the partition of outer cycle C.

The problem with this approach is that it is very sensitive to the structure of the pieces of the outer cycle. For example when let's say piece G_- has a protruding vertex v_-, most traffic sent towards G_- will actually be sent towards v_- leading to high load imbalance. Another idea which is inspired by Skraba et el. [14] is to compute potential functions g and h on nodes whose gradient define the producers' and consumers' paths respectively, and which are not as sensible to protruding vertices. In [14], the gradient field g to define producers' path is set up by fixing the temperature of nodes in G_- to 0 and to 1 for nodes in G_+, and iteratively setting the temperature of the other nodes to average their neighbors until conversion (analogously for H_+/H_-). This process effectively solves a Laplace's equation with Dirichlet boundary conditions over the network. As noted in [14], iteratively solving Laplace equation discretized on N grid-points is well-studied problem and it takes $\mathcal{O}(N)$ time in 2D [16]. They also observed similar running times for graphs with constant degree as usually encountered in wireless networks. To speed-up the convergence, we set the initial values of g and h to be the relative distance to G_+ and H_+ respectively before starting the iterative process as it was already suggested in [9] to speed up the computation of a similar function.

While the initial computation requires some effort, we want to emphasize that in terms of adaptation to dynamic changes in the connectivity structure of the network, this approach of iteratively averaging requires almost no effort apart from periodically checking the temperatures of the neighbors. Small changes in the network topology are also not expected to change the gradient field drastically, so stabilization can be achieved quite quickly.

3   Simulations

To evaluate the performance of the algorithms described above, we performed a set of computer-simulated experiments considering four variants of our approach, distinguishing between a hop-count-based gradient field and a temperature-based gradient field (**Hop/Temp**) as well as shape aware or unaware outer boundary partitioning (**SHA/SHU**). Main objective was to analyze the differences in terms of load balancing, but we also compared our best variant (Temp/SHA) with GHT in terms of responsiveness for the consumers (note, though, that the routing protocol GPSR [7] which GHT is based upon requires geographic location information which our approach does not). The Gabriel graph was used for face routing in GPSR.

Our simulator [12] is not packet-based, and thus it does not take into account some issues that occur in practice (i.e. medium access and message loss). However, we feel that these factors would have similar impact on all algorithms, and thus would not significantly affect the relative performance.

3.1   Load Balancing

Let us first consider the amount of load (number of packets) received by nodes during information exchange between producers and consumers. We compare the distribution of load on nodes for different variants of our scheme.

The simulations in this section are performed on a network containing 12000 nodes distributed uniformly at random with average degree of 29. Every node

Table 1. Inverse cumulative frequencies of load for various double ruling schemes. (a) Hop-Count/Shape aware, (b) Temperature/Shape aware, (c) Temperature/Shape unaware (d) Hop-Count/Shape unaware.

Load		250	450	650	850	1050	1250	1450	1650	1850
Inverse	Hop/SHA	826	366	192	104	61	25	10	5	2
Cumulative	Temp/SHA	694	95	16	4	2	2	0	0	0
Frequency	Temp/SHU	605	190	93	30	23	17	14	4	3
	Hop/SHU	626	268	138	90	69	48	33	22	17

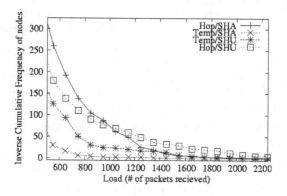

Fig. 6. Plot of Table 1

Fig. 7. Load distribution for various simulations presented in Table 1: Hop/SHA (top-left), Temp/SHA (top-right), Temp/SHU (bottom-left) & Hop/SHU (bottom-right)

in the network is both producer and consumer of information of single data type. This will essentially flood the network where every node is interested in getting information from every other node. Table 1 shows the inverse cumulative frequencies of nodes i.e. number of nodes having load greater than certain values. Figure 6 presents the same as plot. Our scheme clearly outperform others by exhibiting fewer nodes with high load.

Figure 7 shows the actual load distribution in the network during our simulations of Figure 6. We observe that, using the hop count for the double ruling gives nodes with high loads scattered all over the network, while choosing a bad boundary partition yields high loads near the breakpoints between G_-, G_+, H_-, H_+. This is due to many paths being attracted towards these corners. The scheme with bad boundary partition and shortest path gives the worst results in all of the schemes considered. Employing both, a temperature-based gradient field as well as shape aware partitioning of the network delivers the most balanced load distribution with only few exceptions near the corners.

Table 2. Inverse cumulative frequencies of load for various double ruling schemes on a network with a hole in the middle (a) Hop-Count/Shape aware, (b) Temperature/Shape aware, (c) Temperature/Shape unaware (d) Hop-Count/Shape unaware

Load		250	450	650	850	1050	1250	1450	1650	1850
Inverse	Hop/SHA	608	273	153	113	71	42	22	13	5
Cumulative	Temp/SHA	538	82	17	8	5	4	2	0	0
Frequency	Temp/SHU	461	180	95	41	20	11	8	2	1
	Hop/SHU	475	208	114	84	57	28	18	15	15

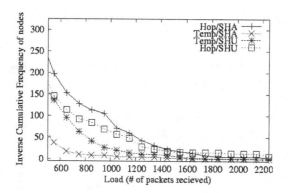

Fig. 8. Plot of Table 2

3.2 Load Balance for Non-regular Network Topologies

Let us now examine how robust our approach is in the presence of holes or communication voids in the network. Table 2 shows the simulation results on a square network containing single hole in the middle, while Figure 8 shows the same as plot. The schemes using shortest paths to define producers' and consumers' paths exhibit rather high load since shortest paths tend to 'hug' the hole boundaries quite closely, the temperature based variant with shape aware boundary partition again performs best.

Also for comparison we consider the simple geometric double ruling scheme, where producers (consumers) simply send their messages along horizontal (vertical) paths. Although this simple scheme gives good results for rectangular (axis-parallel) shaped networks, our scheme (Temp/SHA) exhibits better load balancing properties for networks with holes and more complicated shapes. Figures 9 and 10 gives two example of such cases.

3.3 Distance Sensitivity

Our double ruling scheme has the property that the distance consumer needs to travel to get information from a producer is proportional to the actual distance between them. While in GHTs, a consumer need to travel to hashed location to

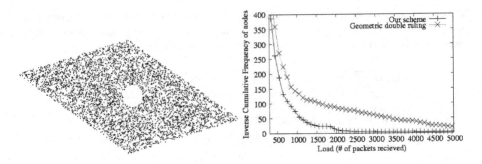

Fig. 9. Comparison with simple geometric double ruling

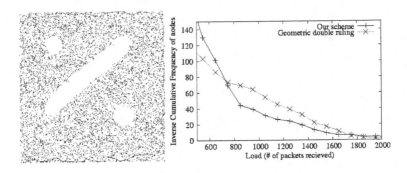

Fig. 10. Comparison with simple geometric double ruling

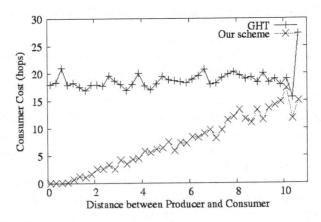

Fig. 11. Average hop-distance consumer need to travel for varying distances between producer and consumer

get information even when the actual distance between producer and consumers is relatively short. We simulated both schemes and consider the average hop-distance a consumer needs to travel for varying distances between producer and consumer. In a network of 10000 nodes distributed uniformly at random, we define 1000 data terms and randomly pick a pair of producer and consumer for each of the type. Figure 11 presents the simulation results. Our scheme exhibits a direct relation between the actual distance of producer and consumer versus the cost incurred by a consumer to get the information, while the GHTs shows no such relation (since GHTs hash to a random location in the network which is typically quite far away).

4 Conclusion

In this paper we have presented a location-free double ruling scheme for information brokerage within a wireless sensor network. The absence of geographic location information requires some novel ideas how to carry over the simple intuition of the basic geometric double ruling scheme. The resulting approach exhibits quite good load balancing properties and also performs favorably in terms of distance-sensitivity from the consumer's point of view when for example compared with popular schemes like GHTs. On the downside, our approach – since based on a boundary recognition phase and the computation of the respective gradient fields – requires some time for startup until it can fully operate. Once in operation, it can rather easily deal with network volatility, though. Directions for future research include simplification of the startup phase as well as a more thorough investigation of the properties of the temperature gradient field to further decrease load imbalance.

References

1. Das, S.M., Pucha, H., Hu, Y.C.: Performance comparison of scalable location services for geographic ad hoc routing. In: 24th Conference of the IEEE Communication Society (INFOCOM) (2005)
2. Fang, Q., Gao, J., Guibas, L.J.: Landmark-based information brokerage in wireless sensor networks. In: Proc. 25th Conference of the IEEE Communications Society (InfoCom), Barcelona, Spain (April 2006)
3. Fekete, S.P., Kröller, A., Pfisterer, D., Fischer, S.: Deterministic boundary recognition and topology extraction for large sensor networks. In: Proc. of SODA (2006)
4. Funke, S.: Topological hole detection in wireless sensor networks and its applications. In: DIALM-POMC '05: Proceedings of the 2005 joint workshop on Foundations of mobile computing, Cologne, Germany, pp. 44–53. ACM Press, New York, USA (2005)
5. Funke, S., Guibas, L.J., Nguyen, A., Wang, Y.: Distance-sensitive routing and information brokerage in sensor networks. In: Gibbons, P.B., Abdelzaher, T., Aspnes, J., Rao, R. (eds.) DCOSS 2006. LNCS, vol. 4026, pp. 234–251. Springer, Heidelberg (2006)
6. Funke, S., Klein, C.: Hole detection or: how much geometry hides in connectivity? In: SCG '06: Proceedings of the twenty-second annual symposium on Computational geometry, pp. 377–385. ACM Press, New York, USA (2006)

7. Karp, B., Kung, H.T.: GPSR: greedy perimeter stateless routing for wireless networks. In: MobiCom '00: Proceedings of the 6th annual international conference on Mobile computing and networking, pp. 243–254. ACM Press, New York, USA (2000)
8. Newsome, J., Song, D.: GEM: Graph embedding for routing and data-centric storage in sensor networks without geographic information. In: 1st Int'l Conf. Embedded networked sensor systems, pp. 76–88 (2003)
9. Rao, A., Papadimitriou, C., Shenker, S., Stoica, I.: Geographic routing without location information. In: 9th ACM Int'l. Conf. Mobile Computing and Networking (MobiCom), pp. 96–108 (2003)
10. Ratnasamy, S., Karp, B., Yin, L., Yu, F., Estrin, D., Govindan, R., Shenker, S.: GHT: A geographic hash table for data-centric storage in sensornets. In: 1st ACM Workshop on Wireless Sensor Networks and Applications, pp. 78–87 (2002)
11. Sarkar, R., Zhu, X., Gao, J.: Double rulings for information brokerage in sensor networks. In: MobiCom '06: Proceedings of the 12th annual international conference on Mobile computing and networking, pp. 286–297. ACM Press, New York, USA (2006)
12. Schmitt, D.: WNETS - a framework for testing and evaluating algorithms and models for wireless sensor networks. Master thesis, Universität des Saarlandes (2007)
13. Shenker, S., Ratnasamy, S., Karp, B., Govindan, R., Estrin, D.: Data-centric storage in sensornets. ACM SIGCOMM Computer Communication Review 33(1), 137–142 (2003)
14. Skraba, P., Fang, Q., Nguyen, A., Guibas, L.: Sweeps over wireless sensor networks. In: IPSN '06: Proceedings of the fifth international conference on Information processing in sensor networks, pp. 143–151. ACM Press, New York, USA (2006)
15. Wang, Y., Gao, J., Mitchell, J.S.B.: Boundary recognition in sensor networks by topological methods. In: Proc. of the ACM/IEEE International Conference on Mobile Computing and Networking (MobiCom) (September 2006)
16. Young, D.: Iterative Methods for Solving Partial Difference Equations of Elliptic Type. PhD thesis, Harvard University (1950)

Level Set Estimation Using Uncoordinated Mobile Sensors

Gagan Raj Gupta[1] and Parmesh Ramanathan[2]

[1] Department of Computer Science
University of Wisconsin, Madison, USA
[2] Department of Electrical and Computer Engineering
University of Wisconsin, Madison, USA

Abstract. We develop level set estimation algorithms for a novel low cost sensor network architecture, where sensors are mounted on agents moving without an explicit objective of sensing. A level set in a planar scalar field is the set of points with field values greater than or equal to a specified value. We model the problem as a classification problem and evaluate a heuristic to reduce the amount of communication assuming that the base station uses a Support Vector Machine classifier. We then develop a fully distributed, low complexity solution which uses opportunistic information exchange to estimate level set boundaries locally at nodes selected using leader election. We observe that the learning rates of the boundary in a locality is proportional to the complexity. Effectiveness of the proposed scheme is evaluated using simulations with data from both synthetic and measured fields. Random way point mobility model is used for node motion and trade off of accuracy and of coverage with communication costs is studied.

1 Introduction

Wireless sensor networks composed of low-cost miniaturized devices with sensing, processing and wireless communication capabilities are envisioned for a wide range of applications in contaminant tracking in the environment, measuring large scale seismic events; surveillance in military zones and habitat monitoring [1,2]. An important query in such networks is the estimation of level sets which allows us to capture the spatial variation in a planar scalar field even when the field is difficult to express in an explicit mathematical form.

In [10], the authors show that the mean square error to estimate edges (sharp boundaries) with static nodes is bounded below by $O(\frac{1}{\sqrt{n}})$, where n is the number of nodes in the network. This means that a four fold increase in number of nodes roughly reduces the mean square error by half. The level set boundaries are usually not sharp, making the detection even more difficult. Hence, to guarantee coverage and good accuracy, a large number of nodes will have to be deployed. If the region is large, such a network will become infeasible.

An attractive alternative is to build networks with mobile sensors. With mobile sensors, constraints like limited sensing range and limited communication

E. Kranakis and J. Opatrny (Eds.): ADHOC-NOW 2007, LNCS 4686, pp. 101–114, 2007.
© Springer-Verlag Berlin Heidelberg 2007

range are relaxed as each mobile node can visit multiple locations to make measurements. The apparent trade off is the latency needed to complete the measurements sequentially. The challenge is to build a network of mobile sensors which coordinate their movements and collaborate to efficiently identify the level sets [12,13]. This comes with a high cost for motion control techniques.

There are situations however, where one can mount sensing devices [4] on agents moving in the region, e.g. animals moving in their habitat [1,3], vehicles moving in a city etc. for the task of sensing. The vehicles used for excursions in natural habitats may be mounted with sensors which measure critical attributes like oxygen level, nitrate levels etc. in the environment which can be used as gross measures of water quality or to provide an early warning of an excursion from normal conditions because of contamination, illicit discharge, or some measurable ecological or biological change [2]. The sensors can be mounted on these agents and no efforts are made to coordinate the motion of each node, i.e., each node makes measurements independently along its arbitrary path. It is assumed that these nodes are provided location detection services. In a military setting, vehicles and personnel may be equipped with sensors even if they are not dedicated to performing any explicit sensing task. The sensors operate in the background and make measurements on the move at places they visit. This network can be used for example, to make detailed exploratory maps of the region. These networks are attractive due to low cost of deployment and coordination of mobility. In this paper we study networks of such mobile sensors with uncoordinated mobility (UM) which collaborate to estimate the level sets of the field.

Since, every UM node makes measurements only along its path, without information exchange, the scope of estimation is limited. Collaborations among the UM nodes is used to solve the challenges due to noisy measurements, storage and power constraints. If more than one node has made measurements at the same location, the estimations can be robustly fused. Such information exchange is opportunistic, as it occurs only among nodes coming across the same vicinity.

In this paper, we first compare our approach with existing work and then describe the problem formally. We then present a centralized solution using Support Vector Machine (SVM) and describe an approach to minimize communication costs. A practical implementation however, should be distributed and should explore the trade-off between information processing and communication. We explore the various trade-offs in the distributed scheme and discuss results from a few case studies.

2 Related Work

Willet et al. describe an adaptive sampling scheme [10] to estimate sharp boundaries (edges) using static sensor networks. The scheme is energy efficient and theoretical bounds on error and energy costs are derived. In contrast, Singh et al. [13] formulate an approach to estimate boundaries using mobile nodes. They show that adaptive sampling and planning the motion of the nodes reduces the energy cost involved in estimation of the boundary. Estimating level sets is a much more challenging problem because level set boundaries may not correspond to singularities or edges in the underlying function.

We believe that a large class of applications can be designed using level set estimation as a primitive. Hence to evaluate the potential of Uncoordinated Mobile Sensor Architecture, we choose level set estimation problem. [5,6] motivate the use of detecting isolines for efficient continuous monitoring in sensor networks. The problem of detecting level sets with static sensors has been formulated in [7,11], and some theoretical results on error-performance trade-off have been presented. The solution requires high node density and involves high set up costs.

A practical scheme using mobile nodes to track level curves has been described in [12] by Leonard and Zhang. However, in their scheme, motion of the sensors is tightly coupled with the measurements and is sensitive to noise. If the environment is hostile, node movement is restricted,which makes planning very difficult. A network composed of uncoordinated mobile nodes, alleviates the cost of planned motion [14]. Such a solution is also robust to node failures as opposed to static sensor networks and coordinated mobile networks. This motivates us to explore UM model as a practical solution to large scale problems.

The novelty of this work is both in the sensor network architecture and the development of the solution techniques. Although classification using SVM is a standard procedure, we develop new techniques for level set estimation. By sending data only in a band around the boundary, we save on communication costs. This technique is interesting in its own right even for the machine learning community because limiting the training set to only the data around the boundary makes the learning problem even more challenging.

The distributed solution developed in this work is also customized for the problem of detecting level sets. The mobile nodes contend with each other to develop local estimates of the level sets. However, there is opportunistic collaboration amongst the nodes through the exchange of measurements. Efficient and robust classification algorithms are employed at each node to process these measurements and arrive at piecewise linear functions estimating the level set boundary. Low cost communication primitives are then used to fuse these estimates at the base station. We demonstrate the effectiveness of our algorithm with several case studies.

3   Problem Description

Level set estimation is the process of using noisy observations of a d-dimensional function $\mathcal{F}$ defined on the unit hypercube to estimate the region(s) in $[0, 1]^d$ where $\mathcal{F}$ exceeds some threshold z; i.e. $L_z \equiv \{\mathbf{x} \in [0, 1]^d : \mathcal{F}(\mathbf{x}) \geq z\}$

We assume that the field is stationary. This is justified if the field varies slowly as compared to the rate of sensing. A measurement is always associated with its measured time and location, and the measurement made by node u is stored as a 3 tuple, $M_i^u = [v_i^u, t_i^u, p_i^u] \; \forall i = 1, ..., m_u$ where v_i^u is the measured value of the field at time t_i^u and at a location p_i^u. m_u is the number of such measurements currently in the database of node u. As the sensors are not perfect, the measurements are corrupted with noise. Thus, the value v_i^u, measured by the node u, has additive white Gaussian noise n_i^u, i.e.,

$$v_i^u = \mathcal{F}(p_i^u) + n_i^u \tag{1}$$

We assume that the queries for level set estimation within an error threshold arrive at the base station and are propagated into the network. We assume that the network is equipped with location mapping services, through GPS or other location services as proposed in [9].

The level set boundary divides the region into two, and every point in the region can be classified as being present or absent in the level set of a particular value z. This classification function C_z for the value z takes the value 1 when $\mathbf{x}$ lies in L_z and 0 otherwise; i.e. $C_z : [0,1]^d \rightarrow \{0,1\}$

$$C_z(\mathbf{x}) = \begin{cases} 1 & \text{if } \mathbf{x} \in L_z \\ 0 & \text{if } \mathbf{x} \notin L_z \end{cases}$$

Depending on the nature of the field $\mathcal{F}$, these boundaries may be smooth or have discontinuities. These functions could also be defined piecewise. Using these measurements the level sets can be learned, of course with errors. If C_z^* is the estimated classification function, then the percentage classification error is given by

$$E = \sum_{i=1}^{N} \left\| \frac{C_z^*(\mathbf{x}_i) - C_z(\mathbf{x}_i)}{N} \right\| \tag{2}$$

Coverage is the percentage of boundary thus approximated within the error threshold at any given time.

The average latency is defined as the average time spent before achieving the required error threshold for the estimate of the boundary. The amount of energy spent in sensing is assumed to be negligible and mobility comes for free. We assume that the amount of energy spent in communication is proportional to the square of distance between the nodes and consumes the major portion of battery supply.

The objective of this paper is to design, implement and evaluate solutions to estimate the level sets with high level of accuracy, and to understand the trade-offs with latency and communication costs. The trade off between coverage and the amount of communication is also explored.

4   Solution

Since the level set boundary divides the region into two classes, we model the problem as a classification problem. In machine learning literature, the classification problem has been extensively explored and tools of different computation complexity and performance have been proposed. We briefly review some of the fundamentals of Support Vector Machines (SVM) [16]. Subsequently, the distributed algorithm is developed using Linear SVM classifiers which is a low-complexity classifier suitable for implementation on a sensor node.

4.1   Review of SVM

Suppose, we are given some training data $(\mathbf{x}_1, y_1), ..., (\mathbf{x}_n, y_n)$ where $y_i \in \{1, -1\}$. Assume that the classes represented by the subset $y_i = 1$ and $y_i = -1$ can be separated by a decision surface in the form of a hyperplane $\mathbf{w}^T\mathbf{x} + b = 0$; where, $\mathbf{x}$ is an input vector, $\mathbf{w}$ is an adjustable weight vector, and b is a bias. The goal of a support vector machine is to find the *optimal separating hyperplane* (OSH), for which the margin of separation M between the hyperplane and the closest data point is maximized.

The pair $(\mathbf{w}_o, b_o)$ with appropriate scaling, must satisfy the constraint:

$$\mathbf{w}_o^T\mathbf{x} + b_o \geq 1 \quad \forall y_i = +1 \tag{3}$$

$$\mathbf{w}_o^T\mathbf{x} + b_o \leq -1 \quad \forall y_i = -1 \tag{4}$$

The particular data points $(\mathbf{x}_i, y_i)$ for which $y_i[\mathbf{w}^T\mathbf{x}_i + b] = 1$ are called support vectors, that lie closest to the decision surface and are therefore the most difficult to classify. Since the distance to the closest point is $\frac{1}{\|\mathbf{w}\|}$, finding the OSH amounts to minimizing $\| \mathbf{w} \|$ and the objective function is: $min\phi(\mathbf{w}) = \frac{1}{2} \| \mathbf{w} \|^2$ subject to the constraints (3) and (4).

If $(\alpha_1, \alpha_2..., \alpha_N)$, the N non-negative Lagrange multipliers associated with constraints (3) and (4), the solution $\mathbf{w}$ has an expansion $\mathbf{w} = \sum i\alpha_i y_i \mathbf{x}_i$ in terms of support vectors. The classification function can thus be written as

$$f(\mathbf{x}) = \text{sgn}(\sum i\alpha_i y_i \mathbf{x}_i \mathbf{x} + b) \tag{5}$$

If the data is not linearly separable, SVM introduces slack variables and a penalty factor such that the objective function can be modified as

$$\phi(\mathbf{w}) = \frac{1}{2} \| \mathbf{w} \|^2 + C(\sum_{i=1}^{N} \zeta_i) \tag{6}$$

which satisfy the constraints (7)

$$y_i(\mathbf{w}_o^T\mathbf{x} + b_o) \geq 1 - \zeta_i \quad \forall 1 \leq i \leq n \tag{7}$$

The input data can be mapped into a higher-dimensional feature space in which the data may be become separable and hence the OSH is constructed. Using the kernel trick, the dot product required in (5) can be computed efficiently by $k(\mathbf{x}, \mathbf{y}) := (\phi(\mathbf{x}).\phi(\mathbf{y}))$ provided that the kernel k satisfies Mercer's condition [19]. Finally, we obtain the classification function

$$f(\mathbf{x}) = \text{sgn}(\sum i\alpha_i y_i k(\mathbf{x}_i, \mathbf{x}) + b) \tag{8}$$

To test the accuracy of the estimated boundary computed by using the SVM, we use the statistical technique called Cross-validation [17]. In this scheme, a sample of data (measurements made by sensor nodes) is partitioned into subsets such that the SVM algorithm is initially executed on a single subset, while the other subset(s) are retained for subsequent use in confirming and validating the estimate. The initial subset of data is called the training set; the other subset(s) are called validation or testing sets. The percentage of points in the testing set that are misclassified is called the cross validation error. SVM has strong theoretical foundations based on statistical learning theory, and has particular advantages when applied to problems with limited training samples and noisy data and hence is expected to yield low cross validation errors.

4.2   Centralized Scheme

In this scheme, the base station aggregates the measurements made by the mobile nodes to answer the queries regarding level sets. The measurements can be considered as training examples for the problem of learning the level set boundaries. Each training example maps a location in the region to a class. In the light of above discussion about SVMs, it is clear that the training data collected by the nodes can easily be used to estimate the complicated (usually non-linear) level set boundary. Assuming that the base station has enough computational resources, we use kernels like Radial Basis Functions to map the training data to a higher dimensional space so that most of the training examples can be separated with a hyperplane. By observing the learning curve of the SVM, we find that the error decays exponentially with the size of training set and the curve levels off at certain error value. Figure 2(b) shows a typical learning curve.

Recall, that the support vectors or the points near the boundary are the most important for the SVM learning algorithm. We can reduce the amount of communication by trying to send only those points which are likely to be near to the boundary. Assuming that the field will not change abruptly near the boundary, we can define a tolerance limit so that we only communicate

$$(v, p) : \|v - z\| < \tau * z \tag{9}$$

The value of τ must be higher than the noise level and we study the tradeoff of τ with accuracy.

In the centralized scheme, each node transfers its measurements to the base station, independent of the other. We now develop a distributed solution where are the nodes also process the measurements locally to estimate level set boundaries. Such a practical scheme is of interest to us, because these are easy to deploy and the communication costs are low.

4.3   Distributed Algorithm

Assume that the region has been partitioned into non-overlapping cells and a node is selected as the leader for each cell. Mobile node N_j stores measurements made in cell C_k and maintains a list $\mathcal{L}_i$ of cells for which it is a leader. A node receives information from other nodes only for those cells for which it is a leader. A state is maintained with each measurement (a bitmap of all nodes possessing the measurement) to avoid redundant data from being communicated.

While(Status of Node N_i is Active)

        Make measurement at the current location $p = P(N_i)$

        Add (v,p) to the database of cell C_k s.t. $p \in C_k$.

        Increment the count of number of data points in the cell,$\eta_i(k) \rightarrow \eta_i(k) + 1$

        $\forall Nodes\, N_j$ in communication range, $D(P(N_i), P(N_j)) \leq COMM\_THRESHOLD$

            Send(List of cells $\mathcal{L}_i$ for which N_i is the leader)

            Receive(Measurements in cells $\mathcal{L}_i$ possessed by Node N_j)

            Receive(List of cells $\mathcal{L}_j$ for which N_j is the leader)

            Send(Measurements in cells $\mathcal{L}_j$ possessed by Node N_i)

        If (near the base station)

            Compute boundaries in the cells $\mathcal{L}_i$ and the cross validation error.

            Send(Boundaries of those cells in which error <ERR_THRESHOLD)

        End

Fig. 1. Distributed Level Set Estimation of L_z

The boundaries of the level set for cells $\mathcal{L}_i$ are computed using Linear SVM. If the cross validation error is less than the threshold specified by the query, the equation of boundary is communicated to the base station. The scheme is formally described in Figure 1.

Partitioning of the region into cells involves the choice of cell size and cell locations. As shown later in Figure 4(a), static cell partitioning leads to poor performance if the boundary is too complicated or is present at the *corner* or the *edge* of the cell. Hence, we develop a dynamic scheme for choosing cell size and position. When a node detects that it has crossed the level set, it selects the region around the current location as the cell and becomes the leader for this cell. To determine the cell size, an adaptive algorithm is used. It begins with a large cell size and splits the cell if the cross validation error is high. Splitting a cell allows the complicated boundary to be approximated by many piecewise linear functions. This improves the performance as shown later in Figure 4(b). Leader election is also used to avoid large overlap between cells. When two nodes are within communication range they exchange information about their cells. If two cells overlap significantly, only one of them is kept and the other is deleted.

Having short range communication among the nodes can be much more efficient as compared to the frequent long range communication with the base station. Thus the scheme presented here allows for a natural trade off between the communication and computational resources at the nodes.

5   Evaluation

In this section we present the results describing the performance of the above schemes on simulated fields. One of the commonly used models for simulating fields due to K point sources is,

$$\mathcal{F}(x) = \sum_{i=1}^{K} \frac{q_i}{D(x, y_i)^\alpha} \tag{10}$$

where y_i is the location of source i in the d-dimensional region $[0,1]^d$. The intensity decays exponentially with distance with parameter α. For simplicity of exposition, we begin with a field produced by random placement of $K = 4$ point sources in the 2-dimensional region which is normalized to $[-1,1]^2$. The value of alpha is chosen to be 1.

There are 10 nodes equipped with homogeneous sensors (having the same characteristics) moving on a 100x100 grid. Each node moves one step on the grid after every time unit based on the *random way point mobility model*. We vary the noise level, sensing rate and level set. As time progresses, the nodes start communicating the measurements to the base station, which continuously runs the classification algorithm.

5.1   Centralized Scheme

To understand the trade-off of accuracy with the communication cost we test the classifier thus learned with the testing set generated using equation 10. Figure 2(b) shows that the classification error for L_5 decays rapidly with the number of measurements communicated.

As is frequently observed with learning algorithms, the Figure 2(b) shows that the error, levels off after sometime. If the required error threshold is not met, measurements will have to be made on a finer grid. We also see in Figure 3(a) that the scheme only suffers *graceful degradation with addition of noise*. The noise level indicated is the ratio of the standard deviation of AWGN to z, value of level to be estimated.

As explained in Section 4.2, we can save on the communication cost by sending only the data when the measured values are within a certain tolerance. In other words, we communicate only the data lying in a band of certain size, around the level curve. As we increase the size of the band, more data will be communi-

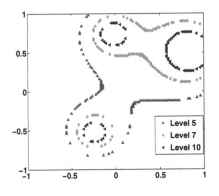

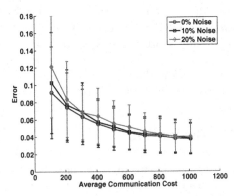

(a) Level Curves of the Simulated Field

(b) Learning Curves with Centralized Scheme for L_5

Fig. 2. Basic Centralized Scheme

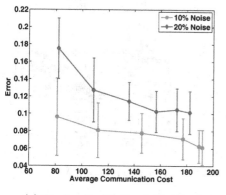
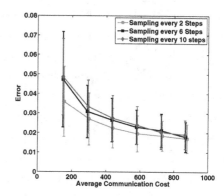

(a) Trade-off with Band Size for L_7

(b) Learning Curves for different sampling rates

Fig. 3. Experiments with the Centralized Scheme

cated. Figure 3(a) shows that in general, *increasing band sizes gives diminishing returns*.

So, the band size can be chosen appropriately. In general, it was observed that decreasing the band size beyond this limit, increases the time taken for the overall classification error to become less than the threshold.

Figure 3(b) shows the effect of changing sampling rate on accuracy. A high sampling rate makes measurements highly correlated to the mobility model while low sampling rate is very similar to random sampling of the field. We plot the cross validation errors with time for different sampling rates. We observe comparable learning curves and accuracy levels for different sampling rates. From this, we can indirectly infer that the mobility model does not influence the effectiveness of this scheme, because when the sampling rate is low and the number of nodes is large, the measurements will be sampled almost uniformly randomly over the region.

5.2 Distributed Scheme

For the distributed case we use the same field, given by Equation 10. The cells where the error is less than the threshold (5%) are sent to the base station. The learning curves in each of cell is similar to those shown earlier in the centralized case. Thus, we focus on the coverage, which is the percentage of boundary estimated within the given error threshold.

Figure 4 shows a snapshot of the cells possessed by various nodes at different times where the cross validation error is less than the threshold. Training examples and the boundary is shown in each of the cells. Figure 4(a) shows the case where cell size and location is fixed. Since, linear classifiers perform poorly in cells having sharply curved boundaries these cells are not communicated to the base station. In contrast, choosing cell location and size dynamically

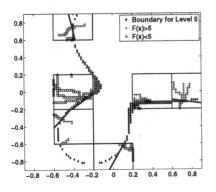

(a) Fixed Cell position, Fixed Cell Size

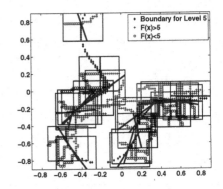

(b) Dynamic Cell position, Adaptive Cell Size

Fig. 4. Comparison of Static and Dynamic Cell Partitioning Schemes

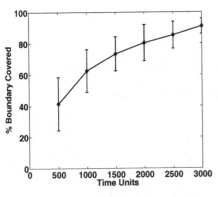

(a) Coverage of the boundary with time

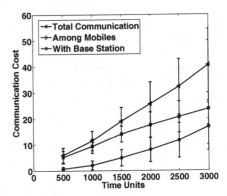

(b) Cumulative Communication cost with time

Fig. 5. Experiments with the Distributed Scheme

improves coverage (Figure 4(b)). These cells also *overlap considerably* and hence we implement *leader election* to reduce the number of cells as discussed before in Section 4.3. For the sake of simplicity, we have only shown cases without any noise. In the presence of noise, thresholds were selected appropriately to ensure that bogus cells without any actual level boundary are not communicated although they may have training examples from both classes (due to noise). Figure 5(a) shows how the coverage of the boundary increases with time for the scheme with Adaptive Cell Location and Adaptive Cell Sizes.

The communication cost is normalized by the energy required to transmit one message, across unit distance. We assume that the cost increases as the square of distance. As the nodes keep on discovering new cells in the region, cost of communication with the base station increases almost linearly as shown in

Figure 5(b). communication threshold among nodes is chosen to be an order of magnitude smaller than threshold for communication with the base station. Thus inter node communication costs are kept low and simultaneously good coverage is also ensured (see Figure 5(a)).

6   Case Studies/Applications

In the previous section we have demonstrated the efficacy of the distributed scheme. In this section we apply the distributed algorithm with dynamic cell size and location in the following case studies. These applications are motivated by the importance of in-situ measurements that could potentially reduce the error in measurement and increase our understanding of large scale physical phenomena like contaminant flow. They are ideal for deployment in adverse settings such as explosion plumes, oil slicks etc. Tracking the illumination intensity levels is important in environmental studies where it is highly correlated with animal activity and plant growth etc. It is also important to monitor illumination intensity in stadiums, public places etc.

6.1   Environmental Pollution

We used a sensor field of varying pollutant concentration generated by a pollutant flow modeling tool, $WQMAP^{TM}$ as described in section 5 of [15]. A typical query for such a network is the estimation of the spatial extent of contour of particular levels of pollution in the ecosystem. Figure 6(a) shows the estimated boundary for two different level curves, L_{36} and L_{120} at $T = 5000$ $units$ corresponding to permissible and very high levels of pollution respectively within 5% error.

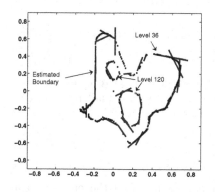

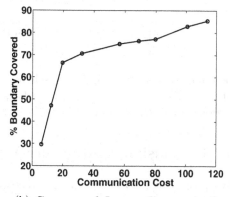

(a) Level sets L_{36} and L_{120} in the Pollution field

(b) Coverage of L_{36} vs. Communication cost

Fig. 6. Results from the Pollution Experiment

Initially, those portions of the boundary which can be easily approximated by lines are estimated with relatively less number of measurements and hence incur low communication cost. The regions with complicated boundaries require cells to be split up and incur larger communication costs. This is responsible for presence of two distinct regions in figure 6(b), where we plot the extent of coverage of pollution level L_{36} with cumulative cost of communication (normalized).

6.2   Illumination Levels in a Room

A light field was set up in a dark room using electric bulbs. We used Crossbow motes MICA2 MPR400CB with MTS300CA sensor boards [21] to make measure light intensities in the range [0-255] on a 150x150 grid and the data was used to simulate tracking of well lit regions, L_{141} within 5% error shown in Figure 7.

Figure 7 shows the snapshot of the boundaries possessed by the nodes at $T = 5000$ $units$ for L_{141}. As we can see, in this case there are two disjoint level

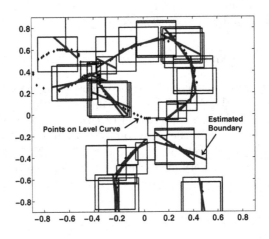

Fig. 7. Boundary of L_{141} in the Illumination experiment

boundaries, but the distributed scheme is competent enough to estimate both of them. Also note that there are a large number of cells of smaller size in the regions where the boundary has a large gradient.

7   Conclusions and Future Work

We proposed a novel sensor architecture and demonstrated the efficacy of a simple decentralized algorithm to carry out level set estimation. Various implementation details have been provided to reduce the communication costs. It is a cost efficient solution to applications in security, environmental studies, civilian application etc. Thus, the scheme has the advantages similar to coordinated efforts to track the level sets with mobile sensors without the cost of mobility. As future work, we would like to understand the trade-offs associated with estimating multiple levels at the same time.

References

1. Szexczyk, R., Osterweil, E., Polastre, J., Hamilton, M., Mainwaring, A., Estrin, D.: Habitat monitoring with sensor networks. Communications of the ACM 47(6), 34–40 (2004)
2. Soreide, N.N., Woody, C., Holt, S.M.: Overview of ocean based buoys and drifters: Present applications and future needs. In: 16th Inter. Conf. on Interactive Information and Processing Systems for Meteorology, Oceanography, and Hydrology (2004)
3. Juang, P., Oki, H., Wang, Y., Martonosi, M., Peh, L., Rubenstein, D.: Energy-efficient computing for wildlife tracking: Design trade offs and early experiences with zebranet. In: ASPLOS, San Jose, CA, October (2002)
4. Sikka, P., Corke, P.I., Overs, L.: Wireless Sensor Devices for Animal Tracking and Control. LCN (2004)
5. Solis, I., Obraczka, K.: Effcient Continuous Mapping in Sensor Networks Using Isolines. Mobiquitous (2005)
6. Meng, X., et al.: Contour maps: monitoring and diagnosis in sensor networks. Computer Networks: The IJCTN 50(15) (October 2006)
7. Liao, P.-K., Chang, M.-K., Kuo, C.-C.J.: A distributed approach to contour line extraction using sensor networks. Vehicular Technology Conference (2005)
8. Buragohain, C., Gandhi, S., Hershberger, J., Suri, S.: Contour Approximation in Sensor Networks. In: Gibbons, P.B., Abdelzaher, T., Aspnes, J., Rao, R. (eds.) DCOSS 2006. LNCS, vol. 4026, Springer, Heidelberg (2006)
9. Tilak, S., Kolar, V., Abu-Ghazaleh, N.B., Kang, K.D.: Dynamic Localization Control for Mobile Sensor Networks. IEEE International Workshop on Strategies for Energy Efficiency in Ad Hoc and Sensor Networks (IEEE IWSEEASN05)
10. Willet, R., Martin, A., Nowak, R.: Backcasting: Adaptive sampling for sensor networks. In: Proceedings of IPSN (2004)
11. Willet, R.M., Nowak, R.D.: Minimax Optimal Level Set Estimation, submitted to IEEE Transactions on Image Processing
12. Leonard, N., Zhang, F.: Generating Contour Plots Using Multiple Sensor Platforms. IEEE Swarm Intelligence Symposium (2005)
13. Singh, A., Nowak, R., Ramanathan, P.: Active Learning for Adaptive Mobile Sensing Networks, IPSN, Nashville, TN (2006)
14. Wang, K.-C., Ramananthan, P.: Collaborative sensing using sensors of uncoordinated mobility, In: Proceedings of International Conference on Distributed Computing in Sensor Systems. LNCS, pp. 293–306 (June 2005)
15. Srinivasan, S., Ramamritham, K., Ramanathan, P.: Contour Estimation using Collaborating Mobile Sensors, DIWANS, Los Angeles (2006)
16. Joachims, T.: Making large-Scale SVM Learning Practical. In: Schlkopf, B., Burges, C., Smola, A. (eds.) Advances in Kernel Methods - Support Vector Learning MIT-Press (1999)
17. Kohavi, R.: A study of cross-validation and bootstrap for accuracy estimation and model selection. IJCAI (1995)

18. Teillet, P.M., et al.: A Framework For In-Situ Sensor Measurement Assimilation in Remote Sensing Applications. In: Proceedings of the 22nd Canadian Symposium on Remote Sensing, Sainte-Foy, Quebec, pp. 111–118 (2001)
19. Cristianini, N., Shawe-Taylor, J.: An Introduction to Support Vector Machines and other kernel based learning methods. Cambridge University Press, Cambridge, UK (2000)
20. Hyyti ä, E., Virtamo, J.: Random Waypoint Model in n-Dimensional Space. Operations Research Letters 33/6, 567–571 (2005)
21. http://www.xbow.com/

The Impact of Delay in Dominating Set and Neighbor Elimination Based Broadcasting in Ad Hoc Networks

Fabián García-Nocetti[1], Francisco Javier Ovalle-Martínez[2,3],
Julio Solano-González[1], and Ivan Stojmenović[3,4]

[1] DISCA, IIMAS, UNAM, Ciudad Universitaria, D.F. 04510, México
{fabian,julio}@uxdea4.iimas.unam.mx
[2] National Institute of Genomic Medicine, Periférico Sur 4809, D.F. 14610, México
fovalle@inmegen.gob.mx
[3] SITE, University of Ottawa, Ottawa, Ontario K1N 6N5, Canada
ivan@site.uottawa.ca
[4] Electronic, Electrical & Computer Engineering, The University of Birmingham, UK

Abstract. In a broadcasting task, a source sends a message to all the nodes of a network. There exist methods for flooding a network intelligently and for scheduling node activities. Dominating sets and neighbor elimination based broadcasting is currently the most efficient broadcasting scheme in terms of the number of re-transmitted messages to complete a broadcast. It provides basis for defining other broadcasting protocols by changing the definition of the delay (timeout) function used to decide how long a dominating node should wait before making a retransmission. In this article, we propose thirteen such variants. They are all reliable, meaning that all the nodes connected to a source will receive the message, assuming an ideal MAC layer. Eight of them are hexagonal based; four are distance-based, giving priority to the neighbors that are further or nearer from the retransmitting node; and one is using a random timeout. Beyond these variants, we propose three different ways to update the timeout values during a broadcasting process. Our experimental data shows that the updating process of the timeout values has no significant impact compared to the selected timeout function. From the thirteen variants we deliberately proposed some worst-case timeout functions to see its impact in the broadcasting process. We confirm by our experimental data that indeed the selected timeout function has an impact in the broadcasting process. Although our experimental data shows that the new further distance-based scheme outperforms almost all schemes in terms of number of messages to complete a broadcast, it also shows that a random function (the way IEEE 802.11 works) is a very good choice.

1 Introduction

Wireless ad hoc networks are formed dynamically by an autonomous system of static or mobile nodes that are connected via wireless links without using an existing network infrastructure or a central base station. The network's wireless topology may change rapidly and without any certain prediction. These type of networks present several challenges due to its intrinsic characteristics; however, they have plenty of actual and future

E. Kranakis and J. Opatrny (Eds.): ADHOC-NOW 2007, LNCS 4686, pp. 115–128, 2007.
© Springer-Verlag Berlin Heidelberg 2007

applications. Sensor networks are among the most important applications in the ad hoc networks field. These networks can be used either to collect data to monitor the weather or to monitor objects in dangerous or in remote environments. Broadcasting consists in sending a message to all the nodes that form the ad hoc network. There exist two models: *one to all* and *one to one*. In the *one to all* model, a node transmits a message to all its surrounding neighbors (nodes that reside within the transmission radius of the sender). In the *one to one* model the transmitter node sends the message to only one receiving node by using directional antennas or by using different frequencies or codification schemes. In this article we will use the first model where all the nodes use omni directional antennas.

Some broadcasting applications include alarming services, configuration of the network and routes discovery. In highly mobile networks, broadcasting can be used as a routing algorithm [HOTV]. One of the most important applications for data broadcasting resides in the sensor networks field, in which such networks send data using the broadcasting paradigm.

Blind Flooding was the first solution for broadcasting. In the first version of it, each node that receives a message retransmits it to all its neighbors. In the optimized version, each node does not retransmit repeated copies of the same message. *Flooding* has some disadvantages, prohibiting its use in networks with a dynamic topology or where requests for information are frequent and must be known by all the nodes.

Intelligent broadcasting and activity scheduling solutions are based on the concept of dominating sets. Some proposed methods; such as multipoint relay (see [SW] for citations) included forwarding the set of neighbors (generating a message overhead). Besides in this method the set of retransmitting nodes depended on the source node. In [SSZ], the authors developed a method, which does not require the inclusion of a forwarding set in the message, and has a fixed set of retransmitting nodes, regardless of the source node. Maintenance does not require more communication overhead, and has competitive performance. It was further enhanced by applying the concept of neighbor elimination.

Hexagonal flooding [DP] is based on selecting forwarding nodes closer to optimal vertices location of a honeycomb network. The authors claimed that their algorithm was beaconless (avoid to use 'Hello' messages), however, their implementation and explanation assumed 1-Hop knowledge.

In this article we propose the idea to apply dominating set based broadcasting, using the generalized coverage criterion and neighbor elimination schemes explained in [SSZ]. We propose thirteen timeout variants of the original timeout function presented in [SSZ]. Our algorithms are reliable, scalable and localized. A broadcasting algorithm is *reliable* if all nodes, connected to the source, are guaranteed to receive the message, assuming ideal medium access layer. We simulated our proposed broadcast algorithms measuring the number of transmissions done for different network densities. We simulated the broadcasting algorithms proposed in [SSZ] to compare them with the ones presented here. Comparing our protocols and the original algorithm [SSZ] the results were very similar and two of them improved marginally the performance of [SSZ].

We present in section 2 the related work of our proposals. In section 3 we explain and show our broadcasting schemes. Section 4 gives performance evaluation of all protocols. Section 5 concludes this paper and discusses relevant future work and open ideas.

2 Literature Review

Ni, Tseng, Chen and Sheu [NTCS] studied the broadcast storm problem. Flooding is usually very costly and will result in serious redundancy, contention, and collisions. They identified this broadcast storm problem by showing how serious it is through analyses and simulations. Several schemes were proposed in [NTCS] (probabilistic, counter-based, distance-based, location-based, and cluster-based) to reduce redundant rebroadcasts and differentiate timing of rebroadcasts to alleviate this problem. These schemes have high percentage of delivery rate with low number of retransmissions. However, they are not reliable.

In [GSS], the authors proposed a broadcasting algorithm based on clustering. In that algorithm, all clusterheads and gateway nodes (nodes common to two or more clusters) will retransmit the message. Each node is assigned a combined id, with the node degree (average number of neighbors) as primary key, and node id as the secondary key to break the ties. Clusterheads and gateways are determined by applying the well known protocol [LG].

In [DP], a flooding protocol based on hexagonal tiling of plane is proposed, with transmission radius defined as the edge length of the hexagons. The authors described a beaconless flooding algorithm. Their protocol is based on selecting forwarding nodes close to optimal vertex location of a honeycomb network. The message contains the location of the previous and the last senders. This enables the receiver to determine the locations of two vertices of the honeycomb network, with side being equal to transmission radius, and with one direction corresponding to the edge direction determined by the last two senders. Upon receiving such message, each node finds its distance to the closer of the two locations, and sets a timeout proportional to that distance. This allows closer nodes to retransmit sooner, which also suppress other retransmissions. The authors failed to prove that their algorithm is reliable. As observed in [KM], the protocol may repeat flooding if the neighbors do not agree on the choice of the node near the common ideal point. [KM] introduced a stopping rule to prevent that, and applied this type of flooding for route discovery. In fact, we found (experimentally) counterexamples, showing that the algorithm is not reliable.

Wu et al described, in a series of articles (the first one being [WL]), a lightweight backbone construction scheme. We will use a modified definition from [SSZ, S-ssz], because of its reduced message overhead. A node is an *intermediate* node if it has two unconnected neighbors [WL]. A node A is covered by a neighbor node B if each neighbor of A is also neighbor of B, and $key(A)<key(B)$. Nodes not covered by any neighbor are *inter-gateway* nodes. A node A is covered by two connected neighboring nodes B and C if each neighbor of A is also a neighbor of either B or C (or both), $key(A)< key(B)$, and $key(A) < key(C)$. An intermediate node not covered by any neighbor becomes an *inter-gateway* node. An inter-gateway node not covered by any pair of connected neighboring nodes becomes a *gateway* node.

Stojmenovic et al. [SSZ], applied the concept of dominating sets to reduce the communication overhead of broadcasting a message; they proposed the use of node degrees instead of node id's as primary key. They also proposed the neighbor elimination method to reduce the number of broadcasts in the network. The basic idea of neighbor elimination is that a node will rebroadcast the message only if it has a neighbor that might need the message. Thus, some of the neighbors are eliminated for re-broadcasting. As a first phase the nodes in the network apply the communication free protocol described in [SSZ] and further explained in [S-ssz] to form a localized dominating set. Nodes marked as gateways will be the only ones that will transmit during the broadcasting process. When a node receives a message to broadcast, if it's a gateway node, it applies the neighbor's elimination scheme found in [SSZ], and sets a timeout before transmitting. If all neighbors are covered the node does not transmits; however, if the node's timeout expires and there are still neighbors uncovered, the node transmits. The authors used only one timeout function for their algorithm:

$$t = \frac{1}{\text{number of neighbors not yet covered}} \text{, and also they compared it with the ones from}$$

[GSS], showing better performance and overall results.

This article studies the impact of various ways of determining the length and type of timeout function used. We keep timeouts as real numbers and artificially avoid the collisions in our study. Some complementary studies have been made in [SSZ]. It was shown in [SSZ] that, when the timeouts are discretized, there is no significant impact of maximal timeout size (maximal backoff) on the performance when random delay is used as timeout (unless the message is long compared to it). It was also shown in [SSZ] that mobility has no notable negative impact, due to the localized nature of the selected protocols.

[GACM] proposed to apply delayed intelligence in a passive clustering protocol. After receiving a packet, each node calculates a delay that is inversely proportional to the distance from the sender and its local energy. Signal strength can be used to find the exact distance if position information is not available.

[ZSL] describes a neighbor elimination scheme that starts from an initiator node (node with lowest ID). Each node sets a timeout inversely proportional to the number of uncovered neighbors. The method corresponds to the one by Peng and Lu, which is not cited. In fact, Peng and Lu add a randomization part in the formula, to avoid collisions when two neighboring nodes, with the same numbers of uncovered neighbors and remaining energy, transmit.

In [KYZS] the authors continued their protocol development from [ZSL] by modifying the random timeout function in the same framework. The timeout is inversely proportional to the product of the number of uncovered neighbors and the remaining energy level.

In [HBSG] the authors proposed a broadcasting protocol called Delayed Flooding with Cumulative Neighborhood (DFCN). DFCN assumed the knowledge of 1-hop neighbors and used a random delay before emitting a rebroadcast. Although this protocol showed a good performance in terms of number of transmissions to complete a broadcast, it is not reliable.

3   Dominating Sets and Neighbor Elimination Based Broadcasting Variants

3.1 Assumptions

All of our proposals assume that the nodes in the network have geographic position information. Our second assumption is that every node in the network has the same transmission radius; and our final assumption is that all nodes use omni-directional antennas.

3.2 Description

The algorithms presented here are based on the algorithm published in [SSZ]. Basically we propose modified versions of the timeout function used to decide if a node transmits or not. Each gateway node, after receiving a message, will set a timeout. We present and describe thirteen variants to define it. These proposals can be classified in three groups: hexagonal based, distance-based, and random-based. In each group we deliberately proposed some "bad" timeout functions to observe the impact of such functions. Beyond defining the timeout functions with these variants, we also propose three different options to update the timeout values during the broadcasting process. To explain this, imagine a gateway node A that has an active timer T. When A receives a message (if there are still neighbors to cover) it updates the value of T with a new calculated value U (obtained by the possible thirteen variants). In order to update T, node A has the following three options:

- Option 1: delete the current value of T and assign it the value of U.
- Option 2: add value U to the current value of T.
- Option 3: calculate the difference D between U and the time that has passed since node A received for the first time this message. If this difference is positive node A sets T to this value D, if not, T remains with the same value.

These three options are actually implemented and measured here. Additional options can be considered in case $D<0$. One option is to transmit at this point. However, if several nearby nodes have the same event, a collision would occur. Another option is to use timeout $min(D, U)$. That is, if the new timeout is shorter than the time left from the remaining 'old' timeout, then replace it with new one. This is not implemented since an estimated low forwarding priority from one neighbor is expected to be more important than high forwarding priority from another neighbor. The node will still retransmit at the end of the timeout if its help is really needed. Figure 1A shows an example of a network after applying the dominating sets concept.

3.2.1 First Hexagonal Variants
When a node A starts a broadcast or retransmits a message, it appends its location to the transmission. Each gateway neighbor B of A receiving the message calculates six imaginary points. The six points are formed by an imaginary hexagon with center at A's position and with hexagon side's length equal to the transmission radius of A. Each node B, receiving the message, calculates its distance from each of the six points of the hexagon and takes the minimum value d of such distances. We will name the process of creating this imaginary hexagon as "Hexagon1". Finally each receiving gateway node B calculates its timeout. For the Hexagon1 topology we propose four timeout variants. These variants are shown in table 1. As we mentioned before we propose two worst-case timeout functions to evaluate its impact. In these first hexagonal variants Hexagon1Max1 and Hexagon1Max2 are the "bad" selected functions.

Table 1. Possible timeout variants (Hexagon1) that a gateway node B can apply

Variant	Timeout(B)	Min value	Max value
Hexagon1Min1	$\dfrac{1}{\text{number of neighbors still not covered}} + \dfrac{d}{R}$	$\rightarrow 0$	2
Hexagon1Max1	$\dfrac{1}{\text{number of neighbors still not covered}} + \left(1 - \dfrac{d}{R}\right)$	$\rightarrow 0$	2
Hexagon1Min2	$\dfrac{d}{R}$	0	1
Hexagon1Max2	$\left(1 - \dfrac{d}{R}\right)$	0	1

In figure 1B, node A starts a broadcast and its neighbor nodes B, C, D, and F receive the message. These receiving nodes know the ideal points $H1$, $H2$, $H3$, $H4$, $H5$, and $H6$. Table 2 shows the values of the timeout functions of receiving nodes using Hexagon1Min1.

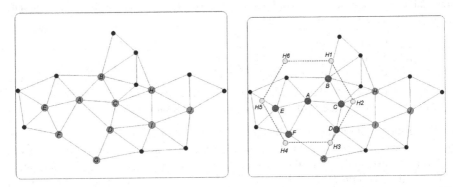

Fig. 1. A) Local dominating set of a network. Gateway nodes A, B, C, D, E, F, G, H, I, and J are the only ones that will transmit if necessary. B) A starts a broadcast. Receiving nodes B, C, D, E, and F, set their timeouts.

3.2.2 Second Hexagonal Variants

For the second variants, we use the characteristics shown in [DP] to calculate two points of a hexagon instead of six. Each node Q transmitting the message now appends its position and the position of the node S that previously sent the message to node Q. Since repeated copies of the message can be received before the timeout expires, we define node S as the node that sent the last copy of the repeated messages before the timeout of Q expires. After node Q transmits the message, each receiving node M calculates two points K and L such that lines QL and QK form 120 degrees and -120 degrees with line SQ. Also, the distances QL and QK are equal to the transmission radius R. Each node M calculates its distance from the two points K and L and takes the minimum value d of such distances. We will name the process of creating these points as "Hexagon2". Finally M calculates its timeout. For the Hexagon2 topology we propose four timeout variants. These variants are shown in table 3. Again, as in the first hexagonal variants we propose two worst-case functions (Hexagon2Max1 and Hexagon2Max2) to observe its impact in the selected timeout function.

Table 2. Example of timeout values (Hexagon1Min1) of nodes from figure 2 after receiving a message

Receiving Node	Closest hexagonal point	Number of neighbors not yet covered	Timeout function
B	H1	3	$\dfrac{1}{3} + \dfrac{distance(B, H1)}{R}$
C	H2	3	$\dfrac{1}{3} + \dfrac{distance(C, H2)}{R}$
D	H3	3	$\dfrac{1}{3} + \dfrac{distance(D, H3)}{R}$
E	H4	2	$\dfrac{1}{2} + \dfrac{distance(E, H4)}{R}$
F	H5	2	$\dfrac{1}{2} + \dfrac{distance(F, H5)}{R}$

Table 3. Timeout variants (Hexagon2) that a receiving node M can apply

Variant	Timeout(M)	Min value	Max value
Hexagon2Min1	$\dfrac{1}{\text{number of neighbors still not covered}} + \dfrac{d}{R}$	$\to 0$	2
Hexagon2Max1	$\dfrac{1}{\text{number of neighbors still not covered}} + \left(1 - \dfrac{d}{R}\right)$	$\to 0$	2
Hexagon2Min2	$\dfrac{d}{R}$	0	1
Hexagon2Max2	$\left(1 - \dfrac{d}{R}\right)$	0	1

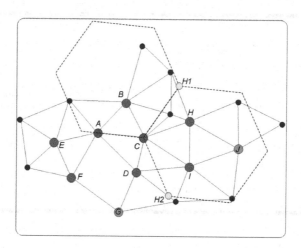

Fig. 2. Node C transmits appending its position and the position of node A. Receiving nodes H and I calculate its distance d the points of the hexagon H1 and H2.

Figure 2 shows an example of how the timeouts are obtained using Hexagon2Min1. Node C transmits the message along with its coordinates and the coordinates of the node from which it received the last copy of the same message, in this case node A. All the new receiving nodes (H, and I) calculate points $H1$ and $H2$ and set their timeouts. Table 4 shows how node H and node I construct their timeouts after receiving the message from C.

Since this variant needs that the transmitter sends its position along with the position of the previous sender, the reader may question how the broadcast can be initiated. In this case when a node wants to start a broadcast to the network, it appends its position two times. One set of coordinates as the transmitter and one set, as it was the previous transmitter. When the receiving nodes notice that both set of coordinates are the same, they realize that the sender has initiated a broadcast. In this case, receiving nodes calculates six positions of the imaginary hexagon shown in variant 1 to set their timeouts. Then the process continues as explained in variant 2.

Table 4. Example of timeout values (Hexagon2Min1) of nodes from figure 3 after receiving a message

Receiving Node	Closest hexagonal point	Number of neighbors not yet covered	Timeout value
H	H1	2	$\dfrac{1}{2} + \dfrac{distance(H, H1)}{R}$
I	H2	3	$\dfrac{1}{3} + \dfrac{distance(I, H2)}{R}$

Table 5. Distance-based timeout variants that a receiving node B can apply after receiving a message from node A

Distance-based variant	Timeout(B)	Min value	Max value
DistPerimeter1	$\dfrac{1}{\text{number of neighbors still not yet covered}} + \left(1 - \dfrac{distance(A, B)}{R}\right)$	$\to 0$	2
DistNodes1	$\dfrac{1}{\text{number of neighbors still not yet covered}} + \dfrac{distance(A, B)}{R}$	$\to 0$	2
DistPerimeter2	$1 - \dfrac{distance(A, B)}{R}$	0	1
DistNodes2	$\dfrac{distance(A, B)}{R}$	0	1

3.2.3 Distance Variants

In these variants instead of using the vertices of a hexagon, we propose to use the distance between the transmitting and the receiving nodes, and the distance between the receiving nodes and the perimeter of the transmission pattern. We propose four distance-based variants that we will call DistPerimeter1, DistNodes1, DistPerimeter2, and DistNodes2. Table 5 shows how a receiving node B calculates each distance-based variant after receiving a message from node A. DistNodes1 and DistNodes2 are the "bad" selected functions for these distance variants.

Figure 3 and table 6 show an example of how the timeout functions are set up when DistPerimeter1 variant is used to perform a broadcast.

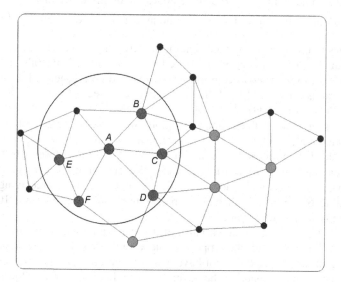

Fig. 3. Node A starts a broadcast. Nodes B, C, D, E, and F set their timeouts applying Dist-Perimeter1 variant

Table 6. Example of timeout values of nodes from figure 4 applying variant 3

Receiving Node	Number of neighbors not yet covered	Timeout function
B	3	$\frac{1}{3}+\left(1-\frac{dist(B,A)}{R}\right)$
C	3	$\frac{1}{3}+\left(1-\frac{dist(C,A)}{R}\right)$
D	3	$\frac{1}{3}+\left(1-\frac{dist(D,A)}{R}\right)$
E	2	$\frac{1}{2}+\left(1-\frac{dist(E,A)}{R}\right)$
F	2	$\frac{1}{2}+\left(1-\frac{dist(F,A)}{R}\right)$

3.2.4 Random Variant

The timeout random variant is as its name suggests. A receiving node B starts a random timeout before transmitting, if its timeout expires and it has neighbors left to be covered it transmits otherwise it abstains to transmit: timeout (B) = random value.

We propose this timeout variant since IEEE 802.11 uses this paradigm. We want to evaluate the viability of applying the concept of localized dominating sets in currently deployed networks that uses 802.11. For our implementation our random value was a random number between 0 and 1.

4 Performance Evaluation

We considered a network of $n = 100$ static nodes, randomly distributed over an area of 100 x 100. In order to control the average node degree d, we sorted all $\frac{n(n-1)}{2}$ (potential) edges in the network by their length, in increasing order. The transmission radius R is equal to $nd/2$-th edge in the sorted array. We used Dijkstra's shortest path algorithm *(SP)* to test whether a graph was connected. We generated a total of 100 connected graphs for each of the following network degrees, $d = 5, 6, 7, 8, 9, 10, 15, 20, 25, 30, 35, 40, 45, 50, 60, 70, 80, 90, 99$.

After creating each of the connected graphs we started a broadcast from a randomly selected node (with ideal MAC layer), using dominating Set Based Broadcasting with each of the proposed timeout variants using the three options to update the timeout values during a broadcasting process. We measured the average number of transmissions to complete a broadcast and the percentage of nodes receiving the message. For all tests and for all schemes the receiving nodes percentage was 100%. This confirms the reliability of our proposals.

In tables 7 and 8 we show the variance of the average number of transmissions using options 1, 2 and 3 for all the proposed timeout functions. It is clear from these results that the option selected did not have a significant effect in the number of transmissions to complete a broadcast; however, among all options, option 2 presented the best performance. This makes us believe that option 2 is the one that has to be used in the broadcasting scheme.

Table 7. Variance of the average number of transmissions using options 1, 2 and 3

Deg	Hex1Min1	Hex1Max1	Hex1Min2	Hex1Max2	Hex2Min1	Hex2Max1	Hex2Min2
5	0.0137	0.0524	0.0513	0.1200	0.0388	0.0208	0.0849
6	0.0170	0.0785	0.1214	0.1588	0.0320	0.0170	0.2046
7	0.0481	0.0440	0.1443	0.3502	0.0581	0.0250	0.1687
8	0.0101	0.1212	0.1252	0.7228	0.1062	0.0485	0.2109
9	0.0171	0.1562	0.2117	0.6686	0.1486	0.0397	0.3449
10	0.0732	0.0940	0.2577	0.7987	0.0809	0.0441	0.2671
15	0.0646	0.2316	0.2163	0.9664	0.0710	0.0694	0.3031
20	0.0526	0.1604	0.1402	0.9000	0.0386	0.0562	0.0944
25	0.0624	0.2404	0.2044	1.3774	0.0704	0.0813	0.2548
30	0.0530	0.2682	0.2817	0.9651	0.0604	0.1385	0.1372
35	0.0230	0.2385	0.2082	1.1676	0.0603	0.1200	0.0842
40	0.0074	0.1332	0.1110	0.5552	0.0132	0.0505	0.0199
45	0.0060	0.0030	0.0394	0.2443	0.0037	0.0625	0.0009
50	0.0005	0.0230	0.0270	0.0702	0.0006	0.0201	0.0008
60	0.0001	0.0009	0.0010	0.0037	0.0000	0.0010	0.0001
70	0.0000	0.0000	0.0000	0.0000	0.0000	0.0000	0.0000
80	0.0000	0.0000	0.0000	0.0000	0.0000	0.0000	0.0000
90	0.0000	0.0000	0.0000	0.0000	0.0000	0.0000	0.0000
99	0.0000	0.0000	0.0000	0.0000	0.0000	0.0000	0.0000

Table 8. Variance of the average number of transmissions using options 1, 2 and 3

D	Hex2Max2	DistPerim1	DistNodes1	DistPerim2	DistNodes2	Rand	Original
5	0.0604	0.0076	0.0757	0.0497	0.1612	0.0660	0.0005
6	0.1434	0.0046	0.0949	0.0905	0.2914	0.2509	0.0041
7	0.1984	0.0273	0.0819	0.0485	0.4683	0.2386	0.0016
8	0.1371	0.0156	0.1849	0.0547	0.5352	0.4508	0.0040
9	0.3236	0.0140	0.2443	0.2377	0.9300	0.2730	0.0081
10	0.3852	0.0331	0.2577	0.1204	0.9292	0.4693	0.0040
15	0.3050	0.0156	0.2058	0.1270	1.0566	0.2910	0.0040
20	0.2100	0.0462	0.3046	0.2439	0.9730	0.3388	0.0041
25	0.4377	0.0566	0.3232	0.2811	1.3132	0.3969	0.0028
30	0.2626	0.0403	0.4736	0.4156	1.3326	0.3682	0.0021
35	0.2810	0.0247	0.3997	0.3600	1.2544	0.2990	0.0009
40	0.1452	0.0325	0.1308	0.1606	0.7760	0.1276	0.0009
45	0.0541	0.0036	0.0073	0.0567	0.2330	0.0530	0.0004
50	0.0442	0.0016	0.0276	0.0162	0.1080	0.0074	0.0005
60	0.0026	0.0001	0.0012	0.0002	0.0091	0.0043	0.0000
70	0.0000	0.0000	0.0000	0.0001	0.0000	0.0042	0.0000
80	0.0000	0.0000	0.0000	0.0000	0.0000	0.0000	0.0000
90	0.0000	0.0000	0.0000	0.0000	0.0000	0.0000	0.0000
99	0.0000	0.0000	0.0000	0.0000	0.0000	0.0000	0.0000

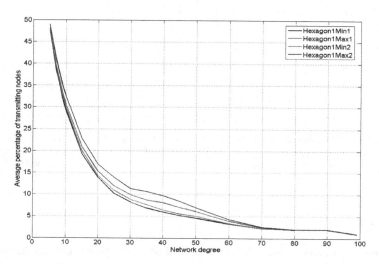

Fig. 4. Average percentage of transmitting nodes using hexagon1 schemes with option2

In figure 4 we show the average number of transmissions to perform a broadcast using some of the proposed timeout functions. These patterns were the same when the broadcast was performed using options 1, 2 and 3.

In figure 5A we show the three options applied to the original DS+ NE. It is clear that the differences between the three option's results are ineligible. In figure 5B we show the three options applied to the random variant. Again the differences between them are almost inexistent. Figure 5C shows the Min and Max values of the average number of transmissions that were registered in all the tests. Finally figure 5D shows the percentage of the max difference between the thirteen variants using the three different options. This figure clearly shows that there exists dependence on the selected timeout function. The difference was over 70% in some cases.

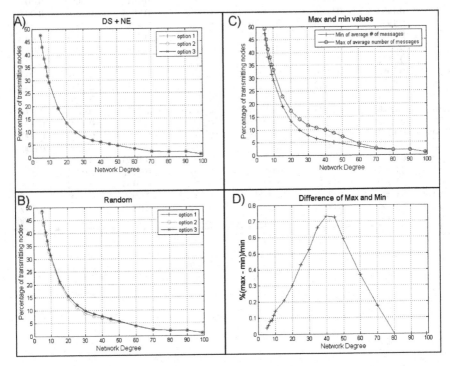

Fig. 5. A) Mean number of transmissions of DS + NE using the three options. B) Random variant using the three options. C) Min and Max values of the average number of transmissions registered in all tests. D) Percentage of the difference between the max and min values founded in all tests.

5 Conclusion

Dominating sets and neighbor elimination based broadcasting is currently the most efficient broadcasting scheme in terms of the number of retransmitted messages to complete a broadcast. Beyond this original protocol we proposed and tested thirteen new broadcasting protocols by changing the definition of the timeout function found in the original scheme. Also we proposed three different options to update the values of the timeout functions during the broadcasting process. The results obtained show that these options did not have a significant impact in the performance of the broadcasting protocols, all tree options performed in a very similar way. However, among the three options tested for updating the timeout functions during a broadcast, option 2 yielded the best results.

For the worst-case timeout functions the number of messages to complete a broadcast was affected in a negative way. This showed that indeed the selected function has an impact in the DS + NE broadcasting protocol. By using option 2, DistPerimeter1 gave the best results overall, followed very closely by the original DS + NE timeout function. Also the random timeout function (option 2) gave very close results to the best timeout function differing at most in 26 % of the average number of messages to complete a broadcast. Another important issue is that the nodes that currently operate under IEEE 802.11, use random delays in order to transmit. This fact and the results obtained in this article clearly show that energy savings can be achieved by using dominating sets plus neighbor elimination based broadcasting in 802.11 networks. However, as an open issue, it has to be tested how real MAC and Physical layers affect these variants.

Acknowledgements

This research is supported by Mexico CONACYT project 37017-A, Canada NSERC Collaborative Research and Development grant CRDPJ 319848-04, and the UK Royal Society Wolfson Research Merit Award.

References

[DP] Durresi, A., Paruchuri, V.: Geometric Broadcast Protocol for Sensor and Actor Networks. In: IEEE 19th International Conference on Advanced Information Networking and Applications, AINA 2005, 25-30 March 2005, vol. 1, pp. 343–348 (2005)
[GACM] Gosnell, M.R., Albarelli, R., Cheng, M.X., McMillin, B.: Energy balanced broadcasting through delayed intelligence. In: Int. Conf. Information Technology: Coding and Computing, ITCC 2005, April 4-6, 2005. vol. 2, pp. 627–632 (2005)
[GSS] Garcia, F., Solano, J., Stojmenovic, I.: Connectivity based k-hop clustering in wireless networks. Telecommunication Systems 22(1-4), 205–220 (2003)
[HBSG] Hogie, L., Bouvry, P., Seredynski, M., Guinand, F.: A Bandwidth-Efficient Broadcasting Protocol for Mobile Multi-hop Ad hoc Networks, ICNICONSMCL 06. IEEE Computer Society Press, Los Alamitos (2006)
[HOTV] Ho, C., Obraczka, K., Tsudik, G., Viswanath, K.: Flooding for reliable multicast in multihop ad hoc networks, 3rd Int. Workshop on Discrete Algorithms and Methods for Mobile Computers and Communications DIAL'M (August 1999)
[KM] Kim, D., Maxemchuk, N.F.: A comparison of flooding and random routing in mobile ad hoc network, 3rd New York Metro Area Networking Workshop (September 2003)
[KYZS] Kim, B., Yang, J., Zhou, D., Sun, M.T.: Energy-aware connected dominating set construction in mobile ad hoc networks, 14th Int. Conf. Computer Communications and Networks ICCCN, October 17-19, 2005, pp. 229–234 (2005)
[LG] Lin, C.R., Gerla, M.: Adaptive clustering for mobile wireless networks. IEEE J. Selected Areas in Communications 15(7), 1265–1275 (1997)
[NTCS] Ni, S.Y., Tseng, Y.C., Chen, Y.S., Sheu, J.P.: The broadcast storm problem in a mobile ad hoc network. In: Proc. MOBICOM, Seattle, pp. 151–162 (August 1999)
[SW] Stojmenovic, I., Wu, J.: Broadcasting and activity scheduling in ad hoc networks. In: Basagni, S., Conti, M., Giordano, S., Stojmenovic, I. (eds.) Mobile Ad Hoc Networking, pp. 205–229. IEEE/Wiley (2004)

[S-ssz] Stojmenovic, I.: Comments and corrections to Dominating Sets and Neighbor Elimination-Based Broadcasting Algorithms in Wireless Networks. IEEE Transactions on Parallel and Distributed Systems 15(11), 1054–1055 (2004)

[SSZ] Stojmenovic, I., Seddigh, M., Zunic, J.: Dominating sets and neighbor elimination based broadcasting algorithms in wireless networks. IEEE Trans. on Parallel and Distributed Systems 13(1), 14–25 (2002)

[WL] Wu, J., Li, H.: A Dominating Set Based Routing Scheme in Ad Hoc Wireless Networks. In: Proc. DIAL M: Int'l Workshop Discrete Algorithms and Methods for Mobile Computing and Comm., pp. 7–14 (1999)

[ZSL] Zhou, D., Sun, M.-T., Lai, T.-H.: A Timer-based Protocol for Connected Dominating Set Construction in IEEE 802.11 Multihop Mobile Ad Hoc Networks, IEEE SAINT, pp. 2–8 (2005)

Building a Trusted Community for Mobile Ad Hoc Networks Using Friend Recommendation

Shukor Abd Razak[1], Steven Furnell[2], Nathan Clarke[2], and Phillip Brooke[3]

[1] Faculty of Computer Science and Information Systems
Universiti Teknologi Malaysia
shukorar@utm.my
[2] Network Research Group, School of Computing, Communications & Electronics,
University of Plymouth, Plymouth, United Kingdom
info@network-research-group.org
[3] School of Computing, University of Teesside, Middlesbrough, United Kingdom
pjb@scm.tees.ac.uk

Abstract. The success of authentication mechanisms, intrusion detection and response systems, and other security measures to protect MANET from attack are very much dependant on nodes' cooperation. Ensuring a node's cooperation is very challenging in a MANET environment because of the nature of such a network; consisting of a set of anonymous nodes that operate independently, without the existence of a central authority to enforce them to cooperate. One of the solutions for this problem is by creating a trust chain between nodes to create a trusted community. This paper discusses some of the existing studies that have been proposed to overcome this issue and proposes a novel trust framework based upon a friendship concept. Results from simulation experiments proved that the proposed trust framework is capable of creating a trusted community in MANET, whilst at the same time addressing limitations of existing trust frameworks.

Keywords: MANET, Friend Mechanism, Trusted Community.

1 Introduction

MANET is a peer-to-peer mobile computing technology that utilises the wireless medium for communication. Although inheriting most of the other wireless networking characteristics (e.g. WLAN and WPAN), it has its own unique characteristics. For instance, it operates in a fully distributed manner, has no fixed network topology, and relies upon nodes cooperation for the success of its operations. Such characteristics create more challenges, especially with respect to security. Trustworthiness plays an important role in MANET security due to the issue of node anonymity. For example, a source node might be interacting with a set of unknown intermediate nodes to facilitate communication with a destination node. Such a scenario puts node communication at risk as the intermediate nodes might be malicious or misbehaved during their participations. Users' concern on this issue could be eased by establishing security associations between nodes in the network.

E. Kranakis and J. Opatrny (Eds.): ADHOC-NOW 2007, LNCS 4686, pp. 129–141, 2007.
© Springer-Verlag Berlin Heidelberg 2007

Initial trust provides a basis for establishing security associations between two or more nodes. For instance, two nodes that have established a mutual trust between them in an offline mode (e.g. via a secure side channel) might agree to become each other's trusted entity when participating in MANET operations. The more nodes that establish mutual trust between them in an offline mode, the more security associations will exist in the network. Security associations will ensure that only legitimate nodes are permitted to participate in network operations, thus reducing the probability of interacting with malicious nodes. In addition, security association could eliminate the problem of anonymous nodes in the network, thus each misbehaving nodes could be identified and penalised.

Several solutions have been proposed to establish a security association between nodes in a MANET environment. However, most of them require expensive cryptography, and sometimes need to assume the existence of a central authority to create, manage, and distribute system secret keys [1]. This paper proposes an alternative method based upon a friendship concept to establish trust and further create security association between nodes. To the best of the authors' knowledge, there are only two studies addressing the concept of friendship to establish trust in MANETs [2, 3]. However, the implementations of a friendship concept in those studies are different from the one being proposed in this paper. Such differences will be discussed further in the next sections.

The paper is structured as follows. Section 2 outlines background issues related to this study, which include discussion of existing solutions, their advantages, and limitations. Details of the proposed trust framework are presented in section 3. Results from simulation in evaluating the proposed trust framework are presented in section 4. Finally, section 5 outlines the future direction of this study and concludes the paper.

2 Background Issue

Fulfilling security requirements in MANET environment is very challenging due to fundamental characteristics of the network (e.g. limited nodes resources, an absence of CA). This paper proposes a friendship mechanism as an alternative solution to the problem. A pair of friend nodes, which are assumed to have a mutual trust between them before joining the network, are capable of creating a security association between them to participate in MANET operations. In addition, the friendship mechanism is able to speed up the creation process of a trusted community in the network. If each security association that exists in the network is exclusively owned by any two nodes that created it, the development pace of the trusted community will be very slow. Each node needs to meet and establish mutual trust with other nodes, which requires a lot of time and effort. The friendship concept proposed in this study makes this process simpler and faster by providing a secure platform for nodes to exchange their security associations. This ongoing trust exchange process between nodes could without doubt reduce the number of anonymous communications, and thus lead to the creation of a trusted community in the networks.

There are some existing studies suggesting a similar friendship concept for a MANET. For instance, in [2], the authors proposed the same concept to authenticate anonymous nodes in MANET environments. In their system, two nodes are considered as friends to each other if they have physically met in the real world before participating in MANET operations. If a node, lets say node A, wishes to have a trust relation with node B, which it never physically met before, node A needs to have at least one node in node B's friends list, lets say node C, to authenticate its identity. If there is no node in B's friend list that has physically met node A before, the recommendation request will then be forwarded to the next hop in the same manner. Once a node that knows the identity of node A is found, the information is sent back to node B to complete the authentication process. However, if no one in the chain knows about node A's identity, node A then must name at least one node, lets say node D, that it has met before to act as a reference node. Node B then will do the same process to authenticate node D's identity. If the identity of node D is known by any of node B's friends in the chains, the identity of node A then is considered authenticated. The introduction of referee nodes in their framework is very useful to speed up the security association's establishment process especially in a situation where only a few trust relationships exist in the system. However, their proposal requires strong encryption to be deployed in the system to avoid identity thefts when the recommendation packets travel across the networks. Their proposed framework also seems to cause extra overhead from the increase number of node's activities in the network because of the complicated process in searching recommendations and referees.

Researchers in [3] also proposed the same approach, but they dropped out the reference mechanism to minimise overhead caused by their proposed framework. They suggested that friendships in the real world could be used to establish trust between two or more mobile nodes that had never met each other in MANET environments. They divided the process to establish trust relationships into two phases. In a first phase, two nodes will establish a mutual trust between themselves by providing their personal information via a secure side channel (e.g. Infrared). Those two nodes will then exchange appropriate security keys, to enable them to communicate with each other using encrypted messages. With the encryption facility installed, these two nodes will be able to recommend their friends to each other, which will be much faster than communication via a secure side channel as they are not required to get physically close to each other to ensure secure communication. They claimed that the mobile characteristic of MANET nodes help in their proposed friendship concept, which is important for the overall performance of their authentication mechanism. However, there is one thing missing in their proposed friendship concept. There is no collaborative effort from each node to create a trusted community in MANET environments. Since the recommendation process will only take place when there is a need to authenticate an anonymous node, the process of creating a trusted community is a responsibility of each node itself.

The trust framework proposed in this paper is based on the two earlier studies mentioned above. It is designed to best suit the MANET environment by considering several aspects such as resources constraints, self-organisation, security, scalability, and the simplicity of the process. It also provides a platform for nodes to exchange their security associations with other trusted nodes in the system, which then leads to the creation of a trusted community. The detailed design of the proposed framework is discussed in the next section.

3 Trust Framework

Security association in autonomous networks such as MANET could be established based upon nodes' initial trusts. Initial trusts between nodes exist via several ways, including based upon the friendships of the bearer (i.e. human) in a real world, or based on the good reputation of other nodes through experiences [4]. Each method has its own advantages and limitations. For instance, initial trust based on a real world friendships are more relevant than that established based on nodes' experiences at the early stages of the proposed framework implementation. This is because in such a situation, each node is very unlikely to have sufficient knowledge/experience about other nodes, thus will not be able to rate other nodes' reputations. Initial trust that based on reputation is more suitable at the later stages when sufficient experiences have been gathered. Perhaps the combination of the two methods could result in a better performance. However, for simplicity, only initial trust based on a real world friendship is implemented in this study to show how a trusted community could be established in MANET environments. This section discusses how initial trust could be exchanged in MANETs, as well as the important concepts behind the proposed friendship framework.

3.1 Initial Trust

In most existing MANETs trust frameworks, researchers claimed that security associations between nodes could be established based on the initial trusts that have been setup beforehand. Although they mentioned how such setup could take place (e.g. via a secure side channel when two nodes are adjacent to each other), in most cases they did not address what actually motivates the nodes to create such relationships. The proposed trust framework in this paper suggests a human (node's bearer) relationship is one of the factors that could motivate initial trust establishment between nodes.

People do not live in this world alone. They socialise, making friends, live in a neighbourhood, and have family. Some people find that their family members are the group of people that they can trust the most. Some others might think differently. In some cases, friends could be the one that are more trustworthy. The issue of trustworthiness is very subjective and it depends upon how the relationship is developed [5]. However, the issue of how the trust relationship is being developed between individuals is not the main focus of this study. What needs to be highlighted here is that everyone has their own sets of friends. Within a set of friends, there might be a few of them that could be trusted, and vice versa. The above statements are supported by a series of surveys conducted in 1986 and 1995 by a group of researchers in Great Britain [6]. The surveys, which were conducted for the British Social Attitudes Survey Series (Britsocat), revealed that the average Briton has 14 close friends. One might thinks that the figure is obsolete because the latest survey was conducted 10 years ago. However, the number might be slightly higher than 14 as communication nowadays is far simpler, with e-mail, instant messenger, and mobile phones technologies [7]. This real world friendship could be a very good basis to setup initial trust between nodes. In this study, it is assumed that each mobile node inherits all the friendship relations established by its bearer and uses them as a basis to establish initial trust with other nodes. Figure 1 illustrates how this process could happen in MANET environments.

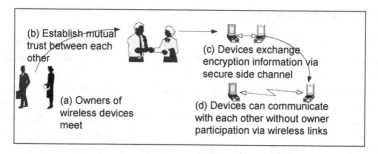

Fig. 1. Trust Establishment between Nodes based on Real World Friendship

In Figure 1, it is shown that initial trust is established bidirectionally between two nodes. However this does not always happen in a real world situation. Initial trust is not always necessarily bidirectional. In some cases, it is established unidirectionally. For instance, person A might be a friend to person B, and vice versa. In a unidirectional trust relationship, A might believe that B is trustworthy, but not the other way round. This is because trustworthiness is a very subjective issue and it depends upon each individual whether to trust or not to trust any other person. This subjective trust relationship creates an advantage in the proposed framework as it increases the number of initial trusts that could exist in the networks. For instance, as shown in Table 1, node A could establish a unidirectional trust relationship with node B without node B's approval, as it is its right to trust node B.

Table 1. Unidirectional vs. Bidirectional Trust Establishment

Nodes	Two Trusted Individuals	Unidirectional Initial Trust	Bidirectional Initial Trust
A	B & C	A-B & A-C	A=C
B	C & D	B-C & B-D	B=D
C	D & A	C-D & C-A	C=A
D	A & B	D-A & D-B	D=B
	Total Relationships	8	4

However, such a scenario will not happen in a bidirectional trust establishment as both node A and node B need to agree on the relationship prior to its establishment. For the case in Table 1, each node is assumed to have two initial trusted friends. However, in a real MANET implementation, it is difficult to estimate the number of trusted friends owned by each. One of the reasons is because the MANET could exist in several environments. Each environment has its own characteristics (i.e. different kind of users, variation in network density, and coverage), which leads to various numbers of trusted friends each node could have. In the proposed framework, it is assumed that this number could vary between 0 and 14 (i.e. the average number suggested in Britsocat survey [6]) depending upon which environment the network is deployed within. For instance, users in a university campus environment might have more trusted friends than users operating in a city environment due to the fact that more friendships could be established between course mates.

3.2 Trust Chain and Recommendation Concepts

Considering a MANET with 4 nodes, and having unidirectional trust relationships as shown in Table 1, there is a possibility for each node to add another node to its trusted lists. In such case, node A could add node D to its trusted lists, node B could add node A, node C could add node B, and node D could add node C to its trusted lists. Node A might not consider to add node D to its trusted list in the first place because it needs more time to ensure node D's trustworthiness. This is a case for 4 nodes, which does not require much time and effort for nodes to build their own trusted lists. However, in a wider and/or denser MANET environment, each node might require a little help from other nodes in the networks to build its own trusted list. The proposed framework in this research suggests the concept of a trust chain via friends' recommendations to help each node build its own trusted list. Based on a scenario as illustrated in Table 1, with a trust chain concept in place, node A this time could rely upon a recommendation from node B and/or node C to establish a security association with node D. This without doubt could save node A time and effort in the process of building its own trusted list. Table 2 shows how more initial trusts between nodes could be established via the trust chain and recommendation concepts.

Table 2. Trust Sharing between MANET Nodes

Nodes	Two Trusted Individuals	Unidirectional Initial Trust	Self + Shared Established Initial Trust
A	B & C	A-B & A-C	A-B , A-C & A-D
B	C & D	B-C & B-D	B-C , B-D & B-A
C	D & A	C-D & C-A	C-D , C-A & C-B
D	A & B	D-A & D-B	D-A , D-B & D-C
	Total Relationships	8	12

The concept of trust sharing in this paper is motivated by a research finding published in [8]. The author introduced a small world phenomenon concept, which suggests any two individuals selected randomly from almost anywhere in this world, are connected via a chain of no more than six acquaintances (often referred to as six degrees of separation [9]). The author brought this concept into a discussion in 1967 with an experiment in which he sent 60 letters to various recruits in Wichita, Kansas who were asked to forward the letter to the wife of a divinity student living at Cambridge, Massachusetts. The letters could only be forwarded by hand to personal acquaintances (directly or through a friend of a friend) who they thought might be able to reach the recipient. He claimed that he has proved the concept when 3 out of 60 letters that he sent reached the recipients but neglected to say about the low (i.e. 5%) chain completion percentage. However, his experiment has motivated other researchers to investigate more on this concept, such as in the Internet context, as observed in [10]. In that study, the author suggested that the World Wide Web is a 'small world' in a sense that all the sites are highly clustered yet the path length between them is small.

The concept of small world phenomenon has been brought into discussion in a wireless network by [11]. This study was based on findings from research in [12], where the authors proposed that by adding a few random links in the system, the average path length between nodes could be reduced dramatically. These few random links could be made available in the ad hoc networks by adding a few 'short cut' nodes in the system. Simulation results from this study proved the hypothesis. However, one question emerging from this study is how to select the few 'short cut' nodes in an autonomous, fully distributed, and self-organised ad hoc network. The author proposed the concept of *contacts*, which will act as short cuts to transform the wireless network into a small world. However, the author did not discuss how these *contacts* can be made available in the system, and this problem remains an open issue. That is the reason the friendship concept is being introduced in this study. It acts as a useful *contact* to create a relationship between two or more anonymous nodes, thus enables more interactions/communications in the networks.

3.3 Key Features of the Proposed Framework

As mentioned earlier, this study is not the first to suggest the concept of friend's recommendations to establish trust in a MANET environment. A similar concept has been proposed in [2, 3] as discussed earlier in this paper. The aims of this study are to provide solutions to several unaddressed issues in previous studies, as well as to suggest some enhancements that could be made to provide a reliable trust relationship framework for a MANET. The subsections that follow discuss key features of the proposed framework.

3.3.1 Light Weight
One of the main concerns in MANET operations is the node's limited resources [13]. The proposed framework in this study minimises the problem by not incorporating heavy computational mechanisms that might increase the network's operational overhead. For instance, friend lists could only be exchanged between nodes in a single hop communication, which eliminates the need for complex authentication mechanisms that are required if the exchange is permitted over multi-hop links. Complex authentication mechanisms utilise more network resources, especially for computational purposes. In another case, although it is true that a reference concept as suggested in [2] could speed up the trust establishment process, such a concept is being avoided in this study as it could cause more nodes' activities that could increase network's operational overhead. On the other hand, the trust framework as proposed in this study is based upon a trust chain and recommendation concept, which could offer a better trade off between trust establishment speeds and network operational overhead.

3.3.2 Self Organisation
Researchers in [3] used a CA to improve the performance of their trust framework. However, as MANET operates in a self-organised manner without any central point to perform the administration, it is very useful if the assumption of CA existence is avoided. This study introduces a constant friend recommendations concept, which could lead to a trust chain establishment to improve the performance of the proposed trust framework. A pair of friend nodes will exchange their trusted friend list whenever they are in range to each other (i.e. one hop away). Although this approach could not match the performance of a trust framework assisted by a CA,

the results from simulations presented later in this paper show that the number of trust relationships could be significantly improved.

3.3.3 Security Issues

Operating without any CA creates challenges for each node in MANET to deal with security issues. Each node is responsible to protect itself from various passive and active attacks [14]. The trust framework proposed in this study is designed with this issue in mind; to ease the burden for each node in dealing with security issues. Since two nodes are required to have a physical meeting before they could establish trust relationships with each other, an issue about identity theft could be minimised. Another important security issue that could be eased by the friend concept introduced in this study is the problem of false accusation/alarm advertised by a blackmail attacker [15]. With an assumption that all the trusted nodes are behaving properly, the friendship concept could act as a filter for nodes to avoid receiving false accusations from the blackmail attackers.

3.3.4 Scalable

Scalability is not an issue in the proposed trust framework. Unlike in [3], no meeting point is required in the proposed trust framework. A meeting point might boost the establishment of trust relationships, but only in a small network size. For instance, in a large network environment, nodes might need to travel a long distance to reach the meeting point, and in a dense environment, the meeting point might be crammed full. Besides, a CA also might be required to ensure security at the meeting point, which if not taken seriously could be abused by the attackers.

4 Evaluating the Performance of the Proposed Framework

This section provides details of a series of simulation experiments that have been conducted in order to evaluate the performance of the proposed trust framework. Evaluating a new MANET framework is not easy. One of the big challenges, especially in the case of this study, is to set up a huge amount of mobile nodes to represent a standard MANET environment. The applicability of the trust framework cannot be seen with the interactions of a small number of mobile nodes. With such a requirement, although it is always a desire to use an implementation-based approach, a simulation-based approach is often more practical. Moreover, a simulation has been chosen as a preferred approach by previous researchers, as it is useful to support and demonstrate nodes' mobility patterns in MANET environments [16]. There is a concern in the research community about the credibility of MANET simulations. Researchers in [17] brought this issue into a discussion and suggested that errors in simulation models or improper data analysis often produce incorrect or misleading results. However, this is not a big issue in this study, as no critical simulation attributes (e.g. traffic patterns, data transfer rate, or communication disruptions) are directly involved in the simulations. In fact the main reason for using a simulation to evaluate the performance of the proposed framework is because it is more practical considering the huge amount of nodes that are involved in this study. The following sub-sections describe simulation experiments, which aim to evaluate the performance of the proposed trust framework in creating a trusted community in MANET environments.

4.1 Simulation Setup

This study investigates the performance of the proposed trust framework in three different MANET settings, which differ from each other based on the density of the network. The differences in network density and coverage provide some means to investigate the applicability of the proposed framework in several MANET environments. For better understanding, these three MANET settings are classified into two open MANET environments, namely university campus and city network. The first setting, which is the densest setting, is represented by the university campus environment. The other two settings, with x2 and x4 less density than the first settings, are represented by the city-1 and city-2 network environments, which usually have a wider network terrain. The differences in network density level represent one of the factors that might have an impact on the proposed framework's overall performance. This factor is combined with the other two factors, namely the size of nodes' initial trust relationships and simulation time (representing network's age), in two separate simulation setups.

4.2 Results

This section presents the results and observations obtained from the simulation experiments, which were carried out to investigate the effects of different factors over the proposed trust framework's overall performance.

4.2.1 The Effects of Nodes' Initial Friendships
Figure 2 presents the results from simulation experiments conducted to investigate the effects of nodes' initial friendships towards the proposed friendship framework's

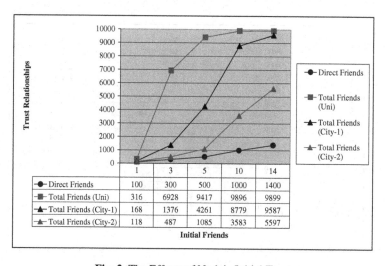

Fig. 2. The Effects of Node's Initial Trust

overall performance. From the results, it can be seen that the number of trust relationships between nodes is increased significantly according to the initial friends owned by each node. For instance, the number of trust relationships established with the help of 1 initial friend is increased from 100 to 316 in the university campus environment. A similar scenario also occurs in the city network environment, where the trust relationships were increased from 100 to 168 and 100 to 118 in both x2 and x4 less dense settings. The significant increase of trust relationships in 3, 5, 10 and 14 initial trusts simulation sets confirmed the effects nodes' initial friendships towards the overall performance of the proposed framework.

Apart from providing a platform for nodes to expand their trust relationships, the proposed friendship framework is also capable of virtually creating a trusted community in the network as described earlier in this paper. Results from the experiments justify this claim. As illustrated in Figure 2, the percentage of a trusted community in the network could be increased significantly with the help of friends. For instance, in a case of 5 initial friends in university environment, the percentage of a trusted community (T) is increased from 5% (without help of friends) to 95% (with the help of friends). The results are even better with 10 and 14 initial friends, where the percentage of a trusted community established in the university environment almost reaches the total 9900 trust pairs (i.e. 100%). In the case of the city environment, although the number of trust relationships created is not as high as in the university environment, the percentage of a trusted community established in both city-1 and city-2 settings are much better than solely depending upon a direct trust establishment. It is important to mention here that the results in this experiment sets were obtained based on a fixed 200 seconds simulation time. The results are expected to be better (especially in the city environment settings) if a longer simulation time were used in the experiments. The next subsection discusses this issue.

4.2.2 The Effects of Network Age (Simulation Time)

Figure 3 justifies that the network's age, which in this study is represented by the simulation time, is one of the important factors that could affect the overall performance

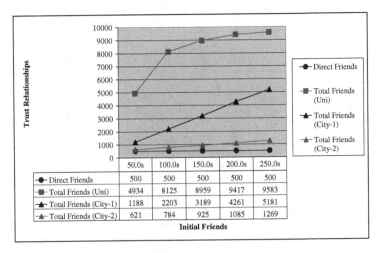

Fig. 3. The Effects of Network Age

of the proposed trust framework. From the results, it can be seen that the number of trust relationships established between nodes were gradually increased over a period of time. By using a fixed number of initial friends (i.e. 5), the total trust relationships established in a university campus environment were gradually increased from 500 to 9583 initial trusts within 250 seconds simulation time – more than a 19 times increase. A similar scenario also occurs in the city network environment, where the total trust relationships were increased to 5181 and 1269 in both 2 times and 4 times less dense settings. In terms of the overall performance (i.e. $T\%$ trusted community), results from the experiments shown that with a sufficient amount of time, the percentage of a trusted community established in the network could be gradually increased towards the 9900 total trust relationships. For instance, the percentages of a trusted community created in both university and city network environments were increased from 5% (without any help from friend) to approximately 96% in university environment, and 52% in city-1 environment (with the help of friends) within just 250.0 seconds simulation time. In addition, a higher $T\%$ value could be expected if more initial friendships were used in the experiment sets. For instance, 10 initial friends instead of 5 are very likely to increase the percentage of a trusted community for both university and city network environments.

4.2.3 The Effects of Network Density

The reason for having three different MANET settings (i.e. represented by a university campus and two city network environments) in the previous experiments is to investigate if the proposed trust framework could perform well in various MANET environments. Each setup represents a different network density level based on the number of nodes and the size of the terrain. The network density level (D) for each environment is obtained using the following formula:

$$D = \frac{n}{km^2} \quad \text{where } n = \text{total nodes; } km^2 = \text{terrain size} \quad (1)$$

Therefore, network density levels for the university and the two city network environments are 0.2n/m2, 0.1n/m2 and 0.05n/m2 respectively. Although with such a difference (i.e. 2 times and 4 times less dense), it can be seen that the total trust relationships between nodes in the two city network environments are still much higher than the trust relationships that have been established solely on each node's efforts (i.e. direct trust). For instance, in the experiment to investigate the effects of nodes' initial trust as discussed in the previous section, although the percentage of a trusted community in the city-1 network is approximately 40% less than the percentage of a university environment (i.e. in a case of 5 initial friends), an approximately 40% increase from initial 5% direct trust is still a good performance.

4.3 Discussion

Based on the observations of the simulation results, it is apparent that the proposed friendship mechanism is capable of expanding node's initial trust towards the creation of a trusted community. Although the true potential of the proposed trust framework is dependent upon various factors (such as network age, the number of nodes' initial trust, and network's density level), its performance is still very good in the least optimum scenarios. For instance, although with only a single trusted friend for each node, the number of trust relationships in the network still could be increased to at least 18% for the case of city-2 environment (refer to Figure 2). Moreover, this 18% is achieved from a very immature network (i.e. only 200 seconds lifetime) in a less dense environment. The percentage is expected to be higher if the network's lifetime is longer than 200 seconds, as shown in Figure 3. It is important to mention here that the simulation time (e.g. 50, 100, 200 seconds, etc.) used in the simulation do not represent a real time in real MANET implementation. Various simulation times were used in this study to show that a longer MANET operational time will have an impact upon the overall performance of the proposed framework. From the observation, the number of direct friends owned by each individual node has a significant effect upon this overall performance. The more direct friends owned by each node means the better the performance of the trust framework will be. However, as mentioned earlier, this direct friendship relation is a very subjective issue. Some people might have as many as 5, 10, or even more individuals or friends that they could trust. On the other hand, some people might have no friends that they could rely upon. However, the later case is almost never true since all people in this world live with friends [6], especially with the current advancements in todays telecommunication technology [7].

There is also another important factor that was not investigated in the experiment but has a major impact upon the overall performance of the proposed framework. The factor is known as the inconsistent value of nodes' initial trust. It is always possible for each node to establish a new mutual trust with another node in the network. This situation mimics a real world friendship scenario, in which each person could establish a new friendship as he/she desires with any newly met person. Such a situation gives an advantage towards the overall performance of the proposed framework. This factor's influence upon overall performance was not experimentally investigated in this study because the nature of direct trust establishment process is very subjective. Apart from that, the inclusion of this factor will not decrease performance but will only serve to the number of indirect trusts in the obtained results.

Results from the simulations also suggested that the proposed framework is applicable in various network density levels. Compared with a denser environment (i.e. university campus), the trust growths in the city network environments (i.e. 2x and 4x less dense) are still significant. This suggests that the framework is scalable although with a slight decrease on the overall performance.

5   Conclusion

This paper has proposed a friendship mechanism, which includes a discussion on its background and how it could help in MANET operations, particularly as an alternative solution in dealing with security issues. The applicability of the proposed framework in various MANET environments and scenarios has been justified via a set of simulation experiments. The main conclusion derived from the experiments is that the proposed framework, with reasonable assumptions, is proven capable of establishing trust between nodes in fast and secure manner. However, the maximum potential of the framework is dependent upon the number of direct friends owned by each node. The more direct relationships established in the network the better the performance the proposed friendship mechanism. In addition, this paper also addressed the issue of network maturity and density levels toward the overall performance of the proposed framework. Having justified the applicability of the framework via simulation, the next step of this study is to investigate its performance in a real implementation. In order to achieve that, a practical prototype must be developed utilising real world friendships.

References

1. Zhou, L., Haas, Z.J.: Securing ad hoc networks. IEEE Network 13(6), 24–30 (1999)
2. Weimerskirch, A., Thonet, G.: A Distributed Light-Weight Authentication Model for Ad-Hoc Networks. In: Kim, K.-c. (ed.) ICISC 2001. LNCS, vol. 2288, pp. 341–354. Springer, Heidelberg (2002)
3. Capkun, S., Hubaux, J.-P., Buttyan, L.: Mobility Helps Security in Mobile Ad Hoc Networks. In: Proc. of MobiHoc'03, Annapolis, Maryland, USA, pp. 46–56 (2003b)
4. Walsh, K., Sirer, E.G.: Experience with an Object Reputation System for Peer-to-Peer Filesharing. In: Proc. of NSDI '06: 3rd Symposium on Networked Systems Design & Implementation, CA, pp. 1–14 (2006)
5. Castelfranchi, C., Falcone, R.: Principles of Trust for MAS: Cognitive Anatomy, Social Importance, and Quantification. In: Proceedings of the 3rd International Conference on Multi Agent Systems, Paris, France, pp. 72–79 (1998)
6. Britsocat. Number of close friends (1995), http://www.britsocat.com/ BodySecure.aspx? control=BritsocatMarginals&var=PALS&SurveyID=224
7. Frean, A.: Friends like These: The relationships that shape our lives. ESRC Society Today - The Edge 13, 8–10 (2003)
8. Milgram, S.: The Small World Problem. Psychology Today 1, 61–67 (1967)
9. Guare, J.: Six Degrees of Separation: A Play. Vintage Book, New York (1990)
10. Adamic, L.: The Small World Web. In: Abiteboul, S., Vercoustre, A.-M. (eds.) ECDL 1999. LNCS, vol. 1696, pp. 443–452. Springer, Heidelberg (1999)
11. Helmy, A.: Small Worlds in Wireless Networks. IEEE Communications Letters 7(10) (2003)
12. Watts, D.J., Strogatz, S.H.: Collective dynamics of 'small-world' networks. Nature 393, 440–442 (1998)
13. Salem, N.B., Buttyan, L., Hubaux, J.-P., Jakobsson, M.: A Charging and Rewarding Scheme for Packet Forwarding in Multi-hop Cellular Networks. In: Proc. of 4th ACM International Symposium on Mobile Ad Hoc Networking and Computing (MobiHoc'03), Annapolis, USA, pp. 13–24 (2003)
14. Al-Jaroodi, J.: Security Issues in Wireless Mobile Ad Hoc Networks at the Network Layer. In: Technical Report TR02-10-07. Computer Science and Engineering, University of Nebraska-Lincoln (2002)
15. Zhang, Y., Lee, W., Huang, Y.-A.: Intrusion Detection Techniques for Mobile Wireless Networks. Wireless Network 9(5), 545–556 (2003)
16. Kurkowski, S., Camp, T., Colagrosso, M.: Manet Simulation Studies: The Incredibles. SIGMobile Mobile Comm. Rev. 9(4), 50–61 (2005)
17. Andel, T.R., Yasinac, A.: On the Credibility of Manet Simulations. IEEE Computer Magazine 39(7), 48–54 (2006)

Dependable and Secure Distributed Storage System for Ad Hoc Networks

Rudi Ball, James Grant, Jonathan So, Victoria Spurrett, and Rogério de Lemos

Computing Laboratory
University of Kent, UK
r.delemos@kent.ac.uk

Abstract. The increased use of ubiquitous computing devices is resulting in networks that are highly mobile, well connected and growing in processing and storage capabilities. The nature of these ubiquitous systems, however, also increases the risk of building systems that are undependable and potentially insecure. This paper investigates the use of autonomous agents combined with an intrusion tolerance technique for providing secure and dependable storage for ad hoc networks. The proposed approach is based on the fragmentation-redundancy-scattering (FRS) technique that is able to tolerate both accidental and intentional faults by fragmenting confidential information into insignificant fragments, and by scattering these fragments in a redundant fashion across a network. Two algorithms that are able to maintain a constant number of fragments replicas were developed for this study: one based on the game of life, and the other based on roaming ants. Both algorithms were simulated in NetLogo, and simulated further in a cluster of computers for evaluation of their scalability.

1 Introduction

This paper presents an approach for a decentralized storage system for ad hoc networks. The proposed approach allows users to store information in a dependable and secure manner in networks that are constantly changing. The storage system is dynamic, self-organizing, and self-adaptive to changes that occur in the network, and from the perspective of dependability it does not rely on the existence of any particular device, or set of devices of the network.

Work on secure computing systems has focused mainly on intrusion prevention, that is, the means for preventing the occurrence of intrusions [6], and which is based on forecasting and preventing, as far as possible, the different intrusions that could affect overall system security. Such approaches become infeasible in the context of open and decentralized systems containing a large number of components. Instead of attempting to prevent any type of intrusion, which in the context of ever changing environments may be very costly to achieve, some contributions have already been made in tolerating them. The basis for such approach is that, in case intrusions are successful, the whole system's security will not be compromised since the intrusions will be handled in the same way faults are tolerated. Although the term "intrusion tolerance" has been introduced in the past [5], only recently has there been an

E. Kranakis and J. Opatrny (Eds.): ADHOC-NOW 2007, LNCS 4686, pp. 142–152, 2007.

increased interest in this area. Several major projects, such as, MAFTIA, OASIS and ITUA, have produced groundwork into concepts, mechanisms and architectures for intrusion tolerant systems.

In this paper, we present a solution for a dependable and secure storage system based on intrusion tolerance that can be scalable to large pervasive systems. The approach presented in this paper is based on the fragmentation-redundancy-scattering (FRS) technique [4][5], which itself has been one of the approaches used for building intrusion tolerant computing systems. Since in dynamic ad hoc networks it is not possible to make precise assumptions about the number of node failures, we have investigated non-conventional approaches based on swarm techniques for implementing the scattering and retrieving of information fragments. The results reported in this paper refer to two algorithms: a variant of the game of life, and swarm-based ant algorithm. In these two algorithms, the information fragments are autonomous agents that manage their replicas locally within their neighbourhood, while the overall number of fragment replicas in the system becomes an emergent property.

The rest of the paper is structured as follows. In Section 2, we present some background about the fragmentation-redundancy-scattering (FRS) technique. Section 3 presents the overall design of a dependable and secure storage system for ad hoc networks. In Section 4, we present two of the algorithms implemented for the scattering and retrieving of fragments. In Section 5, the scalability of the proposed algorithms is evaluated in the context of a system containing 28K nodes. Related work is presented in Section 6. The last section presents some concluding remarks, and identifies future directions for research.

2 Background

The dependability of a system is the ability to avoid service failures that are more frequent and more severe than is acceptable [1]. A failure can be characterized as a security failure when there is the violation of a security policy. A security policy is normally expressed through properties, which are related to the absence of unauthorized access to, or handling of, system state. System failures are caused by faults, and an intrusion is a malicious external fault that might lead to a security failure. Thus intrusion tolerance is one of the means for the provision of security, and is regarded as the process for tolerating the violation of a security policy.

The fragmentation-redundancy-scattering (FRS) technique is an example of intrusion tolerance, which uses fault tolerant techniques as a means to avoid system failures and the violation of security policies [4][5]. The aim of FRS is to tolerate both accidental and intentional faults/intrusions by fragmenting confidential information into insignificant fragments, and randomly scattering these fragments in a redundant fashion across nodes of a network. Fragments contain no significant information, so any intrusion into some part of the system only gives access to unrelated fragments, thus maintaining the confidentiality of the information. By increasing the number of fragments a file is broken into, we can reduce the usefulness of a fragment, thereby improving the security of the system. Before fragmenting, the original information is encoded and signed. Incorporating fragment digests may also protect the integrity of the information. Redundancy is added to tolerate accidental or deliberate destruction, or alteration of fragments. Moreover, in case some nodes suffer denial of service attacks, information fragments can always be retrieved from other nodes, depending on the existing failure assumptions. The complete information that has been fragmented can only be reassembled by an authenticated user in a trusted computing base. The motivation behind the FRS technique is that an intruder attacking an individual node has no access to all fragments. Even if an intruder gets access to all n fragments, $n!/2$ cryptanalysis have to be performed to re-constitute the whole information [4].

3   Fragmentation, Redundancy and Scattering in Ad Hoc Networks

The system comprises a network of devices connected in a peer-to-peer formation, where nodes are only able to communicate with those nodes within their communication range. The system is completely decentralized with all nodes having equal status. Nodes are expected to provide some storage space to the network in exchange for the ability to store and retrieve their own files in the system. We shall use the term *storage node* to refer to any node that offers storage space to the network, and the term *client* node to refer to any node that wishes to either store or retrieve a file from the network.

The client is responsible for the encryption, fragmentation, and scattering of the information that needs to be stored in the system. Files are encrypted as a whole using a symmetric block cipher. The use of asymmetric ciphers is not necessary as data is intended to be confidential and readable by the publisher only. The fragmentation process operates on the ciphertext output of the encryption process. The ciphertext is split into fragments of equal and fixed size. Each of these fragments is associated with a unique identifier. There is no meaningful relationship between the name of the fragment and its contents. Integrity checking in this case is performed solely by the client. Storage nodes are required to hash the content of each fragment and compare it with the name before storing, if the name does not match the content hash, it should reject the fragment. If a fragment has been accidentally or maliciously corrupted, the fragment name will no longer be valid, and therefore cannot be propagated to correctly functioning storage nodes in the system. The client has little control over the information once it has been stored in the system. Over time as more files are stored in the system the number of total fragments, as well as the performance overhead will increase. To avoid this problem, each fragment may be associated with a lifetime value. This lifetime value can be specified by the client, although storage nodes may impose an upper limit on this value.

The client is also responsible for the retrieval, assembly and decryption of the information stored in the system. The first step in the retrieval process is the collection of fragments from the network. The client reads the list of fragment names that were stored during the fragmentation process for a particular file. The client should verify the integrity of fragment data by comparing the content hash to the fragment name and should repeat the collection of any corrupted fragments. Once all the necessary fragments are collected, the fragment data must be re-assembled in the correct order using the ordering information saved during the fragmentation process.

The reassembled data should be identical to the ciphertext used by the encryption process, and can be decrypted using the same encryption key.

As the number of files stored by a particular client increases, the amount of information that must be stored in some trusted computing base also increases. It is possible to reduce the amount of this information through the recursive application of the storage activity process. The secret information stored by a client can be fragmented, named, and stored in the system like any other file. This process can be repeated as many times as necessary until the amount of information stored by the client is reduced to some acceptable size.

4   Algorithms for Scattering and Retrieval

During the design of the system, several candidate algorithms were devised for the purpose of ensuring the availability of fragment replicas in the network. One major constraint considered when designing the algorithms is that devices in the network may fail or act maliciously at any time. Each of the algorithms aimed to fulfil the following criteria: to maintain an acceptable number of fragment replicas, to be resistant to hostile attacks and multiple node failures, and to preserve the anonymity of data holders. During the study, several algorithms were developed and analysed, however, for this paper, we present two of the algorithms that have shown more promising results: a variant of the game of life, and a swarm-based ant algorithm. In the following, for each algorithm, a brief description is presented, together with its pseudo code, and a threat analysis. The last subsection compares the two algorithms, in terms of the outcomes of the simulations on NetLogo [12].

4.1   A Variant of the Game of Life Algorithm

This algorithm was largely inspired by Jon Conway's *Game of Life*. On an iteration of the algorithm, a hostile node containing a fragment replica asks its neighbouring nodes if they hold a copy of the same fragment. Based on the number of responses received, the replica can decide to destroy itself, do nothing, or replicate onto a neighbouring node. Figure 1 displays the pseudo code for the algorithm, written in the context of a fragment replica.

```
ask neighboring nodes if they hold this fragment
   SET count as number of positive responses
   IF count > some upper bound THEN
      destroy self
   ELSE IF count < some lower bound THEN
      replicate onto neighbour that sent a negative response
```

Fig. 1. Pseudo code for the variant of the game of life algorithm

This algorithm is subjected to two main threats. In the first threat, hostile nodes always respond positively to any query to whether they hold or do not hold a replica of a fragment. If this happens on a larger scale, then all fragments within the system may be deleted, causing the fragment population to fail. In the second threat, also known as the Mexican standoff, during an iteration, if a group of connected nodes holding fragments query each other about replicas, and each of the nodes responds positively, thus obtaining the number of replicas above the self-destruction threshold, then every node simultaneously takes the decision to destroy its fragment, which might cause fragment replicas to be destroyed.

4.2 A Swarm–Based Ant Algorithm

The algorithm used by the autonomous agents is largely inspired by the Messor project [7] that models the behaviour of *Messor Sancta ants* in autonomous agents for providing task load-balancing in grid-computing environments. The proposed ant algorithm makes use of mobile agents, or ants, that move around the network trying to replicate and retrieve fragments.

Replication ants are created by storage nodes to maintain the availability of their stored fragments, and also by the client during the scattering process. The ants have a maximum walk length. Each time they move to a new node, their hop count is incremented. If they move to a node they have already visited, they move to their previous node, and decrement their hop counter. If they are at a node where all neighbours have been previously visited, they return along their path, until they reach a node with an unvisited neighbour. Their hop count is decremented along this temporary return journey. Once an ant's hop count equals its walk length, if it has not encountered a fragment of the same type as the one it are carrying, the fragment is deposited on the last node visited. If a fragment of the same type is encountered on the outwards journey, the ant kills itself, and deletes the replica of the fragment that it holds. The originating node does not delete its fragment. Figure 2 displays the pseudo code for a replication ant.

```
SET hops to 0
WHILE (hops not equal to walk-length) {
  IF (any unvisited neighbours) THEN
      choose unvisited neighbour n at random
      visit n
      IF (hops > 1) OR (random value > p) THEN
          hops = hops + 1
  ELSE
      hops = hops - 1
      visit previous-node

  IF (current-node has this-fragment) THEN
      terminate
}
store fragment on current-node
terminate
```

```
WHILE fragment not stored on current node
  IF (any unvisited neighbours) THEN
      choose unvisited neighbour n at random
      visit n
      visit random unvisited neighbour
  ELSE visit previous node
  GET fragment data
  WHILE not at start node
      visit the previous node
```

Fig. 2. Pseudo code for a replication ant   **Fig. 3.** Pseudo code for a retrieval ant

Retrieval ants are responsible for locating and fetching fragments in the system and are generated by the client during file retrieval. Upon visiting a node, a retrieval ant will first check to see if the current node holds the required fragment. If so, the ant returns to the source with the fragment data by following its own trail in the routing records. If the node does not hold the fragment, the retrieval ant examines the routing table to decide which node to visit next. If a retrieval ant encounters the same node twice, it should choose the next best path based on the criteria listed above. The pseudo-code for retrieval ants is shown in Figure 3.

A major threat of the ant algorithm is if the ants are constantly lost, deleted or their fragment contents corrupted. This may prevent an ant from creating further replicas in the network. However, all existing fragments will continue to produce ants at every given time period. A potential attacker who detects an ant with only one hop can gain information about the original position of a fragment: either it is a client or a fragment holder. We reduce the impact of this vulnerability by employing a probabilistic increase of the hop count at the start of the ants walk.

4.3   Comparison of Simulation Results

For evaluating the suitability of the proposed algorithms, each of the algorithms was implemented in NetLogo. This allowed us to investigate the performance of the algorithms under various operational conditions. The simulations were carried out using 100 nodes laid out in 10x10 static configuration in which each node was connected to other 4 nodes. These simulations were useful for tuning the different parameters associated with the algorithms being analysed. For example, for the variant of the game of life, the set up for the kill and replicate thresholds were tested under different operation conditions, and for the swarm-based ant algorithm, the impact of the walk length of an ant in the number of replicas was analysed.

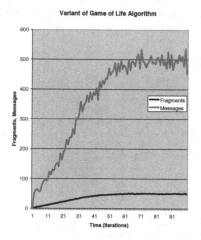
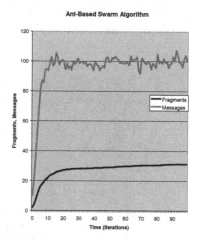

Fig. 4. Number of fragment replicas and messages for the variant of the game of life algorithm

Fig. 5. Number of fragment replicas and messages for the swarm-based ant algorithm

One of the first experiments was to evaluate whether the algorithms were stable in terms of number of replicas, and how many resources would be needed for maintaining a stable number of replicas. Essentially, the objective of these preliminary experiments was to confirm that the emergent behaviour obtained from the localised control in the number of replicas was an effective solution for the provision of a dependable and secure storage system for ad hoc networks. Two of the parameters investigated were the number of fragment replicas in the system, and the number of messages required for maintaining a stable number of replicas. The outcome of the experiments for the variant of the game of life and the swarm-based ant algorithm are shown in Figures 4 and 5, respectively. As it can be observed, the game of life variant requires twice as many messages than the swarm-based ant algorithm for maintaining a stable number of replicas. Although not very clear from the diagram, it was observed that the number of fragment replicas generated by the game of life variant was higher than the swarm-based ant algorithm.

Another set of test performed on the NetLogo simulations was for evaluating the robustness of the algorithms in the presence of several hostile nodes. The results presented in Figure 6 were obtained in the context of a dynamic system rather than a static configuration, which was used for comparing the performance of the different algorithms that were evaluated. In a dynamic configuration, which is a more accurate representation of ad hoc networks, nodes are free to move around at random, connecting to other nodes within some specified range. For these experiments, each simulation started with 1 initial fragment, which was allowed to propagate through the system using one of the algorithms, and after 1000 iterations the number of replica fragments was observed. The number of hostile nodes in the network was varied between simulations until a total of 100 nodes. The outcome of these experiments has shown that both algorithms were capable of building sufficient replicas for ensuring availability of fragments, even in the presence of multiple hostile nodes. Although the algorithms have shown a significant drop in replica populations as the number of hostile nodes increase within the system, they require one third of the nodes to be hostile for causing system failure. However, it was observed that the ant algorithm does not totally collapse because the fragment replica population is often reduced to a single fragment.

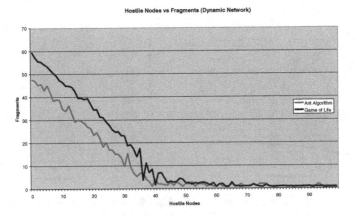

Fig. 6. Algorithms robustness in the presence of hostile nodes

5 Implementation and Evaluation

In this section, we discuss the overall implementation of the developed prototype that was used for evaluating the performance of the two algorithms described above, and the feasibility of the FRS technique as a dependable and secure solution for storage in ad hoc networks.

5.1 Implementation

For the encryption of information, we have employed the Advanced Encryption Standard (AES), which is a symmetric block cipher that uses a secret key encryption. Its combination of robustness, performance, efficiency, low memory requirements, ease of implementation and flexibility, make it desirable to use The algorithm has been designed for achieving great security and speed, and is easily implemented on simple processors. For the provision of the fragment digest, which is used for naming the fragments and to check their integrity, we have adopted the keyed Secure Hash Algorithm-1 (SHA-1), which produces a 160-bit digest providing robustness and improved uniqueness. It should be reiterated that while the components for encryption and integrity were chosen in this prototype to be AES and SHA-1, these components could be easily changed, depending on the desired improvement of confidentiality and integrity required by the users of such a system.

5.2 Prototype Simulation

In order to evaluate the scalability of the proposed algorithms, the prototype of the distributed storage system was simulated on a large number of nodes across a cluster of high-performance computers. The cluster consisted of 28 connected 3.2GHz Intel Pentium 4 computers, each with 1 GB of RAM, running Debian Linux. For the simulations, each computer on the cluster ran 1024 nodes as threads communicating through synchronised message queues, and arranged in a 32x32 grid configuration, with each of the nodes connected with other 4 nodes. Using this configuration, we were able to simulate a system with 28,672 nodes.

The first set of tests compared the NetLogo and the cluster simulations for static configuration of 100 nodes, with each node connected to other 4 nodes. A single

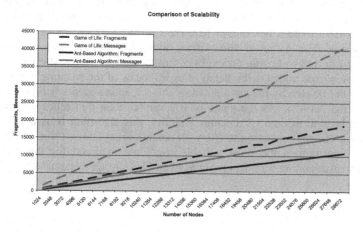

Fig. 7. Cluster scalability simulation of two algorithms - comparison of the number of fragments and messages

fragment was deployed into the network, and the simulation run for 5 minutes, with approximately 0.5 second samples. It was demonstrated from these tests that the fully implementation of FRS approach for tolerating intrusions was equivalent, in terms of number of fragment replicas and messages, to the results obtained from the NetLogo simulations, hence due to space limitations the set of results previously demonstrated are not replicated here.

Another set of tests performed on the cluster was the scalability of the system. Tests were executed with a step increase of 1024 nodes in the system, from 1024 to 28,672 nodes. For each step, the fragment levels were allowed to stabilise and the fragment and message levels were recorded. This process was repeated several times for both algorithms. Figure 7 shows a linear increase in both fragment levels and messages over increasing numbers of nodes. It is clear that the ant algorithm has a lower message overhead than the game of life variant algorithm.

5.3 Evaluation

The purpose of this study was to evaluate whether localised algorithms for replica management could form the basis for a dependable and secure distributed storage for ad hoc networks. The identified criteria for evaluating the feasibility of the proposed solution was whether the total number of replicas it produces at the system level would be stable, whether it would be scalable, and whether it would require an excessive number of resources. The experiences performed provide enough evidence that the first two criteria were achieved, however whether the third criteria was achieved depends on the system under consideration because 100 messages per iteration for a system that contains 100 nodes might be rather resource consuming. An important remark to be made is that no constraints were placed on the number of node failures that were allowed per iteration. However, it is clear from the experiments that after a certain percentage of hostile nodes, the number of fragment replicas in the system are not able to recover.

Comparing the two algorithms, there are a number of other factors that distinguish the game of life variant and the swarm-based ant algorithm. The algorithms take a different approach for restraining and replicating that are required for ensuring a stable fragment population. The game of life variant algorithm relies on communication with its immediate neighbours, where the swarm-based ant algorithm sends out a roaming agent. The game of life variant algorithm has the potential to delete fragments if affected by hostile nodes. In contrast to this, the greatest effect on the ant algorithm by hostile nodes is the deletion of instances of the roaming agents. As the fragment is not deleted by the ant algorithm, these roaming agents will be continually sent out to the network. As the game of life variant algorithm relies on adding and deleting replicas to maintain a dynamic stability, this algorithm is associated with a large amount of communication traffic with neighbouring nodes. Overall, the swarm-based ant algorithm is the most suitable candidate explored for the problem of dynamically maintaining fragment replicas in an FRS storage system.

6 Related Work

The idea of having a file system based on a decentralized solution is not new. File availability, confidentiality, and integrity on large-scale distributed file systems usually rely on the fragmentation of a file, and the scattering of these fragments among different nodes of the system [5][8]. The fragmentation principle is based on splitting an encrypted file into fragments, where each fragment is then distributed. There are two different schemes based on fragmentation and scattering: fragmentation-scattering and replication [4][5], and fragmentation-scattering and threshold [6]. In both schemes, a file read accesses a subset of the fragments, while a file write has to be performed on all fragments. In fragmentation-redundancy-scattering (FRS), fragments of information with little value to a potential intruder are replicated and scattered across a large number of nodes. The fragmentation-threshold-scattering (FTS) scheme is based on the same principles of threshold cryptography [9]. Instead of replication, this scheme relies on the processing of information. Seminal work in this area is the Information Dispersal Algorithm (IDA) [8].

The Information Dispersal Algorithm (IDA) [8], proposed by Rabin, has been considered in the context of Redundant Residue Number System (RRNS) for encoding information [3], which provides uniform coverage of both erasures and errors. The objective of this work is to provide a dependable and secure data storage (DS^2) to mobile wireless networks. It has been shown that this approach and IDA have almost the same performance in terms of code efficiency and complexity [2], even though DS^2 provides richer security features than IDA by exploiting the RRNS codes. An apparent drawback of this approach is that the system has to be reasonably static for it to be feasible. It is difficult to envisage the application of DS^2 to extremely dynamic ad hoc mobile networks where large number of nodes can be joining or leaving the system at the same time. The solution presented for these possible scenarios was for client/user to adopt an appropriate level of redundancy during the creation of the file.

Messor is a grid computing system that employs the use of mobile autonomous agents in order to perform load balancing of CPU intensive tasks in a parallel computing environment [7]. These "ants" wander around the network, picking up jobs from overloaded machines and moving them to areas of the network with less load, resulting in a uniform load balance across the machines. The project is inspired by the emergent behaviour of *Messor Sancta* ant swarms in nature that tend to "group" similar objects in their environment into piles without being told explicitly to do so. Our replication and scattering algorithm has many similarities to the load-balancing algorithm employed by Messor, in particular the way agents are used to distribute objects throughout the network.

7 Conclusion

This paper has proposed an approach for providing a dependable and secure storage system for ad hoc networks. The approach relies on the combined use of the fragmentation-redundancy-scattering (FRS) technique and mobile autonomous agents for tolerating intrusions.

Simulations have shown that the system is effective at maintaining a population of fragment replicas under such conditions as network instability and malicious information destruction. System scalability is shown to be linear in terms of the number of fragment replicas and messages. Perhaps a limitation concerning the proposed approach is the high number of messages that is required to maintain a stable population of fragment replicas. A suggestion for future work in this area is to research a method to limit the growth of fragment replicas, possibly by restricting the movement of replication ants for a particular fragment to some subset of the network. Another approach could be to implement the system on top of an overlay network.

References

[1] Avizienis, A., Laprie, J.-C., Randell, B., Landwehr, C.: Basic Concepts and Taxonomy of Dependable and Secure Computing. IEEE Transactions on Dependable and Secure Computing 1(1), 11–33 (2004)

[2] Chessa, S., Di Pietro, R., Maestrini, P.: Dependable and Secure Data Storage in Wireless Ad Hoc Networks: An Assessment of DS2. In: Battiti, R., Conti, M., Cigno, R.L. (eds.) Wireless On-Demand Network Systems. LNCS, vol. 2928, pp. 184–198. Springer, Heidelberg (2004)

[3] Chessa, S., Maestrini, P.: Dependable and Secure Data Storage and Retrieval in Mobile, Wireless Network. In: Proceedings of the International Conference on Dependable System and Networks (DSN 2003), San Francisco, CA, USA (2003)

[4] Deswarte, Y., Blain, L., Fabre, J.-C.: Intrusion Tolerance in Distributed Computing Systems. In: Proceedings of the IEEE Symposium on Security and Privacy, Oakland, California, USA, pp. 110–121 (May 1991)

[5] Fraga, J., Powell, D.: A Fault- and Intrusion-Tolerant File System. In: Proceedings of the 3rd International Conference on Computer Security (IFIP/SEC'85), Dublin, Ireland, pp. 203–218 (August 1985)

[6] Mei, A., Mancini, L.V., Jajodia, S.: Secure Dynamic Fragment and Replica Allocation in Large-Scale Distributed File Systems. IEEE Transactions on Parallel and Distributed Systems 14(9), 885–896 (2003)

[7] Montresor, A., et al.: Messor: Load-Balancing through a Swarm of Autonomous Agents. Technical Report UBLCS-2002-11. University of Bologna. Italy (2002)

[8] Rabin, M.O.: Efficient Dispersal of Information for Security, Load Balancing, and Fault Tolerance. Journal of the ACM 36(2) (1989)

[9] Shamir, A.: How to Share a Secret. Communications of the ACM 22(11), 612–613 (1979)

[10] Trouessin, G., Deswarte, Y., Fabre, J.-C., Randell, B.: Improvement of Data Processing Security by Means of Fault Tolerance. In: Proceedings of the 14th National Computer Security Conference, pp. 295–304. Washington, USA (1991)

[11] Veríssimo, P., Neves, N.F., Correia, M.: Intrusion-Tolerant Architectures: Concepts and Design. Technical Report DI/FCUL TR03-5. Department of Computer Science, University of Lisbon (2003)

[12] Wilensky, U.: NetLogo. Center for Connected Learning and Computer-Based Modeling, Northwestern University, Evanston, IL (1999), http://ccl.northwestern.edu/netlogo/

An Energy and Communication Efficient Group Key in Sensor Networks Using Elliptic Curve Polynomial

Biswajit Panja and Sanjay Kumar Madria

University of Missouri-Rolla, Rolla, Missouri 65409
{bptfc,madrias}@umr.edu

Abstract. Sensor nodes have limited computation and battery power, and are not very reliable. A sensor network needs to be secure against eavesdrop when it is deployed in hostile environments. In order to provide security at low cost, symmetric key based approaches [11] have been proposed. In [9], an elliptic curve cryptography based approach has been implemented to facilitate the public-key cryptography. However, these schemes become ineffective in terms memory usage, communication time and energy required with the rapidly growing network size. We propose an Energy and Communication Efficient Group key management (ECEG) scheme which reduces the usage of memory, communication and energy in sensors. This scheme designed based on the idea that each node is pre-loaded with a key chain and an elliptic curve. In our scheme instead of computing the key by collaborating with other nodes a point of the elliptic curve broadcasted from the base station is used to compute the key. The simulations conducted using TinyOS [15] shows that ECEG scheme significantly improves the communication and energy usage in computing the key over EccM [9].

1 Introduction

The Sensor networks have applications in military, airport security, habitat monitoring and in robotic toys among others. A sensor network consists of general sensor nodes, cluster heads and relay nodes. All of these nodes are equipped with limited CPU, memory, antenna and battery power [14]. Each sensor node is capable of processing data and communicating with neighboring nodes and in some applications like military, sensor nodes are deployed in a region randomly through a helicopter. In some other applications like border security, they are deployed manually. These sensor nodes communicate with other nodes to form a self reconfigurable network. These nodes fail once the battery is depleted and since in an environment like battlefield, replacing or charging the battery is not possible therefore, the network needs to be reconfigured after the failure of some nodes.

The security of sensor networks is very important when it is deployed in a hostile environment as sensors communicate using insecure radio frequency (RF) channel. Intrusions and attacks have become common threats in such a network. It is difficult to provide confidentiality and authentication of nodes in distributed sensor networks because of high failure rate, limited energy, and frequently changing topology. It is also difficult to prevent an adversary sensor node from compromising the sensor networks because of untraceable sensor nodes, and less physical protection.

E. Kranakis and J. Opatrny (Eds.): ADHOC-NOW 2007, LNCS 4686, pp. 153–171, 2007.
© Springer-Verlag Berlin Heidelberg 2007

Issues such as confidentiality, authenticity and integrity of nodes in Ad-Hoc/sensor networks are achieved using pre-deployed keys in [2, 3, 4, 5, 11, 16]. However, some of these schemes need to store large number of keys not feasible due to restricted memory size of MICA motes. Matt et al. [1] showed that the total number of pre-deployed keys that are to be generated in groups of sensor nodes of different sizes increases exponentially with the network size. In addition, the limited memory in smart sensor nodes [6, 14] makes these approaches impractical.

The lifetime of the sensor nodes can be increased [11, 12, 13] by running the high energy and computation intensive algorithm in the base station. For example, the widely used crossbow sensor motes [14] can handle maximum packet size of 32bytes [6] and according to TinySec [6] and TinyOs [15], one communication between two nodes is equivalent to executing 1000 instructions locally.

Elliptic curve cryptography is an approach of public-key cryptography based on the elliptic curves developed by Koblitz et al [10]. Malan et al. [9] proposed a key distribution scheme for sensor networks denoted as EccM (Elliptic curve cryptography Method) which is based on elliptic curve cryptography. Each node in this scheme computes a {private, public} key pair with the collaboration of another node. This scheme has been implemented for Mica2 motes using TinyOs. It uses 63 bit key which can take 5.8×10^{42} years if an intruder tries to decrypt it using exhaustive key search at a decryption rate of $10^6 \, decrytion / \mu \sec$. However, this scheme is not scalable (with respect to memory, communication overhead and energy consumption) with increasing number of nodes in the network which in turn creates too many key pairs. We propose a scheme in which the memory, communication overhead and energy consumption can be reduced by performing some of the steps of the elliptic curve cryptography in the base station. Involving the base station for computing the key chains and later the points of the elliptic curve for computing and updating the key is suitable here as it does not have the energy or memory constraints.

In this paper, we present an **E**nergy and **C**ommunication Efficient **G**roup key management (ECEG) protocol for sensor networks, which elects the cluster head dynamically based on the energy left and reduces the communication overhead by processing key management related operations (computing key chain and broadcasting points of elliptic curve) in the base station. The scheme computes and updates keys dynamically using polynomials of elliptic curve [9]. ECEG provides keys which can be used for encryption/decryption of messages. The elliptic curve cryptography is used to compute the key chains in the base station. These key chains are transferred to the sensor before their deployment. As the base station maintains knowledge of all the key chains and the group, it broadcasts a point to compute the group key from the key chain. As different group of nodes are deployed with different key chain, each group will have a unique group key. We do not use symmetric key from the base station because it is not cost (communication, bandwidth) effective to broadcast a new key from the base station before the expiration of previous key. Oppose to that, using ECEG nodes can compute a new key from the broadcasted message from the base station. This message contains a new point which can be used by the function and key chain. The detail of this scheme is explained in section 3.

To prevent the failure of entire network due to a single compromised node, ECEG provides the key only to the authentic nodes and updates it before the predefined expiration time. The key computed from the key chain and elliptic curve is same for each group hence it is denoted as a group key. Four types of logical levels of nodes are considered in this protocol. They are, general sensor node, *covering sensor, cluster head* and the base station. The covering sensors are a subset of general sensor nodes which covers the region of other sensors. Each group of nodes has a cluster head responsible for communicating with the base station. We have evaluated our scheme through the analysis and simulations using TinyOs and TOSSIM [14] which shows that the group key can be computed much faster and at much lower energy cost than EccM. We also measure the energy consumption for the key setup phase and compare it with the available energy in the network. We observe that ECEG consumes much less energy compared to the total available energy in the sensor nodes.

2 Sensor Network Model

This section describes the system model of a sensor network considered for developing the ECEG protocol. We consider a network composed of sensors which do not have tamper resistant hardware. The deployment of the nodes is dense so that the multiple sensor nodes can sense a similar event in a particular region. It can sense events like tank movement, vehicle movement, chemical leak and nuclear radiation. The detecting sensors collect data and forward to the cluster head through covering sensors. The covering sensors do the data aggregation [8] and forward the data to the cluster heads. The cluster heads communicate with the base station and with other cluster heads.

Figure 1 shows the system model of a sensor network organized in hierarchical structure consisting of five groups. Each group or region consists of many general sensor nodes, some covering sensors and a cluster head organized into logical levels. For example, the covering sensors of Region-1 are A, E, F and G. The algorithm for choosing the covering sensors and the cluster head are explained in section 2.1. Only the cluster head communicates with the base station after the initialization phase whereas in the initialization phase all the sensors communicate with the base station, so that it is aware of the authenticity of the nodes. The role of the cluster head rotates among the *CS* in a cluster. All the messages are broadcasted and the links among the nodes are bidirectional. The base station is fault tolerant. Each cluster has unique cluster identification and each node knows its cluster identification and its own unique identification in that cluster.

The architecture in Figure 1 can be used in a border security application to monitor people or vehicle entering into the country without authorization, or to monitor the activities of drug traffickers across borders, etc. One can deploy sensor networks across the border in clusters or groups as in Figure 2. We can place the base station in a physically protected area to manage the sensor nodes and collect data reported by them. The base station can send queries or commands to the sensor nodes. Also, the general sensor nodes send sensed data in a periodic interval.

We assume that every node has a pre-deployed key chain. The base station has information about all the key chains. The nodes know their one hop neighbors. The base station is capable of authenticating all broadcasted messages. If a node is

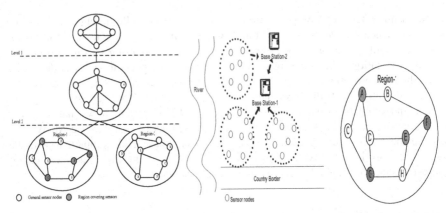

Fig. 1. A hierarchical architecture

Fig. 2. Border security application

Fig. 3. Electing covering
sensors and cluster head

compromised, all its information will be compromised. However, we assume that the
base station is physically secure. Since wireless sensor networks use broadcast based
communication, we assume that an adversary can eavesdrop on all traffic to listen
messages. It can inject false data into the network. It can alter transmitted data, replay
the old data. It can also do resource consumption attack by sending false data or
readings to the sensors. The resource consumption attack can drain the nodes'
batteries and waste network bandwidth. This paper do not address the threat, where an
adversary steals a node, change the security features and put them back in the network
to pretend to be an authentic node.

2.1 Electing Covering Sensors and Cluster Heads

This section provides the proposed schemes for electing covering sensors and cluster
head in each group of sensors.

The reasons for proposing this clustering scheme are as follows:

a) The energy consumption of a group can be reduced by altering nodes for the role
   of cluster head.
b) Most of the nodes will get a chance to become covering sensor or cluster head
   which can create confusion to the intruder because it is difficult to predict who
   will be the next cluster head. As a reminder, after the initialization process only
   the cluster head communicates with the base station for updating the group key.
c) The communication links of the nodes can change with the failure of the nodes,
   so it crucial to find the nodes with maximum links because it can reduce
   bandwidth overhead by minimizing the number of hops for routing packets.

For choosing the Covering Sensors (CS), ECEG finds the links of the sensor
nodes with other sensors in that region. We assume that there are S sensors in the
region R. The link between two nodes is denoted as L. If an instance of a region
R is $R_1 = (S, L)$, where S are the nodes in the region R_1. For example, in

figure 3 the sensor node A in the region S can be denoted as S_A and the link between nodes A and B is L_{AB}.

ECEG uses the following algorithm to get a subset of nodes $S' \subseteq S$ which covers all the links.

Algorithm:

1. $CS \leftarrow \{\}$
2. $L' \leftarrow L \ of \ R_1$
3. While $(L' <> empty)$ do
4.   Let (L_A, L_B) be the arbitrary link of L'
5. $CS \leftarrow CS \cup$ node with the link (L_A, L_B)
6.   Remove from L' every link incident on either L_A or L_B
7. Return CS

In Figure 3, the CS are A, E, F and G. After choosing the CS, the ECEG selects one of the CS as cluster head CH. For choosing the CH an election algorithm is followed, where the CS communicate among them to know the energy left and the CS with highest energy left is chosen as CH. When executing this algorithm they consider what will be the energy left after communicating with other CS. For transmission or reception of messages each node spends energy. Some of the CS are multi-hop away from the other CS, in that case the spent energy also depends on the number of hops for communication. For proposing the algorithm the identification of each CS is denoted as ID_{CSi}. For example, in figure3 the value of i is 4. If in each node the transmission, reception and the total energy is E_{Tx}, E_{Rx}, E_{total}, respectively, and the number of hops is h and the energy left at each node is E_{left}, then the algorithm is as follows:

Algorithm:

1. $n \leftarrow number[CS]$
2. *for* $i \leftarrow 1$ to n
3.   do While (E_{left} of $ID_{CSi} > 0$)
4. $E_{left} = E_{total} - (E_{Tx} + E_{Rx}) \times h$
5.   if $((E_{left}$ of $ID_{CSi+1}) > (E_{left}$ of $ID_{CSi}))$
6. $CH \leftarrow ID_{CSi+1}$
7.   else
8. $CH \leftarrow ID_{CSi}$
9. Return CH

The CS nodes collaborate with each other to find out who is left with the maximum energy. The highest energy sensor among CS is then selected as the cluster head CH. To balance the energy in the network, the cluster head is rotated after a certain period of time. In LEACH [7], it has been shown that sensors can save energy by communicating in hops instead of directly communicating with the base station. In our scheme, the responsibility of cluster head is rotated among the sensor nodes to balance the energy as the cluster head consumes more energy than any other nodes. After proposing the group key computation approach in section 3, we explain the scheme for secure organization for selecting covering sensors and cluster heads.

3 Group Key Using Pre-distribution, Collaboration and Elliptic Curve

In this section, we first present the basic idea of ECEG scheme and then provide the details of this protocol.

Overview of ECEG: Before the deployment of the nodes in the network, the base station generates a key chain for each node using an elliptic curve. All the nodes (N) in a particular group (G), $N \in G$ are pre-deployed with the same key chain $K_{ID_G}(j)$, where ID_G is an identification of the group and j is the index associated with the key chain broadcasted for the key discovery process instead of sending the actual keys. After deployment of the network, each group decides the covering sensors CS and the cluster head CH. The cluster head communicates with the base station to get the point generated from the elliptic curve polynomial. The base station chooses a point for each group for the computation of the group key. After receiving the point, the cluster head broadcasts the point and the associated index to its group members. All the sensor nodes in that group use this point to compute the group key by applying it to the key chain. With every broadcasted message, the cluster head attaches one plain text containing the information about the targeted nodes so that other nodes can drop the message if it is not for them.

Notations	Explanations	Notations	Explanations
E	Elliptic curve	$f_{ID_G}(x)$	Function for E, where ID_G is the identification of group and x is the characteristics of base field
$K(j)$	Generated key chain in the base station and j is and index	S_B	Random integer generated at base station
S_B	Generated key at base station from S_B	E_S	Elliptic curve base point
K_S	Random integer computed at sensor node	IC_S	Certificate generated at the base station
ID_S	Identification of sensor	t_S	Certification expiration date
H	Hash function	I_S	Derived integer by hash function
ID_B	Identification of base station	ID_G	Identification of group
ID_{CS}	Identification of a covering sensor	ID_{CH}	Identification of a cluster head

Detailed description of ECEG Method: 1) Pre-distributing the key chain: Before the deployment of a group of nodes, each node is preloaded with a link layer level key chain. The base station decides the number of groups it will be communicating with so that it can compute key chains for them. For each group G_i, a unique elliptic curve E is computed by the base station and this curve is used to assign the key chains in the sensor nodes. For each G_i, E is defined over a function $f_{ID_G}(x)$ where ID_G the identification of each group and x is the characteristic of the base field. For example, the function can be $f_{ID_G}(x) = x^3 + 17x + 12$. The $f_{ID_G}(x)$ used to compute the key chains $K(j)$ is composed of polynomials derived from E, where $(j \geq 1)$.

For the initialization of key chain computation, the base station selects a random integer s_B. It computes the key S_B using x where $S_B = s_B \times x$. The sensor nodes also select a random integer K_S and send it to the base station. The base station computes the elliptic curve base point E_S for the sensors using K_S and S_B, where $E_S = K_S + S_B$. The base station also generates a certificate IC_S using identification of sensors ID_S, elliptic curve base point E_S and the key of base station S_B, where $IC_S = (ID_S, E_S, S_B, t_S)$, where t_S is the expiration time. It applies a hash function H on IC_S and generates an integer I_S which is in the range of the polynomial of the elliptic curve. The key chains $K(j)$ of the sensors are computed in the base station using the formula $K(j) = S_B I_S + s_B \pmod{n}$. After that the sensors receive the key chains and certificate from the base station. The algorithm is shown in Figure 4, in which s_B can be in the range of $[2, n-2]$ because of the property of the elliptic curve, where the elliptic curve falls in a range points. The other important property of it is the operations such as addition, complement etc that falls in the same integer range. As the key is computed using the function $H(IC_S) \rightarrow I_S \in [2, n-2]$, different key chains can be computed for different groups. The reason for keeping the relation between the computed key chain and the elliptic curve is that a relation can be derived from the old key and the updated key, which is explained later in this section. The computed keys in the key chain have a one way relation as they are computed using the hash function. These keys are not directly used to compute the group key; they are used as raw keys. The intruders can not get all the future keys by attacking only one node. The pre-distribution of the key chain is done before the deployment (offline) of the network, so the communication and computation cost can be minimum and is ignored in the analysis. For example in crossbow sensor motes [14], the key chain can be distributed using MIB510 board which does not have power or computation constraint.

We do not need to know which node will be chosen as the cluster head. The cluster head can have the same key chain as other nodes in a particular group. As the general nodes only with the cluster head after the initialization phase, CH gets the point from the base station which it sends to the general nodes for computing the group key.

2) *Secure communication between nodes and the base station:* After the deployment of sensors in groups, nodes communicate with the base station for the initialization phase. Each node holds an identification set $\{ID_S, ID_G, ID_B\}$ which provides the unique identification to each sensor in a group. A message authentication code (MAC) is generated using a pre-deployed shared key K_{SB} with the base station. A

message of the form:
$$Msg_S = [(ID_S, ID_G, ID_B)_{K_{SB}},$$
$$MAC(ID_S, ID_G, ID_B, K_{SB})]$$

is computed in each sensor node. It contains the encrypted identification and a MAC used to handle insider attacks in case a node pretends to be authentic. The base station B verifies the authenticity of the sensor node S by decrypting the identification information and replies it with a message Msg_B. The $Msg_B = \{(ID_B)_{K_{SB}}, MAC(ID_B, K_{SB}) \mid REP\}$ where REP is used for the acknowledgement.

$$S \rightarrow B : \{(ID_S, ID_G, ID_B)_{K_{SB}}, MAC(ID_S, ID_G, ID_B, K_{SB})\}$$
$$B \rightarrow S : \{(ID_B)_{K_{SB}}, MAC(ID_B, K_{SB}) \mid REP\}$$

After the initialization process, the covering sensors and the cluster heads are elected. After this, the general sensors do not communicate with the base station directly, but only communicate with the covering sensors and the cluster head of the group. Figure 5 shows the communication between four general sensor nodes and the cluster head.

Sensor	Base
$K_S \in [2, n-2]$	$s_B \in [2, n-2]$
	$S_B \in s_B \times x$
$Send(K_S, ID_S)$ $\rightarrow$	$Recv(K_S, ID_S)$
	$E_S = K_S + S_B$
	$IC_S = (S_B, ID_S, E_S, t_S)$
	$H(IC_S) \rightarrow I_S \in [2, n-2]$
	$K(j) = S_B I_S + s_B \pmod{n}$
$Recv(K(j), IC_S)$ $\leftarrow$	$Send \ (K(j), IC_S)$

Fig. 4. Algorithm for computing key chain

The communication cost for this step seems higher, however, since the total processing time of the network is very low as the initialization process is performed only once for each group. After the initialization step, instead of communicating with the base station directly, the general senor nodes only communicate through the covering sensors and the cluster head. Assuming that the communication cost for each node is C_C, for a group of 50 nodes, it is $50 \times C_C$ and for 5 groups it is $5 \times 50 \times C_C$. As the network is deployed manually, the communication cost involved in the self-organization process is minimal. Assuming that the cost for each node for self-organization is C_S and for 5 groups each consisting of 50 nodes, the total cost would be $5 \times 50 \times C_S$. Now, we analyze the C_C and C_S costs. The self-organization process can have multiple steps [6], however, in ECEG the communication cost is involved in only one step where a node sends a message to the base station to authenticate itself, so $C_C < C_S$. In this comparison, we observe that by adopting the manual organization in some applications, the communication cost can be saved.

3) Elliptic curve point computation at the base station for computing group keys in the clusters: As described earlier, the base station generates an elliptic curve E defined over $f_{ID_G}(x)$. Each node is pre-deployed with an elliptic curve and the key chain. Though, every node in a particular group has the same curve but it is not the same for different groups. For example, group 1 can have elliptic curve E_1, group 2 can have curve E_2 and so on. The base station chooses one of the points in the elliptic curve and sends it to the cluster head. The cluster head broadcasts the point P to be used by the general sensor nodes to compute the group key K_G. Before the computation of the group key, a pre-deployed shared key K_{SC} is used for encryption/decryption of messages between the general sensor nodes and the cluster head. Later, the group key K_G is used for this purpose.

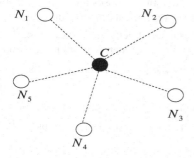
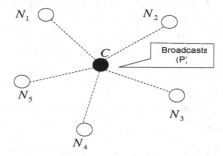

Fig. 5. After distribution of key chain   **Fig. 6.** Point broadcast by cluster head

Figure 6 shows the broadcast of a point by the cluster head. As the elliptic curve for each group is different, a different group key will be generated. The cluster head C broadcasts the encrypted point. The general sensor nodes S as well as the cluster head compute the group key K_G. Then the sensor nodes verifies the K_G by exchanging message M and reply R with the cluster head. The sequence of messages is as follows:

$$C \rightarrow S : \{(P)_{K_{SC}}, MAC(K_{SC}, P)\}$$

$$S : P + f_{ID_G}(x) = K_G$$

$$C : P + f_{ID_G}(x) = K_G$$

$$S \rightarrow C : \{(M)_{K_G}, MAC(K_G, M)\}$$

$$C \rightarrow S : \{(R)_{K_G}, MAC(K_G, R)\}$$

The key chain deployed in the nodes is used as raw keys to compute the group key. After using the point with the raw key, each group can compute its group key.

The advantages of using elliptic cure points are as follows:

1. All the points of the elliptic curve equation are related as they match the left and right hand side of the equation. If some points are lost in the broadcast then still the group key can be computed and updated.

2. The points of the elliptic curve are related with the raw key chain, the function used in the nodes is related with the points.

3. The points can only help to compute the group key from the key chain.

There is a relationship between the points and the keys in key chain K_G, as the key chain is generated from the points of the elliptic curve ECDLP (Elliptic Curve Discrete Logarithm Problem) [10] which can be used to get a group key from the key chain.

4) Key updating: To update the group key, the base station chooses a new point from the elliptic curve. The point is associated with the polynomials of the curve. As the base station communicates directly with the cluster head, it delivers the new point P'. There after, the cluster head broadcasts the new point. It is encrypted with the group key K_G so that the unauthentic nodes can not know the point for computing the new group key.

Figure 7 shows the computation of a new point P' of the elliptic curve from the point P. The general sensor nodes compute the new group key K_G' using P',

$$S : P' + f_{ID_G}(x) = K_G'.$$ After computing the new group key, the sensor nodes communicate with the cluster head to verify it.

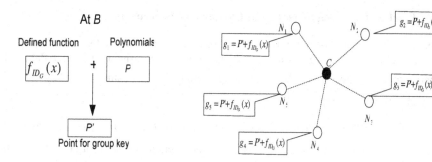

Fig. 7. Point computation at base station for computing group keys in clusters

Fig. 8. Group key computation in each sensor node

Figure 8 shows the new group key computation in each sensor node. The new group keys in five nodes $(N_1, N_2, N_3, N_4, N_5)$ are g_1, g_2, g_3, g_4, g_5 respectively, and $K_G' = g_1 = g_2 = g_3 = g_4 = g_5$. Figure 9 shows the verification of the group key by the cluster head to guarantee that the new group key is same for the group.

$$N_1 \rightarrow C : \{(g_1)_{K_g}, MAC(g_1, K_g)\},$$
$$N_2 \rightarrow C : \{(g_2)_{K_g}, MAC(g_2, K_g)\}.$$

Likewise N_3, N_4, N_5 each sends the group key to C. The cluster head verifies it's K_g' with the received group key from its group members. Then it sends a reply message to the individual nodes. For example, it sends the node N_1 the following message:

$$C \rightarrow N_1 : \{(M)_{K_G'}, MAC(M, K_G')\}$$

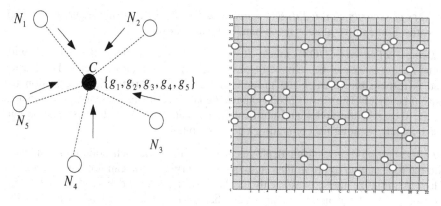

Fig. 9. Verification of group key at cluster head

Fig. 10. 30 points based on the equation $y^2 = x^3 + 17x + 12$

3.1 Electing Covering Sensors and Cluster Heads Securely

The covering sensors or the cluster head can be elected securely using the group key of each cluster. The organization is done after a pre-defined time to balance the energy in the cluster heads. All the messages are encrypted using group key K_g. The message exchanges are as follows:

1. $CH \rightarrow genral-nodes : (re-organization)_{K_g}$: In this message, the cluster head sends a message to all its group members about the reorganization of the group or the selection of the new covering sensors.

2. $general-nodes \rightarrow CH : (REPLY)_{K_g}$: In this message, the general senor nodes reply to the cluster head to start the process of reorganization or selecting new CS and CH nodes.

3. $general-nodes$: start election of CS and CH: In this step the general sensor nodes start the election of CS and CH nodes. The CS and CH selection algorithm is explained in section 2.1.

4. $CS \rightarrow genral-nodes : (Msg : ID_{CS})_{K_g}$: In this step after choosing the CS according to the covering sensor selection algorithm the CS communicate with the general sensor nodes to let them know that they are CS nodes.

5. $CH \rightarrow genral-nodes, CS : (Msg : ID_{CH})_{K_g}$: In this message, one of the CS elected as the cluster head and the ID of the cluster head is broadcasted to its group members.

After the expiration of a pre-defined time, the cluster head sends a message to its group member for the new election of covering sensors and the cluster head. The nodes securely communicate among themselves using the covering sensors algorithm (explained in section 2.1) to choose the CS. After that, the CS_S selected as one of the CS as a CH using the cluster head selection algorithm.

5) *Types of security claimed in this scheme:* ECEG provides group security in sensor networks. Some of the attacks it can handle are as follows:

a. Denial –of-service (DOS) attack: If some node tries to flood the network with bogus messages, it can be removed from the group by the cluster head. The cluster head is responsible to track the group members and report to the base station for unusual behavior. The cluster head changes the key and create a temporary symmetric key for the authentic nodes and notify the base station, so that the base station can broadcast a news point for computing a new group key.

b. Impersonation attack: After deployment of the network each node communicates with the base station to prove their authenticity. Intruder node can not have the key as it has to initialize the process with the base station. As the base station has the knowledge of all the nodes it can catch the intruder nodes and remove from the network.

c. *Replay attack:* The group key is refreshed before its expiration. If an insider node tries to repeat or delay a message for the key freshness, it can be caught as every node knows the expiration time of the key and new key will never match with the previous key.

4 Analysis of ECEG

This section provides the theoretical analysis of ECEG scheme. First, we study the characteristics of an elliptic curve cryptography then prove that groups can be merged using group keys which can be updated before it can be deciphered by an intruder. The elliptic curve $f_{ID_G}(x)$ used for key chain generation has points ranging from 0 to $x-1$. By the properties of elliptic curve [10], $f_{ID_G}(10)$ composed of integers ranges between 0 and 10. The operations such as addition and complement within the field will result into an integer between 0 and 10. The generalized equation used in elliptic curve is as follows:

$$y^2 \bmod p = x^3 + ax + b \bmod p$$

where p is used as characteristics of the base field. For example for $f_{ID_G}(10)$, $p=10$.

In our ECEG scheme, we consider the following elliptic curve equation:

$$y^2 = x^3 + 17x + 12$$

where the value of $p = 23$, $a = 17$ and $b = 12$. The base station can compute at most 30 points $P = (x_p, y_p)$ which satisfies both left and right hand side of the equation. Figure 10 shows those 30 points computed by the base station used in the group key computaion. These points are used to compute the key chain, which depends on the function $f_{ID_G}(x)$. Figure 4 shows the algorithm for key chain computation. For updating the group key ECEG uses the addition rule [10] in which from the chosen two distinct points $P = (x_p, y_p)$ and $Q = (x_q, y_q)$, a point R is computed for updating the group key as $P + Q = R$. The $R = (x_r, y_r)$ is computed using the following formulas given in [10]: $s = (y_p - y_q)/(x_p + x_q)$, $x_r = s^2 + s + x_p + x_q$, $y_r = s(x_p + x_r) + x_r + y_p$.

Lemma 1: If (x, y) is a rational point on the elliptic curve $y^2 = x^3 + ax + b$, then the group key K_G is same for at least 2 groups. That is, using the same group key K_G two groups can be merged.

Proof: Consider the number of nodes $N_1 \in G_1$, $N_2 \in G_2,......, N_n \in G_n$ where G is the group. $(N_1 \cup N_2) \in (G_1 \cup G_2)$, $(N_1 \cup N_2 ...... \cup N_n) \in (G_1 \cup G_2 .... \cup G_n)$. $\exists K_{G_1} s.t \{ K_{G_1}$ used as group key by $N_1 \}$, $\exists K_{G_2} \{ K_{G_2}$ used as group key by $N_2 \}$. By merge operation $(K_{G_1} \cup K_{G_2}) \in (y^2 = x^3 + x)$ (disregarding the constants). By pigeon hole principle, $K_G \in N_1 \cup N_2 ...... N_n$ s.t. $\{ N_1, N_2$ has a common key $K_G \}$.

Lemma 2: Let R be a rational point on $f_{ID_G}(x)$, and let $\{P1, . . . , Pk\}$ be independent points in $f_{ID_G}(x)$. If $R \in \{P,, . . . , Pk\}$ then the updated group key K_G' from $f_{ID_G}(x)$ does not repeat before the key decryption time t.

Proof: Assume updating time interval of group key $t \le t_{exp iration}$ and maximum decryption rate by intruder $D_r = 10^6$ decryption/ μSec [6]. We know that, $P' + F(q) = K_G'$, and time taken for computing K_G' is less than the expiration time. Total decryption time $D_t \notin t$ as $\{D_t > t\}$.

Theorem: In the ECEG protocol, two groups can be merged without affecting the security.

Proof: Let $N_1 \in G_1$, $N_2 \in G_2,......, N_n \in G_n$ where G is the group. Then $(N_1 \cup N_2) \in (G_1 \cup G_2)$, updating the time interval of group key $t \le t_{exp iration}$ and the decryption time $D_t \notin t$. It is not possible for an intruder to decrypt the key during the time interval of merge.

5 Performance Evaluation

We perform simulations to observe that ECEG scheme improves the performance of secure sensor networks by reducing communication and energy usage in elliptic curve cryptography. We use TinyOS and TOSSIM [15] for the simulation with 2000 sensor nodes distributed uniformly in a region of $1000 \times 800 m^2$. The communication range of the nodes is $40m$. The nodes are divided into groups (clusters) of 10 to 50 nodes. The nodes communicate with each other using multi hoping. We simulate the organization of the network using covering sensors and the cluster head selection algorithm. This helps to measure the communication time and energy consumption before the establishment of secure communication in the networks. The measurement of communication time is important for ECEG scheme because we assume that all the sensor nodes are authentic at the time of deployment. If some nodes are physically captured by intruders and the application layer software is modified to communicate with the unauthentic nodes, and then they can pretend to be authentic nodes. For address this problem, the group key computation should be fast enough so that intruders should not be able to decode the initial shared key used for encryption/decryption of packets before computing the group key.

To measure the initial communication time of general nodes with the base station we performed an experiment by varying the number of nodes in each group. Each group contains 10 to 50 nodes. In this experiment the base station sends *INIT* message and the nodes reply with *REP*. By this, the base station can keep track of the nodes in each group. The nodes are preloaded with a key which is shared with the base station. The size of shared key is 60 bits. This shared key is used for encryption/decryption of packets before computing the group key. Figure 11 shows the time taken by a group of nodes to organize and to communicate with the base station. We observe that a group of 50 nodes take 29.67 seconds. The measurement of this time is important as it is used in the process of first group key computation. As a reminder, we assumed that the keys of the base station must not be forged, and it should be able to authenticate all the messages. If an intruder tries to join one of the authentic groups, then the nodes will be able to detect false messages as it does not hold the shared key. By earlier elimination of the false messages and unauthentic nodes, the overhead of the network can be reduced as the messages will not be sent to the base station. From figure 11, we also observe that a group of 40 nodes takes less time for communication than 50 nodes. This is because of the non-uniform deployment of the groups. The group of 40 nodes was able to communicate faster with the base station as they could find faster route because of caching of routes. From this experiment we also conclude that the communication time of general nodes with the base station may not increase with the increase in number of nodes in each group but this may not be the case all the time. For example, when the nodes need to re-discover routes with addition of nodes in each group, then the communication time increases. We also measure the energy consumption by the general sensor nodes for performing the initial communication with the base station. The following energy usage parameters [15] are used for calculating the energy in both figures 12 and 14.

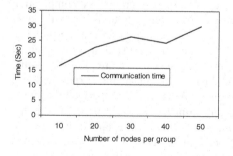

Component	Current drawn(mA)
CPU	
Active	8.0
Idle	3.2
Power save	0.11
Radio	
Reception (Rx)	7.0
Transmission (Tx)	3.7

Fig. 11. Communication of nodes with the base station for initialization

Figure 12 compare the energy consumed for initial communication and the total available energy in each group. It is assumed that each sensor node has two 1.5*volts* AA batteries. We observe that the battery power consumption for initial communication in this scheme (ECEG) is very less compared to the total available energy. The next experiment is to observe the group key computation time to prevent intruders from decrypting the keys. In this experiment, the general sensor nodes first setup a secure communication with the base station using a shared key then a point of the elliptic curve is broadcasted by the cluster head to general sensor nodes. As a reminder, this point is computed by the base station. Figure 13 shows the time taken by groups with different number of nodes for setting up the first group key. Each group consists of 10 to 50 nodes. The graph involves the time the general sensor nodes take to get the point of the elliptic curve and the group key computation time in each individual node. The size of group key computed in this experiment is 60 bits. The summation of secure communication time with the base station of general sensor nodes and group key setup time should be much less than the key decryption time by an intruder. From this experiment we observe that for a group of 50 nodes the key computation time is 37.31 seconds. The summation of secure communication and key setup time for a group of 50 nodes is 67.11 seconds. We conclude that, it is not possible by an intruder to decode the shared key K_{SB} within this time window using the exhaustive key search at a decryption rate of 10^6 decryption/μSec [6]. The intruders do not get enough time to decipher the group key K_G because it is also updated frequently. The updating time is proportional to the size of the group key. The energy consumption for computing the group key observed in this experiment (Figure 14) is very less compared with the available energy in each group.

We compare the time and energy consumption of ECEG protocol with EccM [9]. As a reminder, EccM protocol uses elliptic curve cryptography for computing the {private, public} key pair, but ECEG does not compute the private/public key in the sensor nodes. The base station generates key chain before the deployment of nodes. The individual sensors communicate with their cluster head to compute the group key. Figure 15, 16 shows the time and energy consumption by both the protocols for the number of nodes in a group varied from 10 to 50. It is observed that similar level of security can be achieved in ECEG using less computation overheads and the energy consumption.

Next we simulate three different cases, they are: i) New group key computation from the broadcasted point and the elliptic curve ii) Group key updating with small information sent by the base station iii) Computing the group key from the previous key without help of the base station. In all three cases the base station is aware about the key computation process performed in each group. Figure 17 shows the comparison of the bandwidth usage of the three cases. From this experiment we observe that bandwidth usage of case-1 is highest. We setup the experiment with the different number of nodes in each group varying from 10 to 50. For the first case, the group key is computed from the scratch using all the steps as explained in section 3. The main advantage of this case is, as the process of key computation starts over it is difficult for an eavesdropper to relate the new group key with the previous. For the second case, the base station sends the polynomial to update the function or an updated function. The cluster heads broadcast those to update the group key. The advantage of this case is in reducing the communication overhead and the bandwidth usage. In the last case, the updating of the key is done through the stored polynomial and a function. This case is vulnerable to attack, but the communication between the base station and the cluster head can be reduced.

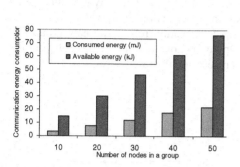

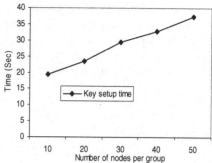

Fig. 12. Energy consumption (milli Joule) by group of nodes compared to available energy (kilo Joule)

Fig. 13. Time taken by each group for initial group key setup

We compared the total cost based on bandwidth usage in all the three cases. We observe that all the three cases have their own advantages and disadvantages. Here we discuss implication of all three cases:

i) New group key computation from the broadcasted point and the elliptic curve: For computing a new group key before expiration of the previous group key, each node in a group needs to go through the steps as explained in section 3. In this case as the process is similar to computing the group key for the first time and the communication overhead and bandwidth usage do not change.

ii) Group key updating with small information sent by the base station: In this case a new group key can be computed using certain information provided by the base station. The information can be a function which can create a new point in the nodes and later that point can be used to compute the group key. The communication overhead and bandwidth usage in this case are lower than the case-1, as it does not need to go through the steps for computing the group key from the point using the elliptic curve.

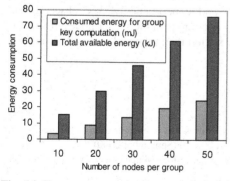

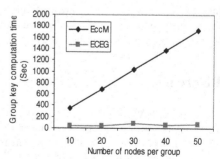

Fig. 14. Energy consumption by each group for initial group key setup

Fig. 15. Comparison between EccM and ECEG for key computation time

iii) Computing the group key from the previous key without any communication with the base station: This is possible by storing information along with the key chain before the deployment of the network. This information relates the new group key with the previous group key. The new group key is not stored in the nodes before the deployment to avoid an intruder to get the present and the future keys.

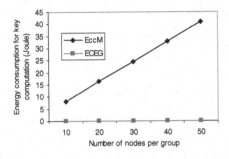
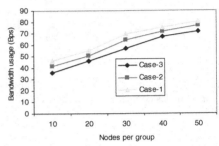

Fig. 16. Comparison between EccM and ECEG with respect to energy consumption for key computation

Fig. 17. Bandwidth usage of three cases

6 Conclusion

In this paper, we proposed an *energy and communication efficient group key management scheme (ECEG)* to improve the group key computation and key updates in sensor networks over EccM algorithm using elliptic curve cryptography. This scheme is based on the idea that energy intensive steps of security algorithms should be processed in the base station. For that, first an elliptic curve is used by the base station to provide the sensor nodes with a key chain before their deployment and then for computing the group key a point is broadcasted from the cluster head. The point is different for different groups and is provided by the base station. For updating the group key, a new point is broadcasted before the expiration of a predefined time. We analyzed the ECEG scheme theoretically and experimentally. We observed that instead of processing all the security related operations in the sensor nodes, if some of the security steps are performed in the base station then time and energy consumption for computing the key can be reduced.

References

1. Carman, D.W., Kruus, P.S., Matt, B.J.: Constraints and approaches for distributed sensor network security. NAI Labs Technical Report 00-010 (September 2000)
2. Chan, A.C., Rogers, Sr., E.S.: Distributed Symmetric Key Management for Mobile Ad hoc Networks IEEE INFOCOM (2004)
3. Du, W., Deng, J., Han, Y.S., Varshney, P.: A Pairwise Key Pre-distribution Scheme for Wireless Sensor Networks. In: Proceedings of the 10th ACM Conference on Computer and Communications Security (CCS), Washington DC, (October 27-31, 2003)

4. Du, W., Deng, J., Han, Y.S., Chen, S., Varshney, P.: A Key Management Scheme for Wireless Sensor Networks Using Deployment Knowledge. IEEE INFOCOM (2004)
5. Eschenauer, L., Gligor, V.D.: A key-management scheme for distributed sensor networks. In: Proceedings of the 9th ACM conferenceon Computer and communications security, Washington, DC, USA (November 18–22, 2002)
6. Karlof, C., Sastry, N., Wagner, D.: TinySec: A Link Layer Security Architecture for Wireless Sensor Networks, Proceedings of the Second ACM Conference on Embedded Networked Sensor Systems (SenSys 2004) (November 2004)
7. Heinzelman, W., Chandrakasan, A., Balakrishnan, H.: Energy-Efficient Communication Protocols for Wireless Microsensor Networks. In: Proc. Hawaaian Int'l Conf. on Systems Science (January 2000)
8. Madden, S.R., Franklin, M.J., Hellerstein, J.M., Hong, W.: TAG: a Tiny AGgregation Service for Ad-Hoc Sensor Networks. OSDI (December 2002)
9. Malan, D., Welsh, M., Smith, M.: A Public-Key Infrastructure for Key Distribution in TinyOS Based on Elliptic Curve Cryptography, IEEE SECON (2004)
10. Miller, V.S.: Use of Elliptic Curves in Cryptography. In: Williams, H.C. (ed.) CRYPTO 1985. LNCS, vol. 218, pp. 417–426. Springer, Heidelberg (1986)
11. Perrig, A., Szewczyk, R., Wen, V., Culler, D., Tygar, J.D.: SPINS: Security protocols for sensor networks. In: Proceedings of Mobicom (2001)
12. Sun, Y., Liu, K.J.R.: Scalable Hierarchical Access Control in Secure Group Communications, IEEE INFOCOM (2004)
13. Steiner, M., Tsudik, G., Waidner, M.: Key Agreement in Dynamic Peer Groups. IEEE Transactions on Parallel and Distributed Systems 11(8), 769–780 (2000)
14. http://www.xbow.com/Products/Wireless_Sensor_Networks.htm
15. http://www.tinyos.net/
16. Ye, F., Luo, H., Lu, S., Zhang, L.: Statistical En-route Filtering of Injected False Data in Sensor Networks, INFOCOM (2004)

A Cooperative CDMA-Based Multi-channel MAC Protocol for Ad Hoc Networks

Yuhan Moon and Violet R. Syrotiuk

Department of Computer Science and Engineering
Arizona State University,
699 South Mill Avenue, Suite 553
Tempe, Arizona 85281, USA
{yuhan.moon,syrotiuk}@asu.edu

Abstract. In this paper we present CCM-MAC, a cooperative CDMA-based multi-channel *medium access control* (MAC) protocol for multi-hop wireless networks. The protocol mitigates the multi-channel hidden and exposed terminal problems through cooperation from overhearing neighbours. By accounting for the multiple access interference obtained through cooperation, it also addresses the near-far problem of CDMA. We provide an analysis of the maximum throughput of CCM-MAC and validate it through simulation in Matlab. A significant improvement in network throughput is achieved over IEEE 802.11 and another multi-channel MAC protocol.

1   Introduction

Most wireless LANs are single channel systems. However, as the number of nodes communicating increases, systems with a single channel suffer declining performance. Contributing to the problem are the well-known hidden and exposed terminal problems. To combat these problems there is growing interest in multi-channel systems. Indeed, the IEEE 802.11 standard [1] already has multiple channels available for use. The IEEE 802.11a physical layer has 12 channels, 8 for indoor and 4 for outdoor use. IEEE 802.11b has 14 channels, 5 MHz a part in frequency. To avoid channel overlap, the channels should have at least 30 MHz guard bands; typically, channels 1, 6 and 11 are used for communication.

In a multi-channel system, the transmitter and receiver must both use an agreed upon channel for communication. This introduces a channel coordination problem. As well, the *hidden* and *exposed terminal* problems remain in the multi-channel setting. Figure 1(a) shows a communication between nodes A and B in progress on channel 1. Suppose that C chooses channel 2 to communicate with D. When A and B complete their transmission, neither has overheard the negotiation of channel 2 between C and D. As a result, a collision might happen if A then chooses channel 2 on which to communicate with B.

Figure 1(b) illustrates the exposed terminal problem in a multi-channel setting. Suppose that there are three channels and two of them are in use by nodes

E. Kranakis and J. Opatrny (Eds.): ADHOC-NOW 2007, LNCS 4686, pp. 172–185, 2007.
© Springer-Verlag Berlin Heidelberg 2007

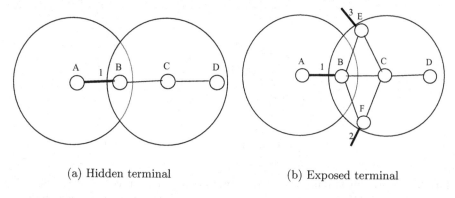

(a) Hidden terminal   (b) Exposed terminal

Fig. 1. The multi-channel hidden and exposed terminal problems

E and F. If nodes B and C want to communicate with nodes A and D, respectively, there is a free channel available. However, both B and C are in the transmission range of E and F. Even though both transmitters could use the same channel, one transmitter will delay its transmission. Without resolving the multi-channel hidden and exposed terminal problems, the optimal efficiency that can be derived from multiple channels can not be achieved.

In this paper, we propose a cooperative CDMA-based multi-channel MAC protocol for *mobile ad hoc networks* (MANETs). It uses *code division multiple access* (CDMA) technology on each channel and a cooperative mechanism to mitigate these problems. The idea of node cooperation is inspired from the CAM-MAC protocol of Luo et al. [2]. It is a simple idea: the reason the hidden and exposed terminal problems happen is because nodes lack knowledge about channel usage. Idle nodes that overhear channel negotiation may help other nodes make informed decisions.

CDMA has a very high spectral efficiency, i.e., it can accommodate more than one user on a channel. While CDMA is widely used in cellular systems there are some difficulties in applying CDMA to MANETs. The *near-far* problem takes place because a signal from a closer source is much stronger than from a source far away. Figure 2 shows a receiver R_2 of T_2 also in the transmission range of T_1. Since T_1 is closer (in terms of distance), its signal drowns out the signal of T_2. In cellular networks this is solved by the base station controlling the power to equalize the signals; this is not viable in MANETs where there is no centralized control. Cooperation is also used to mitigate the near-far problem in our proposed protocol.

The rest of paper is organized as follows. §2 reviews related work on multi-channel MAC protocols, CDMA, and cooperative mechanisms. §3 introduces our CCM-MAC protocol and describes how it mitigates the multi-channel hidden and exposed terminals problems and the near-far problem in CDMA through node cooperation. §4 analyzes the throughput achievable by CCM-MAC. Using

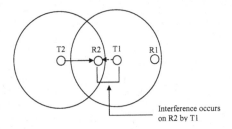

Fig. 2. The near-far problem of CDMA in MANETs

Matlab, we present simulation results comparing our protocol to IEEE 802.11, and the MMAC-CC multi-channel protocol in §5. Finally, in §6, we conclude.

2   Related Work

2.1   Multi-channel MAC Protocols

Some multi-channel MAC protocols use a control channel to coordinate channel selection. Of those without a control channel, [3,4] are equipped with single-transceiver. Lo et al. [3] uses CSMA on multiple-channels. N nodes compete to select one channel from M available; a channel is randomly chosen from the free channel list acquired by sensing at the transmitter. Zhou et al. [4] propose a multi-frequency MAC protocol for wireless sensor networks. It uses multiple frequencies to transmit or receive, and senses the carrier signal on all of frequencies rather than using a handshake. For nodes equipped with multiple transceivers, Nasipuri et al. [5] propose a multi-channel MAC protocol for multi-hop wireless networks that uses power-based channel selection. Both transmitter and receiver monitor all channels and select one that has the lowest signal power. In [6], a "soft" channel reservation is made, meaning a node prefers a channel on which it was last successful. This protocol uses as many transceivers as channels, and is therefore expensive in terms of the hardware required. A power-saving multi-radio multi-channel MAC protocol for WLANs is proposed by Wang et al. [7]. This protocol divides time intervals into three phases so that it can estimate the number of active links, negotiate channels, and then transmit data.

Some multi-channel MAC protocols using a dedicated control channel use one transceiver. Shi et al. [8] introduce the asynchronous multi-channel coordination protocol for WLANs. The control channel uses IEEE 802.11 DCF. Each node maintains a channel table and a variable indicating the channel it prefers. So et al. [9] uses a beacon signal to make periodic transmissions and give contention window time to all nodes which hear a beacon. Nodes then negotiate with each neighbour for a channel. The dynamic channel assignment with power control protocol, proposed by Wu et al. [10], expects the best channel to be the one for which another transmitter located the farthest distance from the transmitter is in use; they check signal power on transmitter side only.

Other protocols that use a control channel use multiple transceivers. Jain et al. [11] include free channel information in the handshake in a protocol that uses as many transceivers as channels. Wu et al. [12] propose a multi-channel MAC protocol with on-demand dynamic channel assignment. Each node uses two half-duplex transceivers, and each transceiver is used on a dedicated channel.

2.2 CDMA in Multi-hop Wireless Networks

Garcia-Luna-Aceves et al. [13] propose a method to assign codes in a dynamic multi-hop wireless network. Using neighbour information embedded in the handshake a unique orthogonal code is found. This protocol does not address the exposed terminal problem, or the near-far problem.

Muqattash et al. [14] propose a CDMA-based power controlled MAC protocol for MANETs. They address the near-far problem by using power control among the nodes. To obtain information about the power strength of the neighbour nodes, each node is equipped with multiple transceivers.

2.3 Node Cooperation

Cooperative mechanisms are becoming increasingly important in wireless networks with the potential to enhance system performance. More common in cellular networks (see, for example, [15]), cooperation is still largely unexplored in MAC protocols for MANETs. From the system point of view, since a node has limitations in terms of its antenna, power, cost and hardware, it is infeasible to use MIMO technology. Cooperative communication explores the benefits of multi-user environment by creating a virtual MIMO system.

Liu et al. [16] propose a cooperative MAC protocol for WLANs. The feature of CoopMac is to use a variety data rates on each channel. If the direct path between source and destination has low SNR, then using an intermediate cooperative node that relays the packet may be effective. A cooperative asynchronous multi-channel MAC protocol (CAM-MAC) is proposed by Luo et al. [2]. In CAM-MAC, the transmitter and receiver obtain channel usage information from idle neighbours after the handshake. Many problems remain, such as the hidden terminal problem, cooperative node selection, and control packet collision.

3   The CCM-MAC Protocol

In this section we describe our *Cooperative CDMA-based Multi-channel MAC* (CCM-MAC) protocol. The basic channel selection mechanism is similar to that used in MMAC-CC [17]. There is one control channel and N data channels. Control packets are transmitted on the control channel using a common code; this allows nodes in transmission range to overhear the channel negotiation. Data packets are transmitted on a data channel, with each node using its unique code which is orthogonal to all other codes. Thus, CCM-MAC combines both the advantages from using multiple channels and from CDMA.

3.1   Channel Negotiation in CCM-MAC

CCM-MAC uses a handshake for channel negotiation. In addition to the usual *request-to-send* (RTS), *clear-to-send* (CTS), and *acknowledgment* (ACK) packets, three additional control packets are used: *decided-channel-to-send* (DCTS) is used to indicate the channel selected, *information-to-inform* (ITI) is used by an overhearing node to aid the transmitter or receiver in its decision, and *confirm* (CFM) to inform neighbours of the receiver of the channel selected.

Figure 3 shows an example of channel negotiation in CCM-MAC. In this example node B has a packet to transmit to node C. If the control channel is idle, B transmits an RTS to C. Node C returns a CTS to B containing a list of free channels at B. Suppose that there is a node A that overhears only the RTS, and a node D that overhears only the CTS. Then, node A sends an ITI to B with information about the channel state around B, and simultaneously, node D sends an ITI to C with information about the channel state around C. (In this example, the ITI do not collide).

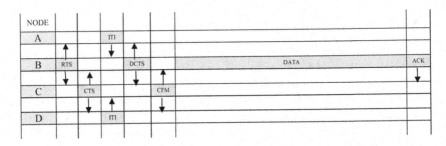

Fig. 3. Example of channel negotiation in CCM-MAC

Using the information contained in the CTS and the ITI, B selects a channel and sends its choice in a DCTS to node C. At the same time, the node A overhears the channel selection and stores this information together with duration information. On receiving the DCTS, if the selected channel is available on the receiver side, C returns a CFM to node B to confirm the choice. In this way, neighbours of the destination also overhear and store the channel selection and duration information. Finally, on receipt of the CFM, B transmits the data packet to C. If the data transmission is completed successfully, C transmits an ACK to node B *on the data channel*.

Recall that all the control packets in CCM-MAC are transmitted using a common code. It is possible that several nodes overhear an RTS/CTS exchange and may want to cooperate in the channel negotiation. Here, we take advantage of the *capture effect* of CDMA. This allows a node to demodulate the strongest signal of those transmitted. Therefore, we assume a node receives the ITI from its closest cooperating neighbour; this node also provides the most accurate information to node x since its transmission range overlaps that of x the most.

For CCM-MAC to mitigate the near-far problem of CDMA, the cooperating neighbours must provide additional information to allow a node to decide

whether it may add another transmission onto a channel with existing communications. This is discussed next.

3.2   Mitigating the Near-Far Problem

There are two factors to consider for the near-far problem in MANETs. One is
the *distance* between nodes.[1] The other is the communication mode (transmit
or receive) of a node. Figure 4 shows two transmitters T_1 and T_2 and their
corresponding receivers R_1 and R_2. In Figure 2, since R_2 is close to T_1 the signal
from T_2 may interfere with the signal from T_1. Therefore, the data transmission
between T_2 and R_2 may fail. However, in Figure 4 the near-far problem does not
occur; each receiver is far enough away from the other transmitter.

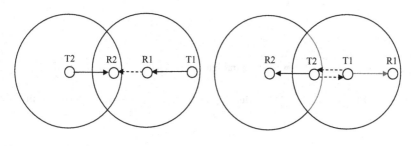

    (a) Interference among receivers   (b) No interference among re-
                                                     ceivers

Fig. 4. Example of interference and no interference between receivers in CDMA

If each node knows the distance to and communication mode of the nodes
around it the near-far problem may be avoided. In CCM-MAC cooperating
neighbours may provide this information. Not only may a cooperating neighbour
help with channel usage information, it can also estimate the distance between
the neighbour and transmitter (or receiver) by checking the signal strength. This
helps in the channel selection. If the distance to a neighbour with an ongoing
transmission is too close, and it is in a different communication mode, by selecting a different channel from that of the ongoing transmission, the near-far
problem can be avoided.

In this way, a transmission may be added to a channel with an ongoing transmission if it does not cause interference. Otherwise, another channel (if available)
is selected. This makes effective use of the multiple channels, and supports high
spatial reuse ratio in the system as more nodes may transmit data concurrently.

3.3   Mitigating Multi-channel Hidden/Exposed Terminal Problems

Figure 5 gives an example scenario to illustrate how CCM-MAC mitigates the
multi-channel hidden and exposed terminal problems. Suppose that nodes A and

---

[1] We assume all nodes use the same signal power.

B are communicating. Node C can initiate data transmission at the same time since it is out of the transmission range of A. But C is hidden to A and could cause a collision at B. This problem is solved by the CCM-MAC handshake. Through the ITI and CFM packets, node C can measure distance information, and obtain channel information to make an informed decision for channel selection to avoid the hidden terminal problem.

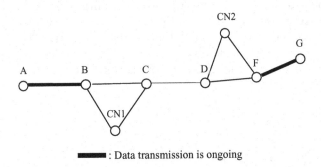

: Data transmission is ongoing

Fig. 5. Example for CCM-MAC mitigating the hidden and exposed terminal problems

There may be some situations in which not enough information is available to make a channel selection. Consider Figure 5 again, and assume that node C and D complete a transmission. Even though C and D may know that nodes B and F are in their respective transmission range, they may not know which channels are in use or the communication mode of each node. As a result, it may cause a hidden terminal problem. In CCM-MAC, cooperating neighbours are again the key to the solution.

In this case, node C sends an RTS to D. When D and any neighbour of C receive the RTS, they estimate their distance to node C by calculating the signal power using

$$\frac{P_t}{P_r} = \frac{(4\pi d)^2}{\lambda^2} \frac{(4\pi f d)^2}{c^2} \tag{1}$$

where P_t is the signal power at the transmitting antenna, P_r is the signal power at the receiving antenna, c is the speed of light, λ is the carrier wavelength, f is the frequency, and d is the propagation distance between antennas.

Meanwhile, D compares the distance between C and D and D and F and determines a free channel list which it includes, along with the distance estimate, in the CTS back to C. The cooperating neighbours CN_1 and CN_2 each transmit an ITI to nodes C and D, respectively, with the following information: the identifier of the node with an ongoing transmission (B and F, respectively), an estimate of the distance between the communicating pair (the distance between A and B, and F and G, respectively), and channel usage information.

In this example, the distance between B and C and D and F is too close, while nodes A and F are relatively far from nodes B and G. Therefore, the communication between C and D should not use the same channel as A and B or F and G to avoid the near-far problem.

The argument for the exposed terminal problem is very similar.

4   Analysis of the Maximum Throughput of CCM-MAC

Unlike technologies such as TDMA and FDMA in which the capacity is fixed and easily computed, CDMA does not have a fixed capacity. A CDMA system can accommodate more users on one channel because it has a very high spectral efficiency. As the number of users increases, the interference increases and the *signal to noise ratio* (SNR) decreases. If the SNR falls below a threshold, the channel is saturated, and no more users are allowed onto the channel. Therefore, the capacity of a CDMA system depends on the number of concurrent users.

In the multi-channel CCM-MAC protocol, there is one dedicated control channel and N data channels; CDMA is used on each channel. A common code is used on the control channel, while each user transmits data packets using a unique orthogonal code on a data channel.

To compute the throughput of the CCM-MAC protocol we use a *transmission frame*. A transmission frame is the time required for the CCM-MAC handshake, the transmission of the data packet, followed by the acknowledgment.

The time $T_{Handshake}$ for a pair of nodes to complete a handshake requires each control packet in the handshake to be transmitted:

$$T_{Handshake} = T_{RTS} + T_{CTS} + T_{ITI} + T_{DCTS} + T_{CFM}.$$

The maximum number of pairs of nodes H_{max} to complete the handshake successfully when noise is not considered is given by:

$$H_{max} = \frac{\frac{D}{B} + T_{ACK}}{T_{Handshake}}$$

where D is the size of the data packet, and B is the bandwidth of each channel. Therefore the throughput of CCM-MAC is given by

$$Throughput(\text{CCM-MAC}) = \frac{H_{max} \times D}{\frac{D}{B} + T_{ACK}}. \qquad (2)$$

However, if H_{max} is more than the channel can support, then it may not be possible for all of the node pairs completing the handshake to communicate. Therefore, we must determine the maximum number of users that can communicate concurrently on one channel.

Similar to Van Rooyen et al. [18] and Turin [19], the received signal Y_{pi} of the ith user in the pth symbol period is given as:

$$\begin{aligned} Y_{pi} &= \sqrt{E_s}(x_{pi} + \eta_i) + \eta_{pi} \\ &= \underbrace{\sqrt{E_s}x_{pi}}_{signal} + \underbrace{(\sqrt{E_s}\eta_i + \eta_{pi})}_{noise}. \end{aligned} \qquad (3)$$

Here, E_s is the energy per symbol, x_{pi} is the data of the ith user in the pth symbol period, η_i is the *additive white Gaussian noise* (AWGN) with zero mean that the ith user experiences from other active users, and η_{pi} is the noise the ith user experiences during the pth symbol period.

The output SNR for the ith user's signal may be expressed by the ratio of signal and noise power from (3) as

$$\alpha_{pi} = \frac{E[\sqrt{E_s}x_{pi}]^2}{E[(\eta_i + \eta_{pi})^2]}$$

$$= \frac{E_s}{E[\eta_i^2] + 2E[\eta_i, \eta_{pi}] + E[\eta_{pi}^2]} \tag{4}$$

since the user's signal $x_{pi} = \pm 1$, i.e., is a data bit denoted by ± 1. The value of $E[\eta_i, \eta_{pi}]$ is zero because the mean of the AWGN is zero. $E[\eta_{pi}^2]$ is $\frac{N_0}{2E_s}$ where N_0 is the noise spectral density [18].

Following Pursley [20], $E[\eta_i^2] \approx \frac{K-1}{3N_c}$, where K is the number of users considering noise and N_c is the number of chips per bit or processing gain. Substituting this approximation into Equation (4) yields an approximate expression for the SNR:

$$\alpha_{pi} = \frac{E_s}{\frac{K-1}{3N_c} + \frac{N_0}{2E_s}}$$

$$\approx \left(\frac{K-1}{3N_c} + \frac{N_0}{2E_s}\right)^{-1} \tag{5}$$

since E_s is a constant.

In this system model, all nodes transmit with the same power level and the received power from each node is also the same. Rearranging Equation (5) to obtain an expression for K, the maximum number of users considering noise, gives:

$$K = 3N_c\left(\frac{1}{\alpha_{pi}} - \frac{N_0}{2E_s}\right) + 1 \tag{6}$$

where N_c, the number of chips per bit or processing gain is T/T_c, where T_c is the duration of the chip pulse.

However, since our protocol is designed for operation in a multi-hop wireless network, it may be that the received power for each receiver is different. The SNR in this case, following Van Rooyen et al. [18], is

$$\alpha_0 = \frac{3N_cP}{\frac{N_0}{T_c} + \sum_{j=1}^{K}\left(\frac{d_{is}}{d_{ij}}\right)^{\beta}P} \tag{7}$$

where the first term of the denominator N_0/T_c is the Gaussian noise power in the chip-rate bandwidth, and the second term is the interfering power component expressed as a sum of the interference induced by all other active nodes. This equation assumes that the transmit power of all nodes is equal, but that each is at a different distance from receiver node i. Here d_{is} is the distance between node i and the source node s, d_{ij} is the distance between node i and active node j, P is the transmit power, and β is the propagation law exponent (normally equal to 4). The inter-node powers are scaled by the distance d_{ij}. Using Equations (5) and (7) a value for K, the maximum number of users with noise, is derived.

Finally, by Equations (2), (3), and (5), the throughput of CCM-MAC in the best case is

$$Throughput(\text{CCM-MAC}) = \frac{M \times D}{\frac{D}{B} + T_{ACK}} \qquad (8)$$

where

$$M = \begin{cases} H_{max} & \text{if } H_{max} < K \times N \\ H_{max} \times N & \text{if } H_{max} > K \times N \end{cases}.$$

5   Protocol Evaluation in Matlab

We use Matlab to simulate CCM-MAC, IEEE 802.11, and the MMAC-CC multi-channel MAC protocol [17].

For IEEE 802.11 the channel bandwidth is $2\,Mbps$. For CCM-MAC, that bandwidth is shared among a control channel and three data channels; the bandwidth of each channel is $500\,Kbps$. For MMAC-CC the bandwidth is shared equally among four channels.

A random grid topology, similar to [14], is used. M mobile users are placed in an area of $1000 \times 1000\,m^2$. The square is split into M smaller squares. The location of a mobile user is selected uniformly at random within each of these squares. The random way point model is used for mobility, with a user speed that is uniformly between zero and $2\,m/s$.

Every user is a source of packets. For each generated packet, the destination is randomly selected from one of the one-hop neighbours. Each node generates packets according to a Poisson process with rate λ, with the same rate used for all nodes. Table 1 shows other parameters of the simulation; these correspond to realistic hardware settings [21].

Table 1. Simulation parameters

Frequency	$2.4\,GHz$
IEEE 802.11 data rate	$2\,Mbps$
CCM-MAC data rate	$1.5\,Mbps$
CCM-MAC control channel rate	$500\,Kbps$
MMAC-CC channel rate	$500\,Kbps$
Transmission power	$20\,dBm$
Processing gain	$11\,chips$
SNR threshold	$15\,dB$
Reception threshold	$-68\,dBm$
Carrier-sense threshold	$-74\,dBm$
Interference threshold	2.78 [18]

5.1   Simulation Results

Figure 6 shows throughput as a function of increasing node density, for increasing packet sizes. The throughput of MMAC-CC is always higher than IEEE 802.11, and the throughput of CCM-MAC is always higher than MMAC-CC. This can be interpreted as the advantage that using a multi-channel protocol brings over a single channel, and then the advantage that CDMA brings over and above using multiple channels.

As the packet size increases, the gap in throughput between CCM-MAC and, IEEE 802.11 and MMAC-CC, increasingly widens. In Figure 6, for 500 *byte* packets and 36 nodes, the throughput of CCM-MAC is 1.3 times higher than MMAC-CC, and the throughput of MMAC-CC is 1.2 times higher than IEEE 802.11. For 1 *Kbyte* packets and 36 nodes, the throughput of CCM-MAC is now 2.5

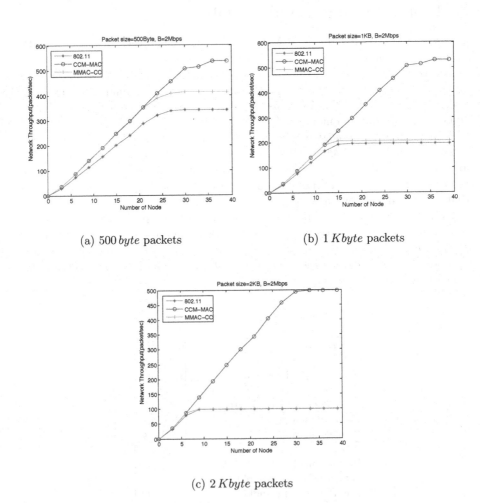

(a) 500 *byte* packets

(b) 1 *Kbyte* packets

(c) 2 *Kbyte* packets

Fig. 6. Throughput as a function of node density

times higher than MMAC-CC, while the throughput of MMAC-CC is now only 1.1 times higher than IEEE 802.11. For 2 *Kbyte* packets, CCM-MAC is 4.9 times higher than the throughput of the other two protocols, which are essentially the same after 12 nodes. As well, the advantage CDMA becomes more pronounced in sparser networks as the packet size increases. For 500 *byte*, 1 *Kbyte*, and 2 *Kbyte* packets the CDMA advantage becomes evident at about 24, 12, and 6 nodes in the network.

In CCM-MAC, the duration of handshake is fixed. However, the total negotiation cycle for all pairs depends on the actual packet transmission time. This is because the first pair of nodes gain access to the control channel for channel negotiation as soon as they complete packet transmission. Hence, the larger the packet size, the more chances for other node pairs to complete their handshake resulting in a larger number of nodes that can be transmitting simultaneously; this increases the overall system throughput.

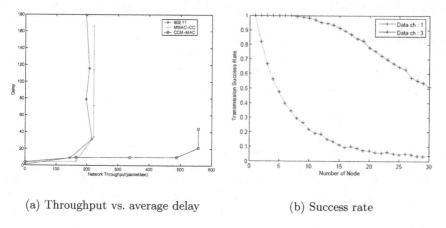

(a) Throughput vs. average delay   (b) Success rate

Fig. 7. Throughput delay curve, and probability of successful packet transmission

We also measured the average packet delay in CCM-MAC, MMAC-CC, and IEEE 802.11. The average delay D is the time elapsed in transmitting one data packet using the entire system bandwidth. Following Kleinrock and Tobagi [22], the delay is given by

$$D = \left(\frac{G}{S} - 1\right) \times R, \text{ where } R = N + 2a + \alpha + \delta \qquad (9)$$

where G is the offered traffic, S is throughput, and N is the number of channels. R is the sum of the packet transmission time, the round trip propagation delay, the transmission time for the acknowledgment (α), and the average retransmission delay (δ). We assume that ACK transmission and propagation delay time is so small that we can ignore their contribution to delay.

We assume that each protocol has same value of δ. Figure 7(a) shows that the average delay for CCM-MAC remains stable at the higher traffic loads. At low traffic loads, IEEE 802.11 and MMAC-CC have a slightly better delay because both procotols have a wider bandwidth than CCM-MAC for each channel. Figure 7(b) shows how the probability of successful packet transmission in CCM-MAC for one and three data channels for increasing node density. It is not surprising that the probability of successful transmission increases as the number of available channels increases.

6   Conclusion

In this paper we presented CCM-MAC, a cooperative CDMA-based multi-channel MAC protocol for ad hoc networks. It addresses the near-far problem of CDMA through multiple access interference, and mitigates the hidden and exposed terminal problems in multi-channel systems, both through cooperation. At high loads, and in denser networks, the protocol shows a significant improvement in throughput as well as lower delay than IEEE 802.11 and MMAC-CC.

Acknowledgments. This research is supported in part by NSF grant ANI-0240524. Any opinions, findings, conclusions or recommendations expressed are those of the authors and do not necessarily reflect the views of NSF.

References

1. Committee, I.C.S.L.S.: Part 11: Wireless LAN medium access control (MAC) and physical layer (PHY) specifications (September 1999)
2. Luo, T., Motani, M., Srinivasan, V.: CAM-MAC: A cooperative asynchronous multi-channel MAC protocol for ad hoc networks. In: Proceedings of the 3rd International Conference on Broadband Communications, Networks, and Systems (BroadNets'06), (October 2006)
3. Lo, F.L., Ng, T.S., Yuk, T.I.: Performance of multichannel CSMA networks. In: Proceedings of IEEE International Conference on Information, Communications and Signal Procesing, vol. 2, pp. 1045–1049 (1997)
4. Zhou, G., Huang, C., Yan, T., He, T., Stankovic, J., Abdelzaher, T.: MMSN: Multi-frequency media access control for wireless sensor networks. In: Proceedings of the 25th Annual Joint Conference of the IEEE Computer and Communication Societies (Infocom'06) (April 2006)
5. Nasipuri, A., Das, S.R.: Multichannel CSMA with single power-based channel selection for multihop wireless networks. In: Proceedings of the 52nd IEEE Vehicular Technology Conference (VTC-Fall'00), vol. 1, pp. 211–218 (2000)
6. Nasipuri, A., Zhuang, J., Das, S.R.: A multichannel CSMA MAC protocol for multihop wireless networks. In: Proceedings of the IEEE Wireless Communications and Networking Conference (WCNC'99), pp. 1402–1406 (1999)
7. Wang, J., Fang, Y., Wu, D.: A power-saving multi-radio multi-channel MAC protocol for wireless local area networks. In: Proceedings of the 25th Annual Joint Conference of the IEEE Computer and Communication Societies (Infocom'06) (April 2006)

8. Shi, J., Salonidis, T., Knightly, E.W.: Starvation mitigation through multi-channel coordination in CSMA multi-hop wireless networks. In: Proceedings of the 7th ACM International Symposium on Mobile Ad Hoc Networking and Computing (MobiHoc'06), pp. 214–225 (May 2006)

9. So, J., Vaidya, N.: Multi-channel MAC for ad-hoc networks: Handling multi-channel hidden terminal using a single transceiver. In: Proceedings of the 5th ACM International Symposium on Mobile Ad Hoc Networking and Computing (MobiHoc'04), pp. 222–233 (2004)

10. Wu, S.L., Tseng, Y.C., Lin, C.Y., Sheu, J.P.: A multi-channel MAC protocol with power control for multi-hop mobile ad hoc networks. The Computer Journal 45(1), 101–110 (2002)

11. Jain, N., Das, S.R., Nasipuri, A.: A multichannel MAC protocol with receiver based channel selection for multihop wireless networks. In: Proceedings the 10th International Conference on Computer Communications and networks (ICCCN'01), pp. 432–439 (October 2001)

12. Wu, S.L., Lin, C.Y., Tseng, Y.C., Sheu, J.P.: A new multi-channel MAC protocol with on-demand channel assignment for multi-hop mobile ad hoc networks. In: Proceedings of the International Symposium on Parallel Architectures, Algorithms and Networks (I-SPAN'00), pp. 232–237 (2000)

13. Garcia-Luna-Aceves, J.J., Raju, J.: Distributed assignment of codes for multihop packet-radio networks. In: Proceedings of the Military Communication Conference (Milcom'97), pp. 450–454 (November 1997)

14. Muqattash, A., Krunz, M.: CDMA-based MAC protocol for wireless ad hoc networks. In: Proceedings of the 4th ACM International Symposium on Mobile Ad Hoc Networking and Computing (MobiHoc'03), pp. 153–164 (2003)

15. Yang, H., Petropulu, A.: ALLIANCES with optimal relay selection. In: Proceedings of the IEEE International Conference on Acoustics, Speech, and Signal Processing (ICASSP'07) (April 2007)

16. Liu, P., Tao, Z., Panwar, S.: A cooperative MAC protocol for wireless local area networks. In: Proceedings of the IEEE International Conference on Communications (ICC'05), pp. 2962–2968 (2005)

17. Nasipuri, A., Das, S.R.: Multichannel MAC protocol for mobile ad hoc networks. In: Boukerche, A. (ed.) Handbook of Algorithms for Wireless Networking and Mobile Computing, pp. 99–122. Chapman & Hall/CRC (2006)

18. Van Rooyen, P.G.W., Ferreira, H.C.: Capacity evaluation of spread spectrum multiple access. In: Proceedings of the IEEE South African Symposium on Communications and Signal Processing, pp. 109–113 (August 1993)

19. Turin, G.: The effects of multipath and fading on the performance of direct sequence CDMA systems. IEEE Transactions on Vehicular Technology 33(3), 213–219 (1984)

20. Pursley, M.: Performance evaluation for phase-coded spread-spectrum multiple-access communication; Part I: System analysis. IEEE Transactions on Communications COM-25(8), 795–799 (1977)

21. Cisco Aironet 350 Series Client Adapters: Data sheet Available: http://www.cisco.com/en/US/products/hw/wireless/index.html.

22. Kleinrock, L., Tobagi, F.A.: Packet switching in radio channels; part 1: Carrier sense multiple access modes and their throughput-delay characteristics. IEEE Transactions on Communications COM-23(12), 1400–1416 (1977)

FDAR: A Load-Balanced Routing Scheme for Mobile Ad-Hoc Networks

XiaoRan Wang, Shigeaki Tagashira, and Satoshi Fujita

Hiroshima University, Japan

Abstract. In this paper, we propose an efficient and practical routing scheme for MANETs based on a new routing metric called the free-degree of nodes. In the proposed routing scheme named FDAR (Free-Degree Adaptive Routing), it is intended to deliver data packets circumventing congested routes, so as to realize a short end-to-end delay and a moderate load balancing of the overall network. In addition, it tries to avoid frequent and unnecessary invocation of route discovery process which would significantly degrade the routing performance of the MANET. The effectiveness of the proposed scheme is evaluated by simulation. The result of simulations indicates that it certainly outperforms previous load-balanced routing schemes including DLAR and LBAR, in terms of the packet loss rate and the average end-to-end delay.

Keywords: Ad-hoc network, routing, load-balancing, metric.

1 Introduction

Mobile Ad-hoc Network (MANET) is a wireless network consisting of several mobile hosts (or nodes), which self-configure to form a temporary network without any centralized administration such as servers and base stations. In recent years, MANET has been deployed in many applications such as the search-and-rescue [13], automated battlefields [15], and disaster recovery [9], and in such networks, each mobile node should seek the aid of the other mobile nodes in forwarding a packet to its destination because of a limited transmission range of wireless communication devices.

Due to several physical constraints such as the mobility of nodes, a narrow communication bandwidth, and a small battery capacity, researches in this field have focused on the development of efficient routing protocols during the past decade. Traditional ad-hoc routing protocols could be classified into two types, i.e., proactive type and reactive type. Proactive routing protocols attempt to maintain consistent, up-to-date routing information in every node by propagating it throughout the network [11,3]. Although it can always find a route to the destination (if any) by using such a globally collected information, a proactive approach generally needs a large number of packet transmissions, which causes a significant power consumption of the mobile nodes. In contrast, reactive protocols find a route to the destination only when it is requested by the source node, without explicitly maintaining a routing information [5,12].

E. Kranakis and J. Opatrny (Eds.): ADHOC-NOW 2007, LNCS 4686, pp. 186–197, 2007.

Most of previous ad-hoc routing protocols, including proactive and reactive ones, adopt the number of "hops" to the destination as the basic metric. Even though it could be easily implemented by using distance-vectors method, it does not take into account the congestion of the traffic load which is not uniformly distributed over the network. In general, a congestion of the traffic load increases the packet loss rate, end-to-end delay, and the battery power consumption even if the number of hops is kept identical [8]. This motivates the study of *load-balanced* routing protocols [7,4,14], which is intended to disperse congestions via selecting an appropriate route in the routing phase. It is reported in the literature that load-balanced routing protocols outperform shortest-path protocols particularly when the traffic load is relatively high [17,6].

In this paper, we propose a load-balanced routing scheme for MANETs based on a new routing metric called the **free-degree** of nodes. In the proposed scheme, named FDAR (Free-Degree Adaptive Routing), it is intended to deliver data packets circumventing congested routes, so as to realize a short end-to-end delay and a moderate load balancing of the overall network. In addition, it tries to avoid excessive overhead of route discovery process, which is frequently invoked by the source node in "unstable" networks such as the MANET. The effectiveness of the proposed scheme is evaluated by simulation. The result of simulations indicates that it certainly outperforms the performance of previous load-balanced routing schemes including DLAR and LBAR, in terms of the packet loss rate and the average end-to-end delay.

The rest of this paper is organized as follows. Section 2 overviews related work. A new routing metric is proposed in Section 3, and a routing scheme based on the new metric is given in Section 4. The result of simulations is shown in Section 5. Finally, Section 6 concludes the paper with future problems.

2   Related Work

This section reviews previous routing protocols for MANETs that utilize the traffic load as the routing metric. Such load-balanced routing protocols can be classified into two types; i.e., traffic-oriented and delay-oriented. Protocols of the first type, such as DLAR [7] and LBAR [4], try to balance the load of the network by distributing the traffic load among nodes. On the other hand, protocols of the second type, such as DOSPR (Delay-Oriented Shortest Path Routing) [14], try to avoid a selection of congested nodes as the intermediate nodes of a route. In the following, we describe the details of DLAR and LBAR, since our proposed scheme is also traffic-oriented. In addition, we will describe the functionalities of DSR [5] since our scheme has the same basic structure with DSR.

2.1   DLAR [7]

In DLAR (Dynamic Load-Aware Routing) protocol, the load of a node is defined as the number of packets buffered in the queue of the node, and the load of a route is defined as the summation of the load of nodes on the route. The definition of the load adopted in DLAR is quite natural, and it seems to work well in actual environments. However, it does not fully reflect the actual load of the network, since buffered packets may vary in size, and it does not consider the effect of access contentions in the MAC layer.

2.2   LBAR [4]

In LBAR (Load-Balanced Ad-hoc Routing), the load of a node is defined as the total number of routes passing through the node and its neighbors. More concretely, by letting P_i be the total number of routes passing through node i, the routing metric of a route r is defined as follows:

$$M_r = \sum_{i \in r} \left(P_i + \sum_{j \in N(i)} P_j \right)$$

where $N(i)$ is the set of immediate neighbors of node i. This metric is more accurate than DLAR, since each node refers to the information on its neighboring nodes; i.e., it could take into account the effect of access contentions caused by the neighboring nodes in the MAC layer. However, it is still insufficient since it neglects the variety of the packet lengths.

2.3   DSR [5]

DSR (Dynamic Source Routing) is a reactive source routing protocol for MANETs, which consists of two basic functions called the **route discovery** and the **route maintenance**.

When a node i wants to find a route to destination j, it first refers to its route cache to check if a route to j is contained in it. If it does, node i uses it to deliver a data packet to j, and otherwise, it starts a route discovery by flooding RREQ packets. More concretely, upon receiving a RREQ packet, each node replies a RREP packet to the source if it is the destination or it has a corresponding route in its route cache. Otherwise, it generates a copy of the RREQ packet and forwards it to its neighbors, after appending its own address to the *route record* in the packet. Note that each intermediate node forwarding a RREP packet to the source (in the backward direction) could acquire the route information contained in the packet and store it in its local route cache.

As for the route maintenance, in DSR, each node is responsible to check the liveness of its incident links. If a node detects that a link is broken while trying a forwarding of a packet to its destination, it returns an error message to the source of the packet. After receiving the error message, the source removes all routes containing the broken link from its route cache, and finds an alternative route to the destination, by referring to the route cache or by conducting a route rediscovery.

3   Proposed Routing Metric

In this paper, we propose a new load-balanced routing protocol called FDAR, which is an extension of DSR described in the last subsection. The proposed scheme is a reactive protocol of traffic-oriented type, and as the routing metric, it adopts the notion of **free-degree** (FD, for short) of routes.

3.1   Free-Degree of Nodes

The FD of nodes is defined as follows.

Definition 1. *The FD of node i, denoted by $\phi(i)$, is defined as follows:*

$$\phi(i) \;\overset{\text{def}}{=}\; \frac{\alpha_i}{(\beta_i)^2} \tag{1}$$

where α_i and β_i denote the transmission and the receiving rates of node i, respectively.

Note that Equation (1) intends to express an availability of a node with respect to the communication activity, in a sense that: 1) there remains an available room if the receiving rate is sufficiently small compared with the maximum bandwidth, and 2) for a fixed receiving rate, the available room becomes large by increasing the ratio of transmissions in its communication activity.

In actual MANET, the transmission and the receiving rates of a node dynamically vary. Thus, we design the proposed scheme in such a way that each node periodically updates its transmission and receiving rates for every T time units, according to the exponentially weighted moving average (EWMA) method, in the following manner:

$$\begin{cases} \alpha_i \leftarrow w \times \alpha_i + (1-w) \times \tilde{\alpha}_i \\ \beta_i \leftarrow w \times \beta_i + (1-w) \times \tilde{\beta}_i \end{cases}$$

where w is a weight with $0 < w < 1$, and $\tilde{\alpha}_i$ and $\tilde{\beta}_i$ are the current values of α_i and β_i, respectively. In the scheme, those rates are measured in terms of the number of bytes instead of the number of packets. More concretely, variable $\tilde{\alpha}$ is calculated as follows ($\tilde{\beta}$ is calculated in a similar way):

$$\tilde{\alpha} = \frac{1}{T} \times (\alpha_{data} + h \times \alpha_{head}) \tag{2}$$

where α_{data} represents the amount of transmitted data during the past T time units (in byte), α_{head} is the number of frames having been transmitted during the past T time units including failed ones, and h is the size of frame header (in byte). Note that Equation (2) reflects the effect of MAC layer contentions by counting the number of failed trials.

3.2   Free-Degree of Routes

In MANETs, conflicts of accessing channels are resolved by using an appropriate MAC protocol. In such congested situations, the traffic load of a node depends not only on the FD of the current node, but also on the FD of the neighboring nodes. In order to take into account such effects, we define the notion of **traffic interference**, in the following manner (note that a similar definition has been proposed in [4]).

Definition 2. Traffic interference around node i is defined as follows:

$$\psi(i) \ = \ \sum_{j \in N(i)} \phi(j). \tag{3}$$

Note that the FD of neighbors can be learned by periodically broadcasting hello message including its up-to-date FD value (see Section 4.3 for the details). By using the notion of traffic interference, we now define the FD of a route r, representing the activity of the route, as follows (as before, a larger FD value indicates a more freedom for the communication):

Definition 3. The FD of route r is defined as

$$\phi(r) \ = \ \sum_{i \in r} (\phi(i) + \psi(i)) \ = \ \sum_{i \in r} \left(\phi(i) + \sum_{j \in N(i)} \phi(j) \right) \tag{4}$$

where the summation in the first term excludes the source and the destination.

After calculating the value of function ϕ for every route connecting the source and destination, it selects a route with a maximum FD as the route to the destination, where a tie is broken by comparing the number of hops.

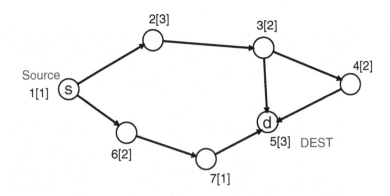

Fig. 1. Example of routing decision

Example: Consider a network shown in Figure 1, in which the number enclosed by a bracket represents the FD of the corresponding node. Let us calculate the FD of a route connecting nodes 1 to 5 through nodes 2 and 3. Since $\phi(1) = 1$, $\phi(2) = 3$ and $\phi(3) = 2$, the "activity" of node 2 is calculated as $\phi(2) + \psi(2) = 2 + 1 + 3 = 6$. Similarly, the activity of node 3 is calculated as 10. Thus, the FD of the route is calculated as 16 ($= 6 + 10$). A similar calculation can be conducted for the other two routes connecting 1 and 5, and in this example, the above route has the maximum FD among them, and is selected by the source node as the delivery route from 1 to 5.

4 Proposed Scheme

The basic structure of FDAR is the same as DSR. In this section, we describe the following points of the proposed scheme in detail: 1) route discovery, 2) route maintenance, and 3) the utilization of hello message.

4.1 Route Discovery

At first, the source initiates a route discovery by transmitting a RREQ packet to its neighbors if it has no valid route to the destination. Each RREQ packet has a field to record the FD value of the corresponding route in addition to the fields used in the original DSR. Before forwarding a copy of RREQ, each intermediate node inserts its address to the node list in the packet, and updates the FD value of the corresponding route by using Equation (4). A key difference to DSR is that in the proposed scheme, no intermediate node generates a RREP packet forwarded to the source node even if it has a valid route to the destination, since cached "FD value" of a route is not accurate in many cases. After receiving the first RREQ packet, the destination waits for receiving succeeding RREQ packets originated from the same source during a pre-designated time period. It then selects a route with a maximum FD, and replies a RREP packet to the source via the selected route.

4.2 Route Maintenance

In order to keep the information on the traffic load of neighbors up-to-date, each node piggybacks its FD value on the data packets transmitted to its neighbors. Thus, in the proposed scheme, a route could be reconstructed before detecting highly congested situations. The process of finding a new route is similar to the route discovery process given in the last subsection, except that the flooding of a RREQ packet is initiated by the destination instead of the source, and the source selects the best route after receiving the RREQ packets. Note that in this route rediscovery, the source does not need to send back a RREP packet to the initiator; i.e., the source may simply start a transmission of data packets.

Here, recall that DSR initiates a route rediscovery upon detecting *any* link failures since it does not distinguish failures due to node mobility from those due to MAC contention loss. Thus, if the traffic load is relatively high, frequent MAC contentions would significantly degrade the performance of DSR. To overcome such problem of DSR, we extend the scheme in the following manner: First, we extend DSR in such a way that it can receive feedback signals from the MAC layer to explicitly detect the occurrence of MAC contention loss. The second point of our extension is that it adopts a tolerance for generating an error message forwarded to the source. More concretely, by using an appropriate threshold θ, the condition for generating an error message is described as follows:

- If it detects a link failure and if it does not receive a feedback signal from the MAC layer.
- If it receives θ successive feedback signals before detecting a success of the packet transmission.

After receiving an error message generated by an intermediate node, the source initiates a route rediscovery as in DSR. Note that "$\theta = 1$" is exactly same as the policy adopted in the original DSR. (In the experiments in Section 5, we fix θ to two.)

4.3   Utilization of Hello Message

In FDAR, each node keeps its neighbors by exchanging hello messages. Each node maintains a list of neighbors and their FD, where FDs in the list will be used to calculate the traffic interference of the node. When a node receives a hello message from a neighbor, it examines its neighborhood list to check whether the neighbor has already been contained in the list. If so, it updates the FD of the node, and if not, it adds the new neighbor to the neighborhood list. On the other hand, if it does not receive a hello message from a neighbor for a predetermined time period, it removes the node from the list. In addition, if hello messages are not received from the next hop along a route, the upstream neighbor sends a notification of link failure to the source, to invoke a route maintenance process.

5   Simulation

We evaluated the performance of the proposed scheme by using NS-2 simulator [10]. We compare the performance of the scheme with three previous schemes described in Section 2, i.e., DSR [5], DLAR [7] and LBAR [4], in terms of the packet loss rate and the average end-to-end delay. In addition, the goodness of our proposed metric is evaluated by comparing it with metrics used in DLAR and LBAR. Finally, we evaluate the effect of the tolerant scheme proposed in FDAR in terms of the total number of routing changes and the aggregate throughput.

5.1   Parameters

Parameters commonly used in the following simulations are as follows. Each node is equipped with a wireless communication device with transmission range

250 m, carrier sensing range 550 m, and the interference range 550 m, where we adopt DCF (Distributed Coordination Function) [2] as the MAC layer protocol. In order to clarify the goodness of the proposed metric, the size of each packet is fixed to 512 Byte, and the size of MAC header is fixed to 30 Byte (recall that those values are used in Equation (2)). The link bandwidth between two adjacent nodes is fixed to 2 Mbps, and each source generates a sequence of data packets in CBR (Constant Bit Rate). We fix the number of nodes to 50, and set the simulation time to 300 seconds.

5.2 Static Case

At first, we consider a static case in which no nodes moves in a given field, where we adopted a field of size 500 m × 500 m. We fixed the sending rate of each source to 50 packets per second, and measured the packet loss rate and the average end-to-end delay by varying the number of source nodes from five to eight.

Figure 2 summarizes the results, where the horizontal axis of the figure represents the number of source nodes. From the figure, we can observe that the

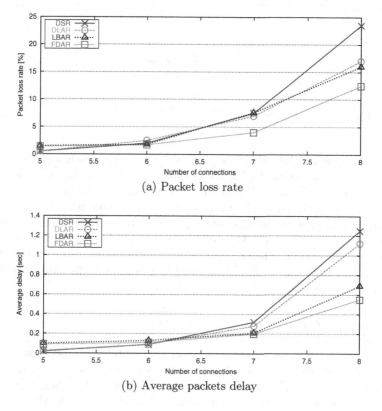

(a) Packet loss rate

(b) Average packets delay

Fig. 2. Results in a static case

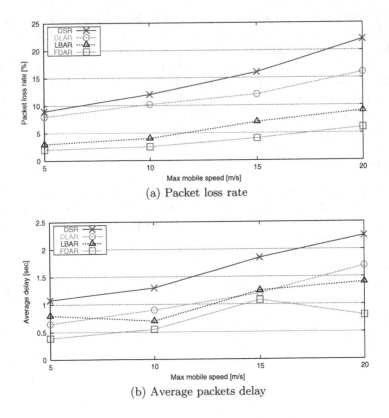

(a) Packet loss rate

(b) Average packets delay

Fig. 3. Results in a dynamic case

proposed scheme exhibits a better performance than the other schemes in almost all cases; e.g., when the number of source nodes is eight, the packet loss rate is reduced to 42% of DSR, and the average packet delay is reduced to 56% of DSR. Although the packet loss rate and the average packet delay are slightly worse than DSR when the number of sources is small, it is because of a relatively high routing overhead in a light traffic load. In fact, as increasing the number of sources, the proposed scheme achieves a good performance, i.e., a high traffic load due to a large number of sources could be evenly and effectively distributed by the proposed scheme.

5.3   Dynamic Case

Next, we consider a dynamic case. As a model of mobility of nodes, we used the *random waypoint model* proposed in [16]. The size of the field is set to 1000 m × 1000 m, and the number of source nodes is fixed to 20. We then measure the packet loss rate and the average end-to-end delay by varying the the maximum speed of nodes from 5 m/s to 20 m/s.

Figure 3 summarizes the results, where the horizontal axis represents the maximum speed of nodes. From the figure, we can observe that the proposed scheme exhibits a better performance than the other three schemes; e.g., the packet loss rate is reduced to 72% of DSR, and the average packet delay is reduced to 63% of DSR. Although DLAR and LBAR also achieve a better performance than DSR, the effectiveness of the load-balancing is not salient compared with our scheme. This result would indicate that the definition of the traffic interference used in the proposed scheme works effectively in a dynamic environment.

5.4 Evaluation of Tolerant Scheme

Finally, we evaluate the effectiveness of the tolerant scheme proposed in FDAR (see Section 4.2). In the experiments, we used a more realistic model for the mobility of nodes known as the Manhattan mobility model [1], which models the mobility of cars in a city area. The size of the field is fixed to 500 m × 500 m, where in the field, six streets are placed at an equal interval (each street has two lanes, i.e., upstream and downstream). We generate 200 mobile nodes in the field, and each node randomly moves along such 12 lanes. The maximum speed

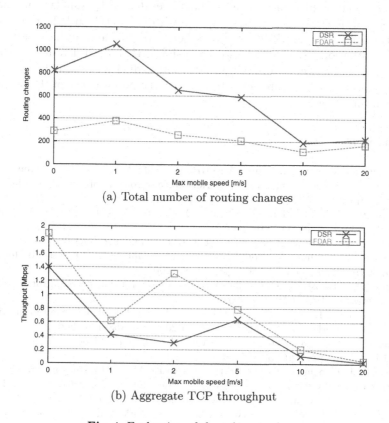

(a) Total number of routing changes

(b) Aggregate TCP throughput

Fig. 4. Evaluation of the tolerant scheme

of each node varies as 0, 1, 2, 5, 10, and 20 m/s, in the simulation. Finally, 20 TCP flows are established in this experiment.

Figure 4 summarizes the results, where Figure 4 (a) shows the total number of the routing changes, and Figure 4 (b) shows the total throughput of 20 TCP flows. The horizontal axis of the figures is the maximum speed of each node. As shown in the figure, a routing change frequently occurs in DSR particularly when the mobility is low, and the total throughput is not as high as our scheme even under a low and moderate mobility, which is because of the unnecessary, frequent invocations of the route rediscovery process. In contrast, the routing performance of our scheme is more stable than DSR, and the number of routing changes reduces to 70% of DSR, and the total throughput improves DSR by at least 35%, under a moderate mobility.

6   Concluding Remarks

In this paper, we proposed a new load-balanced routing scheme for MANETs. The proposed scheme is based on a new routing metric called free-degree of nodes, and it uses several techniques to improve the routing performance in MANETs. The performance of the scheme is evaluated by simulation. The result of simulations indicates that, compared with previous load-balanced routing schemes DLAR and LBAR, it exhibits a better performance in both static and dynamic situations. In addition, we have shown that the tolerant mechanism proposed in the scheme effectively avoids route rediscovery in moderate mobility. As a future work, we are going to evaluate the performance of the proposed scheme in more detail.

References

1. Bai, F., Sadagopan, N., Helmy, A.: IMPORTANT: A framework to systematically analyze the impact of mobility on performance of routing protocols for adhoc networks. In: Proc. IEEE INFOCOM (2003)
2. Crow, B., Widjaja, I., Kim, J., Sakai, P.: IEEE 802.11 wireless local area networks. IEEE Comm. Magezine, 116–126 (1997)
3. Optimized Link State Routing Protocol. IETF RFC 3626 (October 2003)
4. Hassanein, H., Zhou, A.: Routing with Load Balancing in Wireless Ad Hoc Networks. In: Proc. ACM MSWiM, pp. 89–96 (2001)
5. Johnson, D.B., Maltz, D.A.: The dynamic source routing protocol for mobile ad hoc networks. IETF Draft (1999)
6. Jones, E.P.C., Karsten, M., Ward, P.A.S.: Multipath load balancing in multi-hop wireless networks. In: Proc. IEEE WiMob 2005, pp. 158–166 (2005)
7. Lee, S.-J., Gerla, M.: Dynamic Load-Aware Routing in Ad Hoc Networks. In: Proc. IEEE ICC'01, pp. 3206–3210 (2001)
8. Macker, J., Corson, S.: Mobile Ad Hoc Networks (MANET): Routing Protocol Performance Issues and Evaluation Considerations. IETF RFC 2501 (January 1999)
9. Michail, A., Ephremides, A.: Energy Efficient Routing for Connection-Oriented Traffic in Ad-Hoc Wireless Networks. Mobile Networks and Applications 8(5), 517–533 (2003)

10. NS-2 Simulator, http://www.isi.edu/nsnam/
11. Perkins, C.E., Bhagwat, P.: Highly dynamic Destination-Sequenced Distance-Vector routing (DSDV) for mobile computers. In: Proc. SIGCOMM'94, pp. 234–244 (1994)
12. Perkins, C.E., Belding-Royer, E.M.: Ad hoc On-Demand Distance Vector Routing. In: Proc. 2nd IEEE Workshop on Mobile Computing Systems and Applications, WMCSA '99, pp. 90–100 (1999)
13. Rooker, M.N., Birk, A.: Combining Exploration and Ad-Hoc Networking in RoboCup Rescue. In: Nardi, D., Riedmiller, M., Sammut, C., Santos-Victor, J. (eds.) RoboCup 2004. LNCS (LNAI), vol. 3276, pp. 236–246. Springer, Heidelberg (2005)
14. Sheu, S.-T., Chen, J.: A Novel Delay-Oriented Shortest Path Routing Protocol for Mobile Ad Hoc Networks. In: Proc. IEEE ICC'01, pp. 1930–1934 (2001)
15. Taft-Plotkin, N., Bellur, B., Ogier, R.: Quality-of-Service Routing Using Maximally Disjoint Paths. In: Proc. IEEE IWQoS'99, pp. 119–128 (1999)
16. Yoon, J., Liu, M., Noble, B.: Random waypoint considered harmful. In: Proc. IEEE INFOCOM (2003)
17. Zhang, L., Gao, J.: Load balanced short path routing in wireless networks. IEEE Transactions on Parallel and Distributed Systems 17(4), 377–388 (2006)

TOLB: A Traffic-Oblivious Load-Balancing Protocol for Next-Generation Sensornets

Mohamed Aly[1] and Anandha Gopalan[2]

[1] Google, Inc.
[2] Dept. of Computer Science, University of Pittsburgh
maly@google.com, axgopala@cs.pitt.edu

Abstract. The multiple expected sources of traffic skewness in Next-Generation SensorNets (NGSN) will trigger the need for load-balanced point-to-point routing protocols. Driven by this fact, we present in this paper a load-balancing primitive, namely Traffic-Oblivious Load-Balancing (TOLB), to be used on top of any point-to-point routing protocol. TOLB obliviously load balances traffic by pushing the decision-making responsibility to the source of any packet without depending on the energy status of the network sensors or on previously taken decisions for similar packets. We present theoretical bounds on TOLB's performance for special network types such as mesh networks. Additionally, we ran simulations to evaluate TOLB's performance on general networks. Our experimental results show the high benefit (in terms of network lifetime and throughput) of applying TOLB on top of routing schemes to deal with various traffic skewness levels in different sensor deployment scenarios.

1 Introduction

Early sensornet deployments targeted data collection and push-based querying, e.g. [7, 36]. Hence, most current sensornet code bases, e.g. [18, 14], mainly offer tree-based many-to-one and one-to-many routing, query broadcasts, and aggregation during the data collection process [25]. Early sensornet applications were based on this model, e.g. monitoring and surveillance applications [37]. However, researchers and practitioners envision Next-Generation Sensor Nets (NGSN) to be composed of sensors deployed everywhere, together with gateways connecting sensors to Internet users/applications [26, 24]. Gateways may be stationary base stations [10], or mobile ones, such as robots, cell phones, and PDAs [35, 34]. Due to their huge number, sensors will tend to be clustered into geographic areas and geographically addressed *relatively* in each area rather than being assigned GPS-based addresses. In this model, querying loads will mostly be composed of pull-based ad-hoc queries issued by mobile users and/or Internet users. Ad-hoc queries trigger the need for using point-to-point routing [20, 12], a different routing paradigm from the old sensornet data collection model.

The large and continuously varying number of query sources in NGSN highly complicates the task of predicting the query distributions. Furthermore, the possibility of traffic skewness, when the sources and/or the destinations of most

E. Kranakis and J. Opatrny (Eds.): ADHOC-NOW 2007, LNCS 4686, pp. 198–212, 2007.

queries belong to a fairly small subset of sensors, is high. Query hotspots [2], where most queries access a small number of sensors simultaneously, represent one of the major traffic skewness examples. Query hotspots may be in the form of *Data-Centric Storage(DCS) range* queries [2], e.g. many queries asking for a small range of temperature readings stored in one or two sensors, or *geo-centric* queries [26], when many users are simultaneously interested in data generated by sensors in a particular area (e.g. find free parking spots in downtown area). In general, traffic skewness is a major problem that may result in the early death of sensors, network partitioning, and a subsequent reduction in network lifetime. The expected traffic skewness in NGSN introduces the need for robust load-balanced point-to-point routing schemes.

In this paper, we present the *Traffic-Oblivious Load-Balancing (TOLB)* protocol, a load-balancing protocol to be used on top of any point-to-point routing scheme. Our major design goal is *simplicity*. To achieve this goal, we adopt two main concepts: *Traffic-Obliviousness* and *Multipath Routing*. Traffic-Obliviousness means that the route of a packet $p = (s, t)$ is determined independently from the routes of previously issued (s, t) packets throughout the network operation [28, 8, 1]. A *stateless* distributed oblivious routing scheme is one where routing decisions are taken by individual sensor nodes solely based on local information, i.e. with no dependence of the load (energy) status of the remaining network nodes. On the other hand, Multipath Routing means that packets $p = (s, t)$ are routed through different network paths throughout the network operation. Previously presented multipath routing schemes were based on having a *paths enumeration* phase, where a set of network paths P is determined for each packet type (s, t) prior to the network operation. Individual paths of P are interchangeably used to route (s, t) packets based on the load-status of network nodes in a way to balance the energy consumption among all sensors. To blend both concepts together, TOLB substitutes the paths enumeration phase and the dependence on state information in taking routing decisions by *randomization*.

At its core, TOLB is based on a variation of the famous two-stage randomized routing, originally presented by Valiant [33] for bounding congestion in interconnection networks. In plain two-phase routing, an (s, t) packet is first routed to a random intermediate node r before being routed to the final destination t. To maintain obliviousness, TOLB only assumes the ability of each sensor to estimate its location and the approximate boundaries of the network service area. Furthermore, TOLB presents additional optimization heuristics that exploit the power of applying admission-control, making two random choices, and applying partial load-balancing in order to deal with various levels of skewness of both, traffic and node deployments. Through extensive simulations, we show that the major advantages achieved by TOLB are:

– Significantly increasing the network lifetime and throughput against skewed traffic distributions compared to the plain underlying routing scheme. The performance gains achieved by TOLB highly increase when considering typical query semantics such as query-reply pairs and query hotspots. TOLB also exhibits good performance when node densities increase.

- Maintaining a good level of fault tolerance against temporary node failures.
- Maintaining a comparable level of Quality of Service (QoS) and real-time guarantees to that offered by the underlying routing protocol.
- Maintain good performance even under skewed node deployments.

Organization of the paper: The rest of the paper is organized as follows. Related work is presented in Section 2 and the components of TOLB are presented in Section 3. Section 4 presents experimental results and Section 5 concludes the paper and discusses future work.

2   Related Work

In this section, we first provide a quick review of the load-balanced and oblivious routing protocols already presented in literature. Then, we briefly classify the currently available point-to-point routing protocols.

Load-Balanced Routing: Unlike TOLB, all previously presented load-balancing paradigms were embedded in routing protocols. Many of these protocols were based on multipath routing, where a set of paths are determined for each packet type prior to the network operation and paths are interchangeably used afterwords [31, 11]. Directed diffusion [19] presented the idea of finding multiple routes from multiple sources to a single destination while applying in-network data aggregation. Many multipaths routing schemes were later presented based on *Directed Diffusion*, e.g. [13, 29, 6]. Ganesan et al. [13] suggested the use of *braided multipaths* to achieve high resilience and fault-tolerance. Later, Raicu et al. [29] presented a diffusion-based algorithm where load-balancing decisions are made locally using location, power, and load as metrics in order to achieve global energy-efficiency.

Oblivious Routing: The idea of distributed oblivious routing in general communication networks was originally presented by Räcke's seminal paper [28], which later triggered many subsequent improvements, such as [17, 15, 16]. Recently, Busch et al. [8] presented the first theoretical analysis of a valiant-based oblivious routing algorithm on special types of geometric networks like mesh networks and uniform disc graphs. Later, Aly and Augustine [1] addressed the packet admission-control and oblivious routing problem for the first time in sensor networks. The work presented theoretical guidelines for any oblivious routing algorithm to maintain polylogarithmic competitiveness, w.r.t. throughput, against an offline routing algorithm. TOLB uses both, the Valiant paradigm and the admission-control ideas, based on the theoretical guidelines presented by the previous two papers.

Point-to-Point Routing: The need for point-to-point routing has recently increased as many current sensornet applications assume its usage, e.g. data-centric storage (DCS) [32, 4] and muti-dimensional range queries [23, 3, 2]. The first point-to-point routing schemes presented for wireless and sensor networks were based on *geographic routing*, e.g. GPSR [20], where nodes are identified by their geographic coordinates and routing is done greedily. Later, it was pointed that geographic schemes suffer from various limitations, e.g. the inability of current radios to conform with the current planarization algorithms and the unrealistic requirement of GPS-equipped sensors [12, 21]. Driven by these problems, schemes like NoGeo [30] and GEM [27] suggested the use of synthetic *virtual* coordinates assigned by iteratively embedding nodes in a Cartesian plane. Two recent schemes, BVR [12] and Logical Coordinates [9], used a collection of ideas from both geographic and virtual coordinates schemes. The basic idea of both schemes is to let nodes obtain coordinates from a set of landmarks. Routing then minimizes a distance function on these coordinates.

3   The TOLB Protocol

TOLB is designed to run on top of any point-to-point routing protocol to load-balance the traffic in the sensor network. The protocol is composed of a basic load-balancing algorithm and three optimization heuristics that can be set to run based on need. TOLB assumes each sensor node knows its location, the locations of its direct neighbors, and the approximate boundaries of the network service area. Throughout the rest of the paper, we refer to routing any packet $p = (s, t)$ using the underlying point-to-point routing protocol as *routing greedily*. We denote the network service area by A.

We start by presenting TOLB's core load-balancing algorithm.

3.1   The Core Load-Balancing Algorithm

The basic idea of the algorithm is to assign the load-balancing task solely to the source node of any packet. This conforms with the traffic-obliviousness property that TOLB maintains. The algorithm can be described in the following high-level steps:

1. For any packet $p = (s, t)$, the source s selects a location $R = (x_r, y_r) \in A$ at random and routes p *greedily* to R.
2. Let the closest node to the location R be r. Upon receiving p, r routes it greedily to its destination t.

Before describing the implementation details of TOLB, it should be noted that the mechanism for forwarding a packet from a sensor to one of its neighbors depends on the underlying routing algorithm. For example, geographic routing algorithms like GPSR [20] use physical coordinates while algorithms like GEM [27] use logical coordinates. The exact mechanism by which a node determines the location of any sensor in the network is beyond the scope of this paper. Furthermore, the random location selected is based on the assumption that a sensor knows the approximate physical boundaries of the network service area, A (in case of physical coordinates) or the virtual boundaries of A (in case of virtual coordinates).

Upon receiving $p = (s, t)$, s selects a random $R = (x_r, y_r) \in A$. It then sets the *first* destination of p to be R and the *second* destination to be t while setting a *destination flag* to 0 to indicate that the packet should be sent to its first destination. Then, s greedily routes p to its first destination R. Upon receiving p, each intermediate node first checks the destination flag to determine whether to forward p to the first or the second destination. When an intermediate node determines that it is closer to R than any of its neighbors, it sets p's destination flag to 1 indicating that p should be forwarded to its second destination. The process ends when p reaches its final destination t.

The load-balancing effect of the algorithm results from the randomized selection of R. As packets are routed greedily, this results in rotating the use of the different available paths in the network for routing (s, t) packets as time progresses. It is possible to prove theoretical bounds on the performance of the algorithm on special network types, such as mesh networks where each sensor can communicate with its four direct neighbors. The performance is measured in terms of stretch and congestion. Stretch is defined as the maximum ratio of a path length (in hops) of any (s, t) packet to that of the respective shortest path of that packet and congestion is defined to be the maximum number of paths using any node in the network. Let C_{opt} be the minimum congestion of the optimal offline load-balancing algorithm. The following theorem, whose proof directly follows from the proofs presented by Busch et al. [8], describes the performance of our algorithm on mesh networks.

Theorem 1. *TOLB achieves an $O(1)$ stretch and an $O(\log n) * C_{opt}$ congestion on both mesh networks and uniform disk graphs assuming $S = A$ and any underlying greedy routing protocol.*

We now present additional heuristics that would improve the performance of the TOLB load-balancing algorithm in special network settings, e.g. skewed node deployments.

3.2   The Admission-Control Protocol

The importance of the presence of an admission-control protocol in an oblivious routing scheme was raised by the following theorem proved by Aly and Augustine [1]. Let an always-send routing algorithm be the one where a packet p received by any node k is forwarded toward its destination as long as k has enough energy to forward p to one of its neighbors.

Theorem 2. *[1] Given a balanced binary tree $T(V, E)$ and a set of demands D, an always-send distributed oblivious routing algorithm A_{as} cannot maintain polylogarithmic competitiveness if either: 1. D is a set of adversarial demands, or follows a general distribution that is unknown to all sensor nodes; or 2. an adversary sets the tree node capacities (internal nodes or leaf nodes).*

This theorem shows that any distributed oblivious routing algorithm needs a concrete admission-control protocol in order to achieve polylogarithmic competitiveness with respect to throughput in the context of sensor networks. Although our goal is not directly to achieve polylogarithmic competitiveness, however, we use this theorem as an indication showing that the presence of an admission-control protocol improves the performance of any oblivious routing algorithm. The TOLB protocol is intended to maintain obliviousness on top of any routing algorithm. Thus, the combination of TOLB and the underlying routing algorithm can be considered as a distributed oblivious routing algorithm. In light of the above theorem, we append TOLB with a packet admission-control protocol. The basic idea that we present here is that some local information may be available at individual sensors. The usage of this information would improve TOLB's performance without ruling out its obliviousness property.

Before presenting the protocol, we define the *counters-list* as a list of counters maintained by each sensor node and containing a counter corresponding to each of the node's neighbors. Based on this definition, our admission-control technique can be summarized in the following points.

- Initially, all counters are set to zero. Whenever a packet is forwarded to neighbor j from node i, i increments the counter corresponding to j by 1.
- Whenever a packet p, arising in node s, is to be greedily forwarded to neighbor j of s, s compares j's counter value to the values of the counters of the rest of its neighbors. If the ratio of j's counter to the sum of all counters exceeds a threshold c, s reruns TOLB's core load-balancing algorithm to get another neighbor j'. The process is repeated for several times till an unloaded neighbor j_u is selected and p is then forwarded to j_u.

It is important to note that the admission-control protocol presented above did not use any information passing technique. Instead, all information used was inferred by keeping track of the number of packets sent through each direction (and subsequently through each neighbor). The intuition behind this is that the number of packets forwarded to each neighbor can be considered as an approximation of the total number of packets that passed through all the paths on which the neighbor falls. Subsequently, this can show us rough indications about the energy status of nodes along these paths.

The importance of the above admission-control protocol lies in achieving load-balancing for networks with skewed node deployments. A first example of such a setting is a network containing different node densities in different areas. For simplicity, we can think of a network with two sides, a left side with scarce sensor deployment and a right side with dense sensor deployment. We define the *skewness path* to be the path between the source and destination for some *s-t* pair that constitutes a large portion of the total network traffic. Let the skewness path be between the two sides of the network and the source be falling on the intersection line between the two sides. When applying TOLB's core load-balancing algorithm, the source's neighbors falling on the left side will be more loaded that those falling on the right side. Thus, these neighbors would be depleted with a faster rate than the right-side neighbors. In such a case, applying the admission-control protocol would help in improving the level of load-balancing achieved.

A second setting where admission-control shines is when s is the source of some traffic skewness in the network and one of its neighbors is close to death. Note that this information can be easily deduced from the counter value corresponding to this neighbor (thus, the number of packets sent through this neighbor), i.e., without any wireless communication dedicated for such reason assuming all sensors start the network operation with equal amounts of energy. In such a case, the admission-control may decide to take this symptom as a sign showing that most of the nodes on the path that will be followed by the packet are either dead or closed to death. This information can be used to take a decision of not sending any further packets along this path.

3.3   The Two-Choices Paradigm

We now move on to present the two-choices paradigm whose main goal is to enforce load-balancing in another type of skewed deployment. The idea of this paradigm comes from the famous balls-and-bins model. It is well knows that, given n balls that are thrown at random, one at a time, into n bins, the maximum load of a bin is approximately $\log n / \log \log n$ with high probability. Azar et al. [5] showed that in case, for each ball, two random bin selections are made and the ball is thrown in the least loaded bin among the two, the maximum load of a bin drops to $\theta(\log \log n)$, with high probability.

The important implication of the above result is that even a small amount of choice can lead to a significantly improved performance of randomized load-balancing algorithms. Using this intuition, we try to exploit the power of making two choices in TOLB. The paradigm maintains a *counters-list* (already defined in Section 3.2) at each sensor and it works as follows. Whenever a packet p arises in s, s makes two random choices rather than one by selecting two locations R_1 and R_2, both within the boundary of A. Among the two routes s-R_1-t and s-R_2-t, the idea is to try to route p through the route with least loaded nodes (i.e. with higher energy). However, knowing information about the paths' energy status contradicts with traffic-obliviousness. To cope with this problem, s determines the two neighbors j_1 and j_2 that will be involved in greedily routing the packet to R_1 and R_2, respectively. s then picks the location R_i whose j_i's counter has the smaller value and uses this location as the intermediate destination for p.

Like the admission-control protocol, the two-choices paradigm uses the values of the counters, representing messages sent through different neighbors, to get an approximate idea on the energy status of the nodes in the directions (and subsequently areas) corresponding to these neighbors. This idea is exploited by the paradigm to achieve load-balancing for skewed network settings such as *networks with gaps*. As an example, consider a network with randomly distributed sensors, but containing one or more gaps with no sensors in them (due to geographic obstacles, temporary or permanent node failures, etc). Let the source s of some skewness path be existing on the border of one of these gaps. Looking at the neighbors of s, we realize that its direct neighbors falling on the border of the gap will be more loaded than neighbors falling in other locations. For such a setting, applying the two-choices paradigm would be beneficial as it would be unlikely for the two random choices to fall on the border of the gap.

3.4   The Partial Load-Balancing Heuristic

Although load-balancing is an extremely important primitive for any routing protocol, a robust load-balancing protocol should be able to deal with different levels of traffic skewness. Subsequently, we present the option of partial load-balancing in TOLB in order to account for the possibility of regular or slightly skewed traffic loads.

In its high level, the partial load-balancing heuristic can be summarized by applying the TOLB protocol for a subset of the packets injected in the network rather than for every packet arising in any network node. This can be done by defining the *load-balancing factor* $0 \leq \epsilon \leq 1$ which has a unique value for all sensor nodes. Whenever a packet arises in a sensor s, s applies the TOLB protocol with probability ϵ and *greedily* routes the packet immediately to its destination t with a probability $(1 - \epsilon)$. The value of ϵ should be adaptively set in direct proportion to the expected traffic skewness level, thus, high when traffic is expected to be highly skewed and low otherwise. This mixed usage of the two versions of the routing algorithm, with and without TOLB, results in a limited load-balancing effect that is still better than solely using the underlying routing algorithm when a low level of skewness occurs in the network. It is worth mentioning that the traffic skewness level may be determined based on mining the query load history using machine learning techniques that are orthogonal to our concern in this paper.

Now that we have described all the components of TOLB, we move on to experimentally validate its performance in the following section.

4   Experimental Results

In order to evaluate the performance of TOLB, we implemented both TOLB and GPSR using the Glomosim wireless network simulator [22]. We simulated a sensornet cluster of the NGSN. The network is assumed to have *multiple base stations*, both stationary and mobile, acting as sources of the (skewed) traffic. In our experiments, sensors are assumed to be randomly distributed in A (unless otherwise mentioned). Sensors have an equal starting energy amount of $e = 100$ units. Sending or receiving one packet consumes 1 energy unit. Whenever a sensor s sends a packet to one of its neighbors t, only s and t consume energy for sending and receiving this packet, respectively. Furthermore, the wireless medium is assumed to be reliable and does not contribute to any packet loss. Each sensor is assumed to know the approximate boundaries of the service area, A. Also, a sensor is assumed to know its location and the approximate location of any sensor in the network.

To evaluate TOLB's *scalability*, we ran simulations for networks of sizes varying between 1000 and 2000 sensors. The network service area, A spanned a 200x200 square. At the start of every simulation, node locations are picked at random (except for the case of skewed deployments) and multiple stationary base stations are picked at random locations. The number of these base stations is fairly small compared to that of sensors. Initially, each node sends 1 broadcast message to know its neighbor's locations and it receives as many messages as the number of its direct neighbors. No maintenance messages are further sent during the simulation.

Traffic is generated as follows. At the start of the network operation, a small number of stationary base stations are randomly selected. Then, a small amount of destinations are randomly selected. Then, a high percentage of the generated packets are sent between the base stations and the selected destinations. We define this percentage to be the *skewness factor*, x. The rest of the packets are sent from random sources to random destinations. These are meant to be queries issued by mobile nodes, picked by a nearby sensor, and targeting another sensor in the network.

We implemented our protocol on top of GPSR. In measuring TOLB's performance in the different simulations, we focused on two metrics: *Network Lifetime* and *Throughput*. We define the network lifetime to be the time elapsed before the first node death in the network. Throughput denotes the number of successfully sent packets by all network nodes before the network dies. Network lifetime gives an idea of how TOLB load-balances the energy consumption among the different sensor nodes. Throughput on the other hand shows how load-balancing skewed traffic increases the network performance in terms of the number of successfully sent packets.

Simulation results are presented in the following subsections. Note that we only present part of our findings due to space constraints. In each of the graphs below, a point represents the average of 10 runs. It is worth mentioning that we were aware of the standard deviation in all simulation runs and we did not encounter a relatively large variance in any of the simulations.

4.1 Effect of Traffic Skewness Degree

In the first set of simulations, we changed the skewness factor x from 0% to 75% to get an idea on TOLB's performance for different traffic skewness levels. Figure 1 shows results in terms of both throughput and network lifetime. Figure 1(a) shows that the difference in throughput is almost constant between the 0% and the 75% cases. Furthermore, Figure 1(b) shows that the difference in lifetime between both cases decreases with the increase in network size. An important TOLB characteristic that can be deduced from both Figures is that its performance does not highly degrade or depend on the skewness factor.

4.2 Performance Gain Over Greedy

We now present the results that show the effect of our basic TOLB protocol on increasing the throughput when compared to the plain GPSR algorithm. Figure 2(a) demonstrates this comparison for a skewness factor of 70%. The Figure is an example of the big performance gain that TOLB achieves based on load-balancing traffic when compared to plain GPSR.

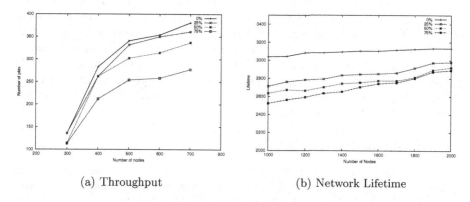

(a) Throughput   (b) Network Lifetime

Fig. 1. TOLB Performance against Different Skewness Levels

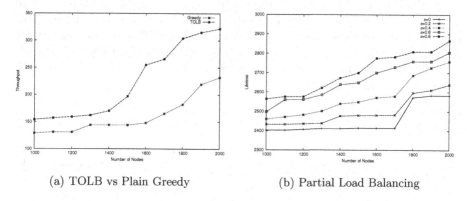

(a) TOLB vs Plain Greedy   (b) Partial Load Balancing

Fig. 2. TOLB-Greedy Comparison and A Study of Different Levels of Partial Load-Balancing

4.3   Benefit of Partial Load-Balancing

We study here the effect of TOLB's partial load-balancing heuristic on increasing network lifetime when traffic is close to regular. We set x to be 30%. For this lowly skewed traffic, we change ϵ from 0 to 0.8. Figure 2(b) shows that the lifetime increases proportionally with ϵ. This implicitly shows that the overheard imposed by TOLB on the different sensors due to using longer paths is totally dominated by the gain achieved by its load-balancing functionality.

4.4   Fault Tolerance

Failures can occur in the sensor network *temporarily* because of environmental conditions or due to the application of a specific energy saving scheme or *permanently* because of node deaths. It is important to test TOLB's performance against the various types of failures. We focused on temporary node failures as they capture the typical sensor network behavior. Thus, we assumed sensors have two modes: On and Off. We then introduced a random distribution to model temporary node failures. For $x = 70\%$, Figure 3(a) shows that the difference in throughput between TOLB and plain GPSR increases proportionally with the network size. This is a direct consequence of the multipath routing that TOLB imposes on GPSR for load-balancing.

4.5   Quality of Service (QoS)

Real-time applications represent an important characteristic of next-generation sensornets. In our application, for example, a user issuing a query can not tolerate waiting a long time without receiving a result. Motivated by this fact, we compared TOLB to plain GPSR in terms of the average time taken by packets to reach their destination. This metric is a twin of the average packet path lengths (Note that the stretch, which is the maximum path length, has been used in the theoretical analysis of TOLB already presented in Section 3.1). Figure 3(b) shows that the difference between TOLB and plain GPSR is almost constant for different network sizes when $x = 70\%$. An important implication of this result is that TOLB does not impose a large or increasing degradation in QoS on the underlying routing algorithm.

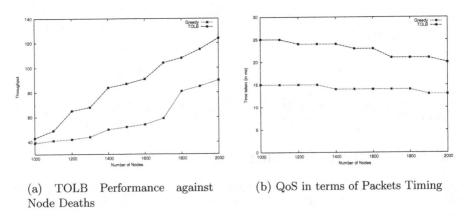

(a)   TOLB   Performance   against   Node Deaths

(b) QoS in terms of Packets Timing

Fig. 3. Fault Tolerance and Quality of Service

4.6   Effect of Correlated Requests

Motivated by our underlying application, a user issuing a query is expected to get an "immediate" answer for such a query. When mapped to requests, this means that whenever an $(s - t)$ is issued, a $(t - s)$ is issued accordingly. Furthermore, if the query spans more than one request, this would mean that multiple requests with the same source s would be issued to many destinations

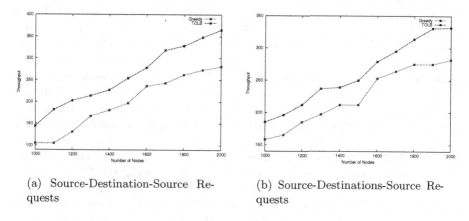

(a) Source-Destination-Source Requests

(b) Source-Destinations-Source Requests

Fig. 4. TOLB vs Correlated Requests

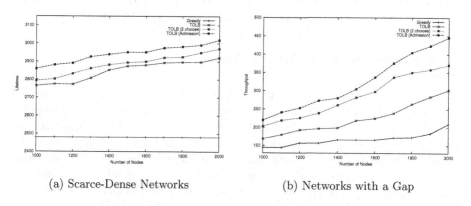

(a) Scarce-Dense Networks

(b) Networks with a Gap

Fig. 5. Skewed Node Deployments

t_j, and replies will be sent back from these destinations to s. Figure 4 presents simulation results modeling these two types of correlated requests. As GPSR acts deterministically, the effect of traffic skewness is magnified. However, load-balancing traffic through multiple $(s - t)$ paths helps in reducing the overhead imposed on individual paths. This is obvious by the relatively large performance gain achieved by TOLB for both types of requests (Figures 4(a) and 4(b)).

4.7 TOLB vs. Skewed Node Deployments

Due to environmental conditions, achieving random or uniform node distributions is difficult. This problem is largely magnified in next-generation sensornets as sensors are stationary and new node deployments are not frequent after the network is initially deployed. For such reason, we simulated skewed distributions of node deployments and networks with *gaps* to compare the effects of applying TOLB's admission-control protocol and two-choices paradigm on top of TOLB with basic TOLB and plain GPSR. Figure 5 shows the results of these simulations for $x = 70\%$. Figure 5(a) shows how applying the admission-control protocol results in increasing the network lifetime achieved by TOLB when compared to the basic TOLB protocol. Of course, the difference between basic TOLB and GPSR is considerably large because of the deterministic behavior of GPSR and the ability of basic TOLB to overcome this problem due to the use of randomization. It can be seen that TOLB with the two-choices paradigm achieves a better performance than basic TOLB. Though this performance is not as good as that of TOLB with admission-control, however, this demonstrates the effect of making two random choices and that it is comparable to using the full knowledge of local information as in the admission-control option. Similar intuitions are valid for Figure 5(b). However, in this case the difference between the effect of the admission-control protocol and that of the two-choices paradigm decreases because of the random distribution of the sensors (outside the gap). Note that we used a threshold $c = 60\%$ for the admission-control protocol. This is just an indicative value and further analysis can provide us with the optimal c.

5   Conclusions and Future Work

In this paper, we presented the Traffic-Oblivious Load-Balancing protocol (TOLB), a load-balancing primitive to be run on top of any point-to-point routing scheme in order to deal with traffic skewness. TOLB is based on the famous Valiant two-phase randomized routing paradigm previously presented used in interconnection networks and communication networks. Additionally, TOLB presents three optimization heuristics that apply admission control, exploit the power of two random choices and partial load-balancing to maintain load-balancing for skewed sensor deployments. We evaluated TOLB theoretically and experimentally to show its ability to load-balance different levels of traffic skewness.

We are currently implementing TOLB on sensornet testbeds to physically test its performance for various network sizes and settings. In the future, we would like to devise traffic skewness detection techniques that can quickly determine the level of traffic skewness and adaptively set the TOLB parameters to deal with the specific skewness level.

References

[1] Aly, M., Augustine, J.: Online packet admission and oblivious routing in sensor networks. In: Asano, T. (ed.) ISAAC 2006. LNCS, vol. 4288, Springer, Heidelberg (2006)
[2] Aly, M., Chrysanthis, P.K., Pruhs, K.: Decomposing data-centric storage query hot-spots in sensor networks. In: Proc. of MOBIQUITOUS (2006)

[3] Aly, M., Morsillo, N., Chrysanthis, P.K., Pruhs, K.: Zone Sharing: A hot-spots decomposition scheme for data-centric storage in sensor networks. In: Proc. of DMSN (2005)

[4] Aly, M., Pruhs, K., Chrysanthis, P.K.: KDDCS: A load-balanced in-network data-centric storage scheme in sensor network. In: Proc. of CIKM (2006)

[5] Azar, Y., Broder, A.Z., Karlin, A.R., Upfal, E.: Balanced allocations. SIAM Journal on Computing (1999)

[6] Baek, S.J., de Veciana, G.: A scalable model for energy load balancing in large-scale sensor networks. In: Proc. of 4th International Symposium on Modeling and Optimization in Mobile, Ad Hoc and Wireless Networks (2006)

[7] Bonnet, P., Gehrke, J., Seshadri, P.: Towards sensor database systems. In: Tan, K.-L., Franklin, M.J., Lui, J.C.-S. (eds.) MDM 2001. LNCS, vol. 1987, Springer, Heidelberg (2000)

[8] Busch, C., Magdon-Ismail, M., Xi, J.: Oblivious routing on geometric networks. In: Proceedings of SPAA (2005)

[9] Cao, Q., Abdelzaher, T.: A scalable logical coordinates framework for routing in wireless sensor networks. In: Proc. of RTSS (2004)

[10] Dai, H., Han, R.: Unifying micro sensor networks with the internet via overlay networking. In: Proc. of LCN (2004)

[11] Dulman, S., Nieberg, T., Wu, J., Havinga, P.: Trade-off between traffic overhead and reliability in multipath routing for wireless sensor networks. In: Proc. of WCNC (2003)

[12] Fonseca, R., Ratnasamy, S., Zhao, J., Ee, C.T., Culler, D., Shenker, S., Stoica, I.: Beacon Vector Routing: Scalable point-to-point routing in wireless sensornets. In: Proc. of NSDI (2005)

[13] Ganesan, D., Govindan, R., Shenker, S., Estrin, D.: Highly-resilient, energy-efficient multipath routing in wireless sensor networks. ACM SIGMOBILE Mobile Computing and Communications Review 5 (2001)

[14] Girod, L., Stathopoulos, T., Ramanathan, N., Elson, J., Estrin, D., Osterweil, E., Schoellhammer, T.: A system for simulation, emulation, and deployment of heterogeneous sensor networks. In: Proc. of SenSys (2004)

[15] Hajiaghayi, M., Kim, J.H., Leighton, T., Räcke, H.: Oblivious routing in directed graphs with random demands. In: Proc. of STOC (2005)

[16] Hajiaghayi, M., Kleinberg, R.D., Leighton, T., Räcke, H.: New lower bounds for oblivious routing in undirected graphs. In: Proc. of SODA (2005)

[17] Harrelson, C., Hildrum, K., Rao, S.: A polynomial-time tree decomposition to minimize congestion. In: Proc. of SPAA (2003)

[18] Hill, J., Szewczyk, R., Woo, A., Hollar, S., Culler, D., Pister, K.: System architecture directions for networked sensors. In: Proc. of ASPLOS (2000)

[19] Intanagonwiwat, C., Govindan, R., Estrin, D., Heidemann, J., Silva, F.: Directed diffusion for wireless sensor networking. IEEE/ACM Transactions on Networking (TON) (February 11, 2003)

[20] Karp, B., Kung, H.T.: GPSR: Greedy perimeter stateless routing for wireless sensor networks. In: Proc. of ACM Mobicom (2000)

[21] Kim, Y.-J., Govidan, R., Karp, B., Shenker, S.: On the pitfalls of geographic face routing. In: Proc. of DIALM-POMC (2005)

[22] Bajaj, L., Takai, M., Ahuja, R., Bagrodia, R., Gerla, M.: Glomosim: A scalable network simulation environment. Technical Report 990027, UCLA (May 1999)

[23] Li, X., Kim, Y.J., Govidan, R., Hong, W.: Multi-dimensional range queries in sensor networks. In: Proc. of ACM SenSys (2003)

[24] Luckenbach, T., Gober, P., Arbanowski, S., Kotsopoulos, A., Kim, K.: Tinyrest - a protocol for integrating sensor networks into the internet. In: Proc. of REALWSN (2005)

[25] Madden, S., Franklin, M.J., Hellerstein, J.M., Hong, W.: Tag: a tiny aggregation service for ad-hoc sensor networks, vol. 36, pp. 131–146. ACM Press, New York (2002)

[26] Nath, S., Liu, J., Miller, J., Zhao, F., Santanche, A.: Sensormap: a web site for sensors world-wide. In: Proc. of SenSys (2006)

[27] Newsome, J., Song, D.: GEM: Graph embedding for routing and data centric storage in sensor networks without geographic information. In: Proc. of SenSys (2003)

[28] Räcke, H.: Minimizing congestion in general networks. In: Proc. of FOCS (2002)

[29] Raicu, I., Schwiebert, L., Fowler, S., Gupta, S.K.: Local load balancing for globally efficient routing in wireless sensor networks. International Journal of Distributed Sensor Networks 1 (2005)

[30] Rao, A., Ratnasamy, S., Papadimitriou, C., Shenker, S., Stoica, I.: Geographic routing without location information. In: Proc. of ACM Mobicom (2003)

[31] Shah, R.C., Rabaey, J.M.: Energy aware routing for low energy ad hoc sensor networks. In: Proc. of IEEE Wireless Communications and Networking Conference (WCNC) (2002)

[32] Shenker, S., Ratnasamy, S., Karp, B., Govidan, R., Estrin, D.: Data-centric storage in sensornets. In: Proc. of HotNets-I (2002)

[33] Valiant, L.G.: A scheme for fast parallel communication. SIAM Journal on Computing 11 (1982)

[34] Vincze, Z., Vass, D., Vida, R., Vidacs, A., Telcs, A.: Adaptive sink mobility in event-driven multi-hop wireless sensor networks. In: Proc. of InterSense (2006)

[35] Westphal, C.: Scaling properties of routing protocols in sensor networks with mobile access. Technical report, Nokia (July 2006)

[36] Yao, Y., Gehrke, J.: Query processing for sensor networks. In: Proceedings of CIDR (2003)

[37] Zhao, J., Govindan, R., Estrin, D.: Sensor Network Tomography: monitoring wireless sensor networks. Computer Communication Review 32(1), 64 (2002)

A Comparative Analysis of Multicast Protocols for Small MANET Groups

Abderrahim Benslimane[1], Cédric Ferraris[2], and Abdelhakim Hafid[2]

[1] Laboratoire Informatique d'Avignon
Université d'Avignon BP 1228
84911 Avignon Cedex 9, France
abderrahim.benslimane@univ-avignon.fr
[2] Network Research Laboratory, University of Montreal
Pavillon André-Aisenstadt, H3C 3J7, Canada
{ferraric,ahafid}@iro.umontreal.ca

Abstract. In this paper, we investigate the performance of multicast protocols for wireless mobile ad hoc networks (MANETs). Recently, there have been several proposals to reduce overhead and enable scalability in respect to the number of multicast members. Explicit multicast protocols, which differ from common approaches proposed for MANET multicast routing, are based on the Xcast scheme and intends to reduce energy consumption. These protocols are best suited for use with Small Multicast Groups operating in dynamic networks of any size. Classical multicast protocols which are tree based or mesh based are more appropriate for large scale multicast group. We evaluate the performance of EM2NET, an Explicit Multicast Protocol for MANET, with other multicast protocols proposed for ad hoc networks via extensive and detailed simulation. The performance differentials are analyzed using varying network load, mobility and multicast group size in order to provide a qualitative assessment of the applicability of the protocols in different scenarios. Results indicate that EM2NET offers better energy conservation and also supports mobility better than existing explicit protocols.

1 Introduction

Many multimedia applications, such as teleconferencing and videoconferencing, involve group communications among hosts. Multicasting is a very useful mechanism which is intended for group-oriented computing and allows reducing bandwidth utilization. It consists of transmitting datagrams to a group of hosts identified by a single multicast address. We consider the problem of multicast routing in Mobile Ad hoc NETworks (MANETs). An ad hoc network is a dynamically reconfigurable wireless network with no fixed infrastructure or central administration. In a typical ad hoc environment, network hosts work in groups to carry out a given task. Applications such as disaster recovery, crowd control, search and rescue, and automated battlefields are typical examples of where ad hoc networks are deployed. Due to nodes mobility in MANETs, the network topology changes frequently and unpredictably. In

E. Kranakis and J. Opatrny (Eds.): ADHOC-NOW 2007, LNCS 4686, pp. 213–225, 2007.

addition to this, bandwidth and battery power are limited. These constraints make routing and multicasting in ad hoc networks extremely challenging.

During recent years, several multicast protocols have been proposed specifically for MANETs (e.g., ODMRP [1], MAODV [2], CAMP [3], and AMRIS [4]). These protocols all follow the traditional multicast approaches, i.e., they maintain multicast state information at intermediate nodes. When applied for use with small and sparsely distributed groups, these approaches may become even less efficient because they suffer from considerable overhead. Moreover, as a result of the constraints we mentioned above, it is difficult to maintain up to date correct multicast state information. A comparative analysis of these protocols has been reported in [5]. To overcome this problem, Explicit Multicast (Xcast) scheme has recently been proposed in order to support a very large number of Small Multicast Groups (SMG, [6]) by explicitly encoding the list of destinations in packets, instead of using a multicast address. It aims to eliminate forwarding states at intermediate nodes. A number of protocols enhance the basic Xcast scheme (DDM [7], SEM [8], and E²M [9]). To the best of our knowledge, E²M is the most recent and the most efficient explicit multicast protocol proposed for MANETs. However, it has a number of limitations which will be elaborated on in Section II. To overcome these limitations, we proposed a new protocol, called EM2NET for Explicit Multicast in MANETs [18], which has the same objectives as E²M with major enhancements. The basic idea behind EM2NET is to forward data only between dynamically determined Intercepting Nodes (IN). EM2NET differs from E²M by taking node mobility into account. Its goal is to reduce energy consumption in general and allow low control overhead and high packet delivery ratio for small multicast groups. In this paper, via detailed simulations, we evaluate its performance and compare it with existing multicast protocols (E²M, ODMRP and MAODV). We chose to compare EM2NET with ODMRP and MAODV in order to show the efficiency of protocols using the Xcast scheme when they are applied to small multicast groups. We intend to determine if explicit multicast methods successfully reduce the control packet overhead and how it impacts on the protocols performance (e.g., energy consumption). We conduct extensive simulations employing a wide range of mobility and traffic load conditions, as well as different multicast group characteristics. We apply metrics that show the efficiency in addition to the effectiveness of the protocols. This paper is the first to conduct a performance comparison study of ad hoc wireless explicit multicast protocols.

The rest of the paper is organized as follows. Section 2 provides an overview of the multicast protocols proposed for MANETs. Section 3 describes the EM2NET protocol. The simulation environment and methodology are described in Section 4 followed by simulation results in Section 5. Finally, Section 6 concludes after highlighting some open problems and possible future work.

2   Related Work

One straightforward way to provide multicast in a MANET is through flooding. It is suggested that in a highly mobile ad hoc network, flooding of the whole network may be a viable alternative for reliable multicast. However, this approach has considerable overhead since a number of duplicated packets are sent and packet collisions do occur in a multiple-access-based MANET [10]. In this section we discuss multicast routing protocols proposed for MANETs. We can classify them into three categories based on how routes are created to the members of the group.

2.1 Tree-Based Approaches

Tree-based multicast is a very well established concept in wired networks. Most schemes for providing multicast in wired networks are either source- or shared-tree-based [11]. However, in the highly dynamic environment of mobile ad hoc networks, the traditional multicast approaches used in wired networks are no longer suitable. Representative tree-based multicast protocols are Multicast AODV (MAODV) and AMRIS [12]. Both protocols are on-demand and construct a shared delivery tree to support multiple senders and receivers within a multicast session.

MAODV routing protocol follows directly from unicast AODV [13] and discovers multicast routes on demand using a broadcast route discovery mechanism employing the same route request (RREQ) and route reply (RREP) messages that exist in AODV. It is based on activation messages to ensure that the multicast tree does not have multiple paths to any tree node. Data packets are forwarded only along activated routes.

Existing studies show that tree-based protocols are not best suited for multicast in a MANET where network topology changes frequently. In such an environment, mesh-based protocols seem to outperform tree-based proposals due to the availability of alternative paths [10].

2.2 Mesh-Based Approaches

In contrast to tree-based approaches, mesh-based multicast protocols may have multiple paths between any source and receiver pair, which allow multicast packets to be delivered to the receivers even if links fail. Representative mesh-based multicast routing protocols include Core-Assisted Mesh Protocol (CAMP) and On-Demand Multicast Routing Protocol (ODMRP). These protocols build routing meshes to disseminate multicast packets within groups. ODMRP uses flooding for mesh building while CAMP uses one or more core nodes to assist in mesh building, instead of flooding.

ODMRP creates a mesh of nodes which forward multicast packets via flooding (within the mesh), thus providing mesh redundancy. In ODMRP, group membership and multicast routes are established and updated by the source on demand. Similar to on-demand unicast routing protocols, a request phase and a reply phase comprise the protocol. This whole process constructs or updates the routes from sources to receivers and builds a mesh of nodes, the *forwarding group*.

Tree- and mesh-based approaches have an overhead of creating and maintaining the delivery tree/mesh with time, especially in a MANET environment which is subject to frequent and unpredictable movement of mobile nodes. To minimize the effect of such a problem, explicit multicast is proposed wherein a source explicitly mentions the list of destinations in the packet header.

2.3 Explicit Multicast

Explicit multicast focuses on Small Multicast Groups (SMG) and assumes the underlying unicast protocol takes care of forwarding the packets. It is a new multicast
scheme that can support a very large number of small multicast groups. Xcast explicitly encodes the destination addresses list in data packets instead of using a multicast
address. Thus, the source encodes the list of destinations in the Xcast header and then
sends the data packet to a router. Each router along the way parses the header, partitions the destinations addresses based on each destination's next hop and forwards a
packet with an appropriate Xcast header to each of the next hops.

   Recently, there have been several proposals, such as Differential Destination Multicast (DDM) and Extended Explicit Multicast (E²M), which extend the basic Xcast
scheme to reduce the overhead involved in processing the Xcast header at intermediate nodes.

   E²M is built on top of Xcast and gives scalability to the number of multicast members. It employs the concept of Xcast Forwarders (XF) which are selected dynamically during the message forwarding procedure, according to the number of group
members: a node decides to become an XF if it serves a number of nodes greater than
a given threshold. This avoids the MEMBER_JOIN implosion problem at the source.

   E²M has three major drawbacks: (a) it does not take effectively into account node
mobility, which is a very frequent phenomenon in MANETs. When an XF moves, the
source will be notified with the help of the network layer (i.e., generation of Route Error
RERR, propagated to the source via AODV or DSR [14] routing protocol). Therefore,
the source obtains the list of destinations served by this XF and puts them explicitly in
its extended header for data transfer. This can cause data loss because routes are not yet
established. Our protocol, EM2NET, takes into account the tree management to overcome this potential problem; (b) with Xcast, the average size of the packet header increases with the number of receivers. This will directly increase energy consumption
because of transmission of longer packets. E²M is more scalable according to the multicast group size but the threshold value for being an XF is a major issue. We will discuss
that during simulations. Moreover, each node in E²M has to intercept every JOIN messages in order to determine if it is a new XF, which is an important wasting of energy.
The addresses discovery process in the header will require both power and energy too.
An XF will consume more energy for encapsulation, duplication, and transmission of
multicast packet copies. This problem is resolved with our method that allows load
balancing between branching nodes; and (c) in ad hoc networks, links are not necessarily bidirectional. Asymmetric routing means that the unicast path from A to B may
differ from the path from B to A (for example due to the radio transmission power).
Contrary to E²M, EM2NET uses the shortest path tree from the source to destinations.
Thus, it overcomes the problem of asymmetric routing (when a source receives join
message, it doesn't use reverse forwarding path for the reply).

3 EM2NET Protocol Description

EM2NET is an explicit multicast protocol introduced for small multicast groups. It
constructs a multicast tree consisting of particular nodes called Intercepting Nodes

(IN) contrary to a classical tree-based multicast protocol. An IN is a member already on the multicast tree or any node that is a branching node (a node explores its unicast routing table to detect if two or more of its next hops are disjoint).So, a branching node is a node which send copies of the same packet on disjoint paths.

3.1 Messages and Data Structures

Each node which is an IN for a multicast session maintains a Multicast Forwarding IN Table (MFIT). An entry in MFIT is indexed by the pair (S, G) where S is the unicast address of the source and G the multicast group address (or the session number). It contains the previous hop IN and the list of the next INs. We can observe that EM2NET reintroduces the concept of multicast state information at intermediate nodes by using MFIT tables. However, it is negligible compared to a classical tree-based protocol since only Intercepting Nodes (IN) maintain small size MFIT tables.

EM2NET uses two types of packets: control packets and data packets. When a node wants to join the multicast session (S, G), it sends a JOIN message directly to the source S, in a unicast way. When S receives the JOIN message, it sends a BRANCH_ACK message in response towards the new member (or the list of new members). This message must be sent hop by hop, with an underlying unicast routing protocol, till reaching the new member. The address of the new member (s) is transported in the message. So, it allows discovering IN nodes. If a node, upon receipt of the BRANCH_ACK message, is already a member of the multicast group, it forwards the message unchanged. In this case, no entry will be created in the MFIT for the (S, G) session. If the node becomes a new IN (a branching node or a new member of the multicast group), then an entry is created in its MFIT if it does not already exist. This entry contains the source address S, the multicast address G, the Previous IN (PIN) address, and the sub-list of destinations per next hop (next INs or NINs). Then the new IN informs the PIN via a BRANCH_UP message about its IN identity and continues to forward BRANCH_ACK, while replacing itself as a PIN, to the next hops until it reaches all the new receivers. BRANCH_UP and BRANCH_DOWN messages are exchanged periodically between adjacent INs (respectively towards its upstream predecessor IN and its successor INs down the sub-tree) to support mobility and failure management. A BRANCH_TIMEOUT is associated with each IN to allow updates. This timer is re-initiated by BRANCH_DOWN and BRANCH_UP messages. EM2NET adds strong link connection check between nodes and hence, preserves tree connectivity. Finally, if a member wants to leave the multicast group, it sends a LEAVE message to the source. After a predefined constant, FAILURE_WAITED_ PERIOD, each member on the downstream sub-tree initiates a JOIN operation. Thanks to the periodicity of transmitted control messages, node mobility management is accomplished [18].

Fig. 1 illustrates the EM2NET packet delivery process from a source S towards receivers (R1 to R6). It also shows the packets headers. We can see that the MFIT table at the source S contains a null entry for the predecessor IN and B2 as the successor IN. For B7, the predecessor IN is B6 (which is a branching node) and successors INs are R3, R4 and R5 (which are members of the multicast group).

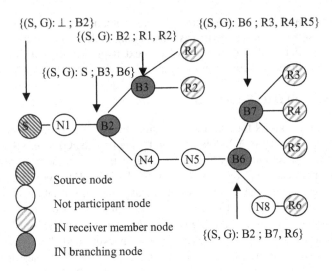

Fig. 1. EM2NET packet delivery and MFIT tables. The content between accolades represents one entry in MFIT.

In this figure, we have a source S which sends data to all group members for session (S, G). Nodes in circles with blue color are IN nodes: B2, B3, B6 and B7. First, S sends a data packet to its group members with a header containing the address of B2. When receiving the data packet, B2 duplicates it and sends one copy to B3 and another to B6. Then, B3 broadcasts the packet to R1 and R2. B6 duplicates the data packet and sends one for B7 and another for R6. Finally, B7 broadcasts the data packet to R3, R4 and R5.

4   Simulation Model and Methodology

4.1   Simulation Environment

We use ns-2 (version 2.29) as the simulation platform [16]. Our simulations model a MANET of 75 mobile hosts placed randomly within a 1000m x 1000m area. When the simulation starts, each node randomly picks a destination and moves towards it at a constant speed. After a node reaches its destination, it stops for a predefined period of time, randomly picks another destination and starts moving towards it. Radio propagation range for each node is 250 meters and channel capacity is 2 Mbits/sec. The initial energy of each node is fixed at 10 joules. Each simulation runs for 600 seconds and there is no network partition during the course of simulation. A free space propagation model is used in our experiments and the IEEE 802.11 MAC with Distributed Coordination Function (DCF) is used as the MAC protocol. In our experiments, we employ RTS/CTS (Request To Send/Clear

To Send) channel reservation exclusively for unicast control packets directed to specific neighbors. All other transmissions use CSMA/CA (Carrier Sense Multiple Access/Collision Avoidance). For the simulations that follow, we have considered CBR traffic with payload size set to 512 bytes.

4.2 Parameters

We alter the simulations by varying three parameters. This is done to clearly identify the impact of changing these parameters on different protocols. The standard case is for a multicast group with 15 members under relatively low traffic load and low mobility (a speed of 1 m/s with no pause time and a delivery rate of 2 pkt/s). When studying the effect of one particular parameter, it is set to different values while others are fixed at the standard case. This way we can focus on the performance changes resulting from the change of the parameter of interest. In particular, 15 nodes were chosen to be part of the multicast group when mobility and traffic load were studied in order to evaluate the performance of the protocols when applied to small multicast groups. We vary the mobility with different pause times: 0, 150, 300, 450, and 600 seconds; the group membership size from 5 to 75 nodes and the traffic load from 1 pkt/s to 40 pkt/s. We averaged the different results from multiple runs.

4.3 Protocols

We have compared the performance of EM2NET with E²M, ODMRP and MAODV. Since E²M improves DDM and thus Xcast in ad hoc networks, we restrict our comparison to E²M. ODMRP was chosen since it has been shown to be the best performer in the comparative study reported in [5]. We chose MAODV as the representative tree-based scheme because it builds and maintains a multicast tree like EM2NET and E²M but contrary to the two explicit multicast protocols, it maintains full state information at intermediate nodes. When implementing E²M, we followed the specifications as defined in the publication literature. We directly asked the protocol designers about details which were not specified in the publications (e.g., various timer values). While ODMRP does not require underlying unicast protocol to operate, we used Ad hoc On-demand Distance Vector (AODV) for MAODV, EM2NET and E²M. Ns-2 already provides the AODV routing agent, while we had to modify it to support Xcast. In order to facilitate the interaction between the Xcast agent and the AODV agent, we have adopted a cross-layer design approach, wherein an Xcast agent can access the AODV routing table to group the nodes which are reachable through the same next hop. In case a route is not available, the Xcast agent simply sends a unicast packet to the AODV agent, which triggers a route request for this destination. When the AODV agent receives a route reply from the required destination, it updates its route table accordingly. An Xcast agent can now directly obtain the next hop information from the routing layer. Table I summarizes properties of the protocols we simulated. The values are the standard values mostly used in previous papers ([5], [15]).

Table 1. Parameter Values for ODMRP, MAODV, E²M and EM2NET

	Parameters	**Values**
	JOIN_QUERY refresh interval	3 s
ODMRP	Acknowledgment timeout for JOIN_REPLY	25 ms
	Maximum JOIN_REPLY retransmissions	3
	GROUP_HELLO interval	5 s
MAODV	HELLO interval	1 s
	Route Discovery timeout	3 s
	XF threshold	8
E²M	Delay after receiving XF_REFRESH at XF	8 s
	MEMBER_REFRESH rate	3 s
	BRANCH_UP and BRANCH_DOWN rate	5 s
EM2NET	BRANCH_TIMEOUT value	10 s
	FAILURE_WAITED_PERIOD value	20 s

4.4 Metrics

We have used the following metrics to evaluate the performance of various multicast protocols. Some of these metrics were suggested by the IETF MANET working group for routing/multicasting protocol evaluation [17]. Others are metrics that characterize explicit multicast protocols.

- **Normalized energy,** energy is a scarce resource in wireless mobile ad hoc networks and its conservation and efficient use is a major issue. It is important that nodes preserve their energy in order to increase the network lifetime. We fairly compare the energy consumption by computing the average energy per data packet (the ratio of the total energy expenditure over all the nodes in the network to the total number of unique packets received at any destination).
- **Packet delivery ratio,** is computed as the ratio of the number of data packets actually received by group members to the number of data packets which should have been received. This number represents the effectiveness of a protocol.
- **Average packet header size,** represents the size needed to include the list of all destinations in the packet header. This number has to be small since intermediate nodes will consume more energy due to transmission of longer packets and power processing which consists of addresses discovery in the header and creation of the multicast copies. This metric is only applicable to E²M and EM2NET.
- **Control bytes overhead,** is the ratio between the number of control bytes transmitted to the number of data bytes sent. It also includes the AODV headers for both E²M and EM2NET. Control packets do not carry any user payload. Accordingly, only the data payload bytes contribute to the data bytes delivered. This value investigates how efficiently control packets are utilized in delivering data.

5   Simulation Results

In this section, we are interested with the study of multicast protocols in the context of small groups and with node mobility. For the lack of space, other simulations related to the multicast group size are given in [18].

5.1   Mobility Speed

5.1.1   Scenarios
Each node moves constantly with a predefined speed of 10 m/s. Moving directions of each node are selected randomly. When nodes reach the simulation boundary, they bounce back and continue to move. The pause times is varied from 0 to 600 seconds. In the mobility experiment, 15 nodes are multicast members and the source transmits packets at a rate of 2 pkt/s.

5.1.2   Results and Analysis
Fig. 2 illustrates the packet delivery ratio of the protocols under different mobility scenarios. ODMRP shows good performance even in highly dynamic situations (pause time of 0 second). The mesh topology provides redundant routes and thus, the chances of packet delivery to destinations remain high even when the primary routes are unavailable. We can say that ODMRP is robust to mobility (as effective as flooding). In MAODV, increased mobility causes frequent link changes since it relies on a single path on the tree. To prevent stale routing information, the protocol has to reconfigure the multicast tree more frequently which result in a packet delivery ratio drop. For both E²M and EM2NET, the number of total packets delivered to destinations decreases as the pause time decreases. There are several factors which contribute to this performance when mobility is introduced. Some of which include packet collision at MAC layer since packets are transmitted without RTS/CTS. However, we can

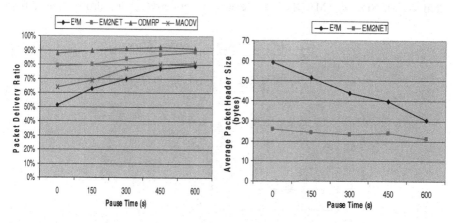

Fig. 2. Packet Delivery Ratio as a function of Pause Time

Fig. 3. Average Packet Header Size as a function of Pause Time

observe that E²M and MAODV have a poor packet delivery ratio compared to EM2NET because they do not take into account node mobility. In addition, the repair procedure employed in EM2NET gives greater importance to a local search first (with a limited TTL) which contributes to a faster tree re-establishment after failure (simulations showed that a node can rejoin the tree after a few milliseconds). By handling node movement, EM2NET improves existing explicit multicast protocols and helps in reducing the packet delivery ratio drop under high mobility.

Fig. 3 shows the impact of node mobility on the average packet header size. As discussed above, when an XF moves in E²M, the source puts the list of all destinations served by this XF in the extended header. This will definitely increase the size needed to include all the destinations. Moreover, it will be difficult for a node to become an XF since there are only 15 nodes which are part of the multicast group. This will not help in reducing the average packet header size. EM2NET is not affected by mobility since nodes use a local search first in order to find an in-tree IN quickly. In fact, the average packet header size remains relatively constant even under high mobility.

We now study the average energy per data packet as a function of pause time. Fig. 4 shows how the packet delivery ratio of the four protocols impacts on the normalized energy consumption. Firstly, we can observe that EM2NET performs better than E²M under all mobility scenarios. In fact, E²M consumes more energy due to its high packet header size as mobility increases (as explained previously). Moreover, it has a poor packet delivery ratio when the pause time is small. On the contrary, EM2NET has a relatively constant average packet header size and does not suffer as much as E²M when mobility is introduced. Another interesting point is the normalized energy consumption of ODMRP. We can see that it still decreases as mobility increases but this reduction is not as severe as under different multicast group sizes. Indeed, E2MNET maintains an acceptable packet delivery ratio even under high mobility because it employs a tree repair procedure. Thus, its average energy per data packet remains relatively constant. ODMRP has also a good packet delivery ratio but nodes consume more energy and hence, its normalized energy remains high compared to EM2NET. For MAODV, the severe packet delivery ratio drop as pause time

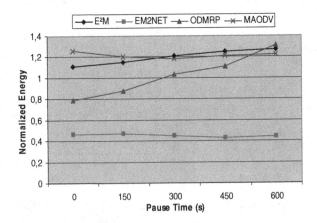

Fig. 4. Normalized Energy as a function of Pause Time

decreases results in an important normalized energy. Indeed, it consumes more energy than EM2NET and E²M while the number of control messages exchanged is greater.

5.2 Network Traffic Load

5.2.1 Scenarios

We now vary the load on the network. The multicast group size is 15 and the node mobility is fixed to 1 m/s with no pause time. Due to the low mobility, we can assume that the packet drops are only caused by buffer overflow, collision, and congestion. The network traffic loads used are between 1 pkt/s and 40 pkt/s.

5.2.2 Results and Analysis

Packet delivery ratios for various traffic loads are shown in Fig. 5. ODMRP shows worse delivery ratios as load grows. Since every data packet is flooded, the number of collisions and buffer overflows grows with the load. However, due to the mesh topology, ODMRP outperforms others multicast protocols. In fact, for both E²M and EM2NET, the packet delivery ratio drops rapidly as the load increases. The degradation is caused by numerous packet collisions. The packet loss rate is less severe in EM2NET however, because it allows load balancing between INs. That means the traffic is not concentrated on particular nodes. In E²M, nodes which are XF will be more subject to buffer overflow because they are responsible for a number of nodes.

Finally, Fig. 6 shows the control bytes overhead for E²M and EM2NET. We have not included ODMRP and MAODV since the control bytes overhead for both protocols remains very high and relatively constant. We can notice that EM2NET has a higher overhead than E²M for low traffic load but it performs better for high traffic load. In fact, EM2NET employs data packets as implicit control packets. A node, upon receipt of a data packet, re-initiates the BRANCH_TIMEOUT timer. Hence, as the load increases, there is less control packets exchanged between INs.

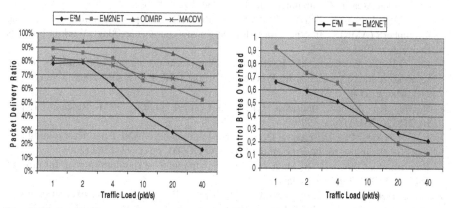

Fig. 5. Packet Delivery Ratio as a function of Traffic Load

Fig. 6. Control Bytes Overhead as a function of Traffic Load

6 Conclusion

In this paper, we were interested within the performance analysis of multicast protocols for large number of small groups. We, mainly, focused on analysing and comparing Xcast protocols, from one hand, to classical multicast based mesh or tree, on the other hand. We have conducted a performance evaluation of four multicast protocols (EM2NET, E²M, MAODV and ODMRP). The detailed simulations have enabled us to perform fair and accurate comparisons of the multicast protocols for a wide range of parameters including mobility and traffic load. Results showed that our proposition, EM2NET, improves the previous methods proposed for explicit multicast protocols in ad hoc networks and also outperforms tree-based schemes. In addition, when compared to a mesh-based protocol (ODMRP), we proved that it leads to better energy conservation, which is an important parameter in ad hoc networks.

A general conclusion is that, in a mobile scenario, mesh-based protocols offer better packet delivery ratios, due to the availability of alternative paths. However, it is done at the cost of a higher control packet overhead. On the contrary, our method is particularly well suited for Small Multicast Groups (SMG). The load balancing between INs makes the protocol more robust to mobility and the average packet header size is reduced. Finally, EM2NET optimizes the nodes batteries and thus, the network lifetime. Currently, we work on developing a new scheme allowing switching between EM2NET and ODMRP.

References

1. Lee, S.J., Su, W., Gerla, M.: Ad hoc wireless multicast with mobility prediction. In: Proceedings of IEEE ICCCN'99, Boston, MA, pp. 4–9 (1999)
2. Royer, E.M., Perkins, C.E.: Multicast operation of the ad hoc on-demand distance vector routing protocol. In: ACM/IEEE MOBICOM'99, Seattle, WA, pp. 207–218 (August 1999)
3. Garcia-Luna-Aceves, J.J., Madruga, E.L.: The Core-Assisted Mesh Protocol. IEEE Journal on Selected Areas in Communications 17(8), 784–792 (1999)
4. Wu, C.W., Tay, Y.C., Toh, C.-K.: Ad hoc Multicast Routing protocol utilizing Increasing id-numberS (AMRIS). Internet Draft, draft-ietf-manet-amris-spec-00.txt, Work in progress (November 1998)
5. Lee, S.J., Su, W., Gerla, M., Bagrodia, R.: A performance comparison study of ad hoc wireless multicast protocols. In: Proceedings of the IEEE INFOCOM 2000, Tel-Aviv. Israel (March 2000)
6. Boivie, R.: A new multicast scheme for small groups, IBM Research Report RC21512(97046) (June 1999)
7. Ji, L., Corson, M.S.: Differential Destination Multicast – A MANET multicast routing protocol for small groups. In: Proc. of the IEEE INFOCOM, pp. 1192–1202 (2001)
8. Boudani, A., Cousin, B.: SEM: A new small group multicast routing protocol. In: Proc. of IEEE ICT2003, Tahiti (February 2003)
9. Gossain, H., Cordeiro, C.M., Anand, K., Agrawal, D.P.: EM: A scalable explicit multicast protocol for MANETs. In: IEEE Int. Conf. on Com. (ICC 2004), Paris, June 20-24, pp. 3628–3632 (2004)

10. Cordeiro, C.M., Gossain, H., Agrawal, D.P.: Multicast over wireless mobile ad hoc networks: present and future directions. IEEE Network, Special Issue on Multicasting: An Enabling Technology 17(1), 52–59 (2003)
11. Ballardie, T., Francis, P., Crowcroft, J.: Core Based Trees (CBT) – An architecture for scalable inter-domain multicast routing. In: Proceedings of ACM SIGCOMM'93, San Francisco, CA, pp. 85–95 (October 1993)
12. Wu, C.W., Tay, Y.C., Toh, C.K.: Ad hoc Multicast Routing protocol utilizing Increasing id-numberS (AMRIS) functional specification. Internet Draft, draft-ietf-manet-amris-spec-00.txt, Work in progress (1998)
13. Perkins, C.: Ad hoc On-demand Distance Vector (AODV) routing. Internet Draft, draft-ietf-manet-aodv-00.txt, Work in progress (November 1997)
14. Johnson, D.B., Maltz, D.A.: Dynamic Source Routing in ad hoc wireless networks. In: Imielinski, T., Korth, H. (eds.) Mobile Computing, ch. 5, pp. 153–181. Kluwer Academic Publishers, Dordrecht (1996)
15. Viswanath, K., Obraczka, K., Tsudik, G.: Exploring mesh and tree-based multicast routing protocols for MANETs. IEEE Transactions on Mobile Computing 05(1), 28–42 (2006)
16. Network Simulator (Version 2), http://www.isi.edu/nsnam/ns
17. Internet Engineering Task Force (IETF) Mobile Ad Hoc Networks (MANET) Working Group Charter, http://www.ietf.org/html.charters/manet-charter.html
18. Benslimane, A., Ferrari, C., Hafid, A.: EM2NET: an Energy-Saving Explicit Multicast Protocol for MANETs. Technical report LIA/University of Avignon (January 2007)

ODCP: An On-Demand Clustering Protocol for Directed Diffusion*

Arash Nasiri Eghbali, Hadi Sanjani, and Mehdi Dehghan

Computer Engineering Department,
Amirkabir University of Technology, Tehran, Iran
{eghbali,sanjani,dehghan}@aut.ac.ir

Abstract. Directed diffusion (DD), uses mechanisms such as data aggregation and in-network processing to suppress the additional data overhead however there is no guarantee that paths from nearby sources join after a few hops. In cases where sensed event is spread geographically, the probability of such combination is reduced. Another problem arises in presence of many source nodes near a single event. In DD for path construction, each source floods a distinct exploratory data (ED) packet through the network, thus a significant amount of network energy is dissipated. The ODCP protocol is proposed to address these two problems: late-aggregation and distinct ED-flooding. In our local on-demand clustering protocol, early aggregation and limited ED-flooding can be achieved by using a virtual sink (VS) near the sources. This node plays the role of sink node and broadcasts local interest messages. Therefore the data packets are sent initially to the VS node and then routed toward destination. Although in simulations we did not consider the improvements gained by early-aggregation, the results show that using this method, connection life-time between source and sink will be increased significantly (up to three times better than directed diffusion).

Keywords: Wireless sensor networks, directed diffusion, energy efficiency, on-demand clustering.

1 Introduction

Wireless sensor networks (WSNs) consist of a number of sensor nodes, scattered in the environment to sense special events such as searching a mobile target or measuring the amount of radioactive radiation in a specified area. These nodes are usually identical with a limited amount of energy. Instead of unique addresses, they are identified by the information gathered by them. They are mostly implemented densely and thus a number of nodes are triggered by a single event. In such situations, each node starts to send the gathered information toward the sink which is usually located far from senders.

Directed Diffusion (DD) [1] is a data-centric routing protocol that uses only local interaction between neighbor nodes. To provide connection between sinks and sources, this work relies on low-rate flooding of events, enabling local re-routing

---

* This work is supported by Iran Telecommunication Research Center ITRC.

E. Kranakis and J. Opatrny (Eds.): ADHOC-NOW 2007, LNCS 4686, pp. 226–236, 2007.
© Springer-Verlag Berlin Heidelberg 2007

whenever the nodes in the primary path, have failed due to energy consumption. In sensor networks, where energy efficiency is of paramount importance, such flooding can adversely impact the lifetime of network.

In DD, data aggregation and in-network processing approaches has been introduced to suppress the additional data overhead. In This protocol, the paths from the adjacent sources often join together after a few hops but there is no guarantee for such combination. In cases where the sensed event is spread geographically, the probability of such combination is reduced. Another problem of DD which arises in presence of many source nodes near a single event is the mechanism used in this protocol for path construction between sources and sinks. By receiving an interest, each source floods an exploratory data (ED). When this packet reaches the destination, the path traversed, is reinforced by the sink and used for later data transmissions. The procedure is repeated for each source separately and thus, a significant amount of network energy is consumed.

Many works has been done recently to improve the energy efficiency of this protocol. In order to decrease the flooding traffic, passive clustering protocol [5] uses a predefined tree with the sink node as root. This tree is constructed at the beginning of routing process. Their protocol aims to omit duplicated packets, transmitted during the flooding period. This can reduce the flooding traffic but this protocol does not solve the problems mentioned above.

In EDDD [10], energy efficiency is obtained through using two kinds of gradients, each one suitable for different types of applications. In order to perform load-balancing between nodes, whenever the delay is of great importance, real-time filter RT is used and the data is forwarded through the shortest path between source and sink. In other situations, best effort BE filters are chosen which lead to longer but more energy efficient paths toward the sink node.

The ODCP protocol is proposed to address two problems: late-aggregation and separate ED- flooding. Early aggregation in our local clustering protocol, can be achieved by using a virtual sink (VS) near the sources which plays the role of sink node and broadcasts local interest messages. Therefore data packets are sent preliminary toward VS node. VS node undertakes the responsibility of sending the data packets toward destination. Hence there is no need for different sources in a single cluster to broadcast separate ED messages. This operation is done only once by the VS node. So the routing protocol overhead is reduced significantly using our ODCP clustering method.

The rest of this paper is presented as following: in section 2, we will introduce on-demand clustering protocol and describe the algorithm in detail. The methodology we used for implementing and testing this protocol is presented in section 3. The simulation results are available in section 4 and a comparison between the original directed diffusion algorithm and our proposed algorithm will be given in this section. Related works are reviewed in section 5. Finally we will conclude the paper and present future works to improve our routing algorithm in section 6.

2 Protocol Description

On-Demand Clustering Protocol (ODCP) is used for collecting information gathered by nearby sensor nodes. Then these data are aggregated and sent toward sink by the

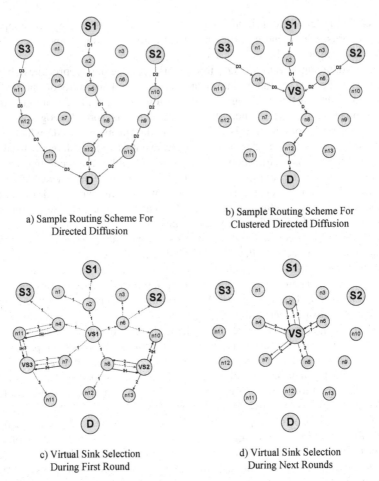

a) Sample Routing Scheme For
Directed Diffusion

b) Sample Routing Scheme For
Clustered Directed Diffusion

c) Virtual Sink Selection
During First Round

d) Virtual Sink Selection
During Next Rounds

Fig. 1. The On-Demand Clustering Protocol. a) Sample routinf scheme for directed diffusion b) Sample routing scheme for clustered directed diffusion c) Virtual sink (VS) selection during first round d) Virtual sink selection during next rounds.

cluster head named as virtual sink (VS). In this section the ODCP clustering protocol will be introduced and all the mechanisms used for cluster head selection and routing within a cluster will be described in detail.

This protocol has four phases. First is the virtual sink selection in which a suitable cluster-head is selected among nodes near the sources. In the second phase, selected VS will explore a path to the sink and in the third phase, after a period of time a new virtual sink is selected among neighbors of prior VS. The fourth phase is considered for situations where the sink node is crashed or lacks enough energy to continue forwarding data packets toward the sink.

Phase I: Virtual Sink Selection
virtual sink selection is the most important and challenging phase in this protocol. As DD is a localized algorithm and each node has only local information, selecting a

single node as virtual sink node between others seems to be a difficult task. The first idea arises in mind is to select one of the sources as VS. Problems will occur when sources are not in radio range of each other. Hence in such case, multiple VSs will be selected and we need to choose one of them as VS. In ODPC one of the nodes in original path between source and sink will be selected as VS. Distance of this node from the source (D_{src}) can be selected, considering density of nodes and spatial properties of sensed event. As we usually prefer to increase the life-time of source nodes, this selection seems to be a better one. In our simulations, D_{src} assumed to be 2 hops. When a node is selected as cluster-head, it broadcasts an *interest* message to all nodes in the cluster. For this purpose, limited flooding method is used. The interest packets are tagged by VS. This tag indicates the time-to-live (*TTL*) of forwarded message and each packet will be ignored after traversing such amount of nodes. The *TTL* value is decremented in each hop until it becomes zero. Therefore the overhead of such flooding is directly related to the size of cluster which is very small in comparison with network size.

In this step, the number of VS selected is the same as number of sources. This selection is shown in the figure 1.c. Firstly VS1, VS2 and VS3 are selected as virtual sink. So one of these VS nodes should be selected as the final VS. A simple approach for such selection is tagging the *VS interest* messages by VS selection time. Cluster nodes (including VS nodes) can select VS with the least time-stamp. So the nearest VS to the sink is selected. Overhead of such flooding increases linearly by the number of sources which is not acceptable especially in high density networks. The *virtual sink inhibit* (IH) message is used to decrease overhead. When a node in cluster that has already received an *interest* message, receives a new *interest* message it simply compare the time-stamp of these messages and send an IH message toward the VS node with the larger time-stamp. By sending IH message, we can both inform VS with the larger time-stamp and also create a path from this VS toward better VS node. In this way, the value of TTL and consequently the overhead of local interest broadcasts can be reduced. However IH messages can be only used when the connections are bi-directional.

In figure 1.c the VS selection mechanism has been shown. Here the VS1 node is the VS with the lowest time-stamp. So the packets from VS2 and VS3 are not flooded in cluster. The paths used by the original DD are depicted in figure 1.a and the paths used by ODCP are shown in figure 1.b.

We used one-phase-pull algorithm to route packets within the cluster. Though the sources can send local exploratory data to the VS and it can reinforce them. However in this case, IH messages cannot be used.

Phase II: Route Discovery

After VS selection phase, we should construct a path from each source toward VS and a path from VS to the sink. Two approaches are used for path construction within the cluster. First is using One-Phase-Pull (OPP) algorithm within the cluster. In this case sources will send data through the path traversed by local interest messages. A more realistic approach is using Two-Phase-Pull (TPP) algorithm in which sources will broadcast a local ED packet in the cluster and VS node reinforces the constructed path after receiving an ED packet from each source.

Then in order to find a path toward the sink, TPP algorithm is used and VS broadcasts an ED packet. When this packet reaches sink, this node will reinforce the path traversed by the ED packet and a path will be constructed from VS toward the sink.

Phase III: Virtual Sink Refreshment
All the traffic generated in the cluster is passed through the VS node. Therefore, the energy of this node will be consumed after a short period of time. So in order to avoid VS failure and perform load-balancing, after the VS expiration period, VS selects one of its neighbors as the next VS. At first the sink will broadcast a *neighbor request* message. Then it waits for a short period of time. During this period, sink neighbors will receive this message and send a *neighbor reply* message which contains energy of each neighbor. So VS can select the neighbor with largest amount of remaining energy. Then it will send a *new sink selection* message to the selected node. This procedure is shown in figure 1.d. In this figure the neighbor request messages are labeled as '1', the neighbor reply messages as '2' and the new sink selection message as '3'.

If a new virtual sink cannot find a path toward the sink after a period of time called PATH_EXPLORATION_PERIOD, it will select another neighbor and the VS refreshment steps will be repeated.

Phase IV: Virtual Sink Expiration
There are two cases that the virtual sink is expired. One is when it does not receive any data packet for a period of time called EXPIRATION_PERIOD. The second case is when the remaining energy of VS and all its neighbors become below a threshold called VS_MINIMUM_ENERGY. In this case the VS node will broadcast a VS_DISABLE message in the cluster. The sources after receiving this message will broadcast an exploratory data toward the sink and send the data directly to the sink node. Additionally each node has an expiration timer. This timer is rescheduled by receiving each VS interest message. So if the VS node fails, sources will themselves take the responsibility of sending data toward the sink.

3   Protocol Description

Performance evaluation experiments for wireless sensor networks is faced with a number of practical and conceptual difficulties. The section summarizes our main choices for the simulation setup.

3.1 Protocol Version

We simulated DD algorithm available with ns-2 simulator, version 2.30. This protocol is implemented for simulator in two versions. We used diffusion3 protocol which is a complete protocol implementation and allows a more realistic evaluation of the protocol.

3.2 Load Model

The traditional DD algorithm, floods an interest message, every 30 seconds and exploratory data floods every 50 seconds. We used these predefined values in our simulations. In this simulation, we used the ping application as the network traffic with the packet rate of 1 packet per second. Also the VS expiration period is assumed to be 60 and 120 seconds.

3.3 Energy Model

In the original directed diffusion, the IEEE 802.11 is used for the MAC layer. For comparability we used the same MAC layer and energy model as in [5] that is the PCM-CIA WLAN card model in ns-2. This card consumes 0.660 W when sending and 0.395 W when receiving.

3.4 Simulation Scenarios

In our approach, for comparison between Directed Diffusion and our On-Demand Proactive Clustering Algorithm, each protocol was tested in a 20*10 grid. For studying the effect of changing number of sources in the efficiency of our protocol we tested two scenarios. In the first case we used 1 sink and 6 sources and in the second case, 1 sink and 6 sources were used. Following parameters were measured during each scenario: mean energy of network nodes, connection lifetime, drop percentage, routing overhead and number of delivered data packets.

For measuring consumed energy, simulations were run for 250 seconds and remaining energy of all nodes was measured in the period of 50 seconds. In order to calculate delay, overhead and drop percentage, these parameters were measured during 100 seconds using variable number of sources. In another scenario connection lifetime was measured using nodes with initial energy of 5 joules. Also effect of changing VS refresh time on connection lifetime was considered using VS refresh periods of 60 and 120 seconds and omitting VS refreshing phase in another scenario.

Ping application was used for all scenarios with the packet rate of one packet per second.

3.5 Energy Calculations

ODCP algorithm aims to decrease the overhead of routing in network nodes by omitting extra ED packet flooding.

For measurement of energy-efficiency of ODCP and comparison between ODCP and DD, we used the scenarios, presented in 0. Initial energy of all network nodes assumed to be 50 joules. In our scenarios, the sources start to send ping data packets towards the sink continually, during simulation period (250 seconds) and average energy of nodes were measured each 50 seconds.

3.6 Connection Time Calculations

For the measurement of energy-efficiency and studying the effect of load-balancing effects on life time of permanent connections between nodes, we used the scenarios, presented in this section. In our simulations, effect of changing VS refresh time using 6 sources and effect of increasing number of sources on connection lifetime was studied. We assumed the initial energy of all nodes in the network to be 5 joules.

In our scenario, the source starts to send ping data packets toward sink continually until connection is broken due to path node failures caused by energy depletion. This period is measured and considered as connection life-time.

3.7 Overhead, Drop Percentage and Delay Computation

In order to compare delay, overhead and drop percentage, between DD and ODCP, these parameters were measured during 100 seconds using variable number of sources. We measured the number of none-data packets, during the connection time and divided it by the number of received data packets to compute overhead. The average delay is also calculated using ping timestamps for each routing algorithm. Drop percentage was measured by dividing number of dropped packets by total number of packets sent.

4 Simulation Results

In this section we will show the simulation results, achieved by implementing the scenarios and assumptions, described in section 3.

4.1 Energy Efficiency

As it can be figured from figure 2.a and 2.b, the energy consumption in ODCP is less than DD. This is due to omitting the ED packets for each source node. Also it is obvious that by increasing the number of sources or increasing network size in comparison to cluster size, the efficiency of ODCP will be increased comparing to DD algorithm. Although by using ODCP algorithm, average energy of all network nodes is increased, average energy of cluster nodes will be decreased due to clustering protocol overhead.

4.2 Connection Life-Time

Connection life-time has been shown in figure 3.a. The results show that the connection life-time will be decreased using the ODCP algorithm using less than 4 sources. This decreasing is due to clustering protocol overhead. However lifetime will be improved when the number of sources grows. In these simulations a fixed VS is used during connection lifetime and VS refresh phase has been omitted.

In figure 3.d, effect of changing VS refresh period on connection lifetime has been shown. In this figure VS(R60) and VS(R120) represents using refresh periods of 60 and 120 seconds respectively and VS(base) stands for ODCP without VS refreshment phase. Result show that by decreasing the VS refresh period, although routing overhead will be increased, connection lifetime will be increased significantly. However as it is shown in figure 3.c number of delivered packet does not increase as connection life time. So in this case after 200 seconds, most of sources are disconnected from the sink but at least one of them is connected and therefore connection is not broken until second 430.

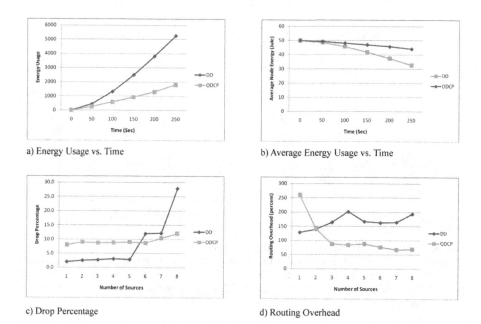

a) Energy Usage vs. Time   b) Average Energy Usage vs. Time

c) Drop Percentage   d) Routing Overhead

Fig. 2. a) Network Energy Usage Comparison between DD and ODCP, b) Average Node Energy Comparison between DD and ODCP, c) Drop Percentage Comparison between DD and ODCP, d) Routing Overhead Comparison between DD and ODCP, simulation parameters: Distance = 100 cm, 6 Sources, Initial Energy = 50 J

4.3 Drop Percentage

Drop percentage will be increased using ODCP approach because during cluster setup, packets will be dropped (figure 2.c) but this value remains constant and is not sensitive to increasing number of sources but in DD by increasing number of sources drop percentage will be increased considerably.

4.4 Average Delay

The delay will be decreased using our on-demand clustering protocol. This decreasing is due to lack of network contention produced by flooded exploratory data. This reduction is shown in figure 3.b. However network response time will be decreased because of time needed for cluster setup and VS selection procedure.

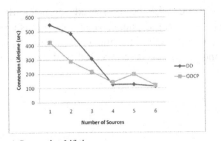

a) Connection Lifetime

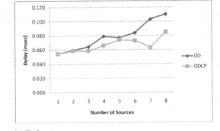

b) Delay

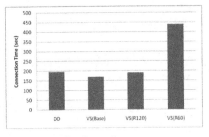

c) Connection Time Comparison

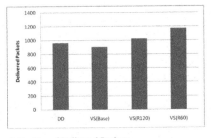

d) Number of Delivered Packets

Fig. 3. a) Connection Life-time Comparison between DD and ODCP, b) Average Delay Comparison between DD and ODCP, c) Connection Lifetime Comparison between DD and ODCP with Different Refresh Periods, 6 Sources, d) Delivered Packet Comparison between DD and ODCP with Different Refresh Periods, 6 Sources. Simulation parameters: Distance = 100 cm, Initial Energy = 5 J.

4.5 Routing Overhead

The reduction of routing overhead caused by ODCP in comparison with DD has been shown in figure 2.d. This reduction is due to omitting extra ED packets.

5 Related Works

Clustering protocols has been proposed for wireless networks in recent years. Some of them will be reviewed in this section.

Heinzelman et al., proposed LEACH [6] protocol which is the base idea used in many other clustering schemes. This protocol consists of rounds in which cluster heads are selected probabilistically in the beginning. The cluster head gathers cluster data using TDMA scheduling and send aggregated data directly to sink. Improvements to [6] are proposed in TEEN [9], PEGASIS [8] and MECH [7]. In TEEN, soft and hard threshold s has been used in each node for sending data in order to decrease the data transmission rate. PEGASIS is a chain-based power efficient protocol that forms the chain toward the base-station greedily using geographical information of each node. The chain leader aggregates the data and sends them toward the sink. Cluster heads are not distributed evenly in LEACH. MECH tries to address this problem by gathering local neighbor information to assign cluster heads. Cluster heads in this protocol routes data among cluster heads to reach the base station instead of sending them directly to the sink.

Also a number of algorithms have been proposed to increase the energy efficiency of Directed Diffusion (DD) [1]. Passive Clustering [5] uses clustering to reduce overhead of flooding packets in DD by constructing a spanning tree. This tree is used for broadcasting interest packets and this way, suppresses overhead of transmission of duplicated packets. In EDDD [10] two kinds of gradients are introduced for different applications. For delay sensitive applications, RT (real-time) filters are used and for the remains, BE (best-effort) filters are used for routing.

To our knowledge, the effect of on-demand clustering on increasing the life-time of connections between sensor nodes especially in data-centric routing algorithms such as directed diffusion has not been explored until now.

6 Conclusion

In this paper on-demand clustering protocol (ODCP) for directed diffusion algorithm is proposed for situations where an event can trigger more than a single sensor node. We used a virtual sink node near the sources. This node gathers information sent by nearby sources and undertakes responsibility of sending them toward the sink. In this way, extra exploratory data packets will be omitted and data from different sources will be aggregated as soon as possible.

Although the improvements gained by early-aggregation have not been considered in our simulations, results show that routing overhead will be decreased significantly (up to three times better than DD). This improvement is gained through omitting the extra exploratory data packets flooded by each source. Also connection life-time between sources and sink will be increased using proper VS refresh periods.

For future work, we suggest using multi-path routing between VS node and the sink to implement load-balancing and further to increase connection life-time.

References

1. Intanagonwiwat, C., Govindan, R., Estrin, D., Heidemann, J., Silva, F.: Directed diffusion for wireless sensor networking. ACM/IEEE Transactions on Networking 11(1), 2–16 (2002)
2. Heidemann, J., Silva, F., Intanagonwiwat, C., Govindan, R., Estrin, D., Ganesan, D.: Building Efficient Wireless Sensor Networks with Low-Level Naming. In: Proceedings of the Symposium on Operating Systems Principles, pp. 146–159 (October 2001)
3. Heidemann, J., Silva, F., Yu, Y., Estrin, D., Haldar, P.: Diffusion filters as a flexible architecture for event notification in wireless sensor networks. Technical Report ISI-TR-556, USC/Information Sciences Institute (April 2002)
4. Heidemann, J., Silva, F., Estrin, D.: Matching Data Dissemination Algorithms to Application Requirements. The First ACM Conference on Embedded Networked Sensor Systems (Sensys'03) , 218–229 (November 2003)

5. Handziski, V., Köpke, A., Karl, H., Frank, C., Drytkiewicz, W.: Improving the Energy Efficiency of Directed Diffusion Using Passive Clustering. Wireless Sensor Networks, First European Workshop, 172–187 (January 2004)
6. Heinzelman, W., Chandrakasan, A., Balakrishnan, H.: Energy Efficient Communication Protocol for Wireless Microsensor Netwroks (LEACH). In: Proc. of 33rd hawaii international conference systems science, vol. 8, pp. 3005–3014 (January 2004)
7. Chang, R.S., Kuo, C.J.: An Energy Efficient Routing Mechanism for Wireless Sensor Networks. In: Proceedings of the 20th International Conference on Advanced Information Networking and Applications (AINA'06), vol. 2, pp. 308–312 (2006)
8. Lindesy, S., Raghavendra, C.: PEGASIS: Power Efficient Gathering in Sensor Information System. In: Proc. of 2002 IEEE aerospace conference, pp. 1–6 (March 2002)
9. Manjeshwar, D.A.: TEEN: a Routing Protocol for Enhanced Efficient in wireless sensor networks. In: Proc. of the 15th international parallel and distributed processing symposium, pp. 2009–2015 (2001)
10. Chen, M., Kwon, T., Choi, Y.: Energy-efficient differentiated directed diffusion (EDDD) in wireless sensor networks. Computer Communications 29(2), 231–245 (2006)
11. Ganesan, D., Cerpa, A., Ye, W., Yu, Y., Zhao, J., Estrin, D.: Networking Issues in Wireless Sensor Networks. Journal of Parallel and Distributed Computing (JPDC), Special issue on Frontiers in Distributed Sens., 799–814 (July 2004)

Quality of Service Support for ODMRP Multicast Routing in Ad Hoc Networks

Amir Darehshoorzadeh[1], Mehdi Dehghan[2], and M. Reza Jahed Motlagh[1]

[1] Computer Engineering Department,
Iran University of Science and Technology Tehran, Iran
[2] Computer Engineering Department, Amirkabir University of Technology, Tehran, Iran
darehshoori@comp.iust.ac.ir, dehghan@ce.aut.ac.ir
jahedmr@iust.ac.ir

Abstract. The primary concerns in ad hoc networks are bandwidth limitation and unpredictable dynamic topology. Therefore, efficient bandwidth utilization is crucial in routing protocols. The On-Demand Multicast Routing Protocol (ODMRP) was designed for multicast routing in ad hoc networks. It is very important to efficiently allocate and consume link bandwidth in this protocol, especially when many groups are working concurrently. In this paper, we propose a new method for estimating bandwidth in multicast protocols and also a new technique for supporting QoS routing in ODMRP by making an acceptable estimation of available and required bandwidth. Simulation results show that using QoS routing for ODMRP improves network performance in presence of mobility, by searching for suitable paths.

Keywords: Ad hoc Networks, Quality of Service, Multicast Routing, ODMRP.

1 Introduction

An ad hoc network is a dynamically reconfigurable wireless network with no fixed wired infrastructure. Each node can function both as a network host for transmitting and receiving data and as a network router for routing packets to the other nodes. Ad hoc networks have numerous practical applications such as military applications, emergency operations, and wireless sensor networks.

Quality of Service (QoS) is the performance level of a service offered by the network to the user. After receiving a service request from the user, the first task is to find a suitable loop-free path from the source to the destination that will have the necessary resources available to meet the QoS requirements of the desired service.

This process is known as QoS routing. After finding a suitable path, a resource reservation protocol is employed to reserve necessary resources along that path [11]. QoS routing protocols search for routes with sufficient resources in order to satisfy the QoS requirement of a flow. Providing QoS support in ad hoc networks is an active research area. Ad hoc networks have certain characteristics that pose several difficulties in providing QoS. Some of the characteristics are dynamically varying network topology, lack of precise state information, lack of central controller, limited resource availability and hidden terminal problem. There are several studies for unicast routing

E. Kranakis and J. Opatrny (Eds.): ADHOC-NOW 2007, LNCS 4686, pp. 237–247, 2007.
© Springer-Verlag Berlin Heidelberg 2007

protocols with QoS in MANET's literature [1][2][3][4][8][12], but QoS support for a multicast protocol should be differently designed from the unicast QoS[9][10].

The On Demand Multicast Routing Protocol (ODMRP) is designed for multicast routing in ad hoc networks [7] The On-Demand Multicast Routing Protocol (ODMRP) was designed for multicast routing in ad-hoc networks. However, ODMRP supports unicast routing too. ODMRP applies on-demand routing techniques to avoid channel overhead. It uses the concept of forwarding group [7], a set of nodes responsible for forwarding multicast data on shortest paths between any member pairs, to build a forwarding mesh for multicast group. In [8] we proposed a new technique that supports QoS routing for ODMRP in unicast mode.

In this paper, we propose a new method for estimating bandwidth in multicast protocols and also a new technique for supporting QoS routing in ODMRP using good estimation of available and required bandwidth. The reminder of this paper is organized as follows. Section 2 describes ODMRP mechanism. Sections 3 describe admission control and our new method for calculating available and consumed bandwidth in multicast protocols. Section 4 describes our new approach for QoS support in ODMRP. Section 5 follows with the simulation results and concluding remarks are made in section 6.

2 ODMRP Mechanism

In ODMRP, group membership and multicast routes are established and updated by the source on-demand. Similar to other on-demand routing protocols, a query phase and a reply phase comprise the protocol (see Fig. 1). While a multicast source has data packet to send, it periodically broadcasts to the entire network a JOIN-REQUEST packet. When a node receives a non-duplicate JOIN-REQUEST, it stores the upstream node ID (i.e., backward route) and rebroadcasts the packet. When the JOIN-REQUEST packet reaches a multicast receiver, the receiver creates or updates the source entry in its Member-Table. While valid entries exist in the Member-Table, JOIN-TABLES are broadcasted periodically to the neighbors.

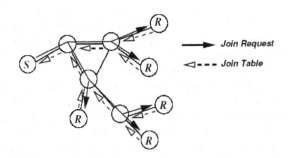

Fig. 1. On-Demand procedure for membership setup maintenance

When a node receives a JOIN-TABLE, it checks if the next node ID of one of the entries matches its own ID. If it does, the node realizes that it is on the path to the source and thus is part of forwarding group, it then sets the FG-Flag and broadcasts its own JOIN-TABLE built upon matched entries. The JOIN-TABLE is propagated by each forwarding group member until it reaches the multicast source via the shortest path. This process constructs the routes from sources to receivers and builds a mesh of nodes, the forwarding group. As shown in Figure 2 the forwarding group is a set of nodes in charge of forwarding multicast packets. It supports shortest paths between any member pairs. A multicast receiver can also be a forwarding group node if it is on the path between a multicast source and another receiver.

After this group establishment and route construction process, a multicast source can transmit packets to receivers via selected routes and forwarding groups. Periodic control packets are sent only when outgoing data packets are still present. When receiving a data packet, a node forwards a packet only if it is not a duplicate and the setting of FG-Flag for the multicast group has not expired.

In ODMRP, nodes do not need to send any explicit control packets to leave the group. If a multicast source wants to leave the group, it simply stops sending JOIN-REQUEST, also if a receiver no longer wants to receive from a particular multicast group; it removes the corresponding entries from its Member-Table and does not send the JOIN-TABLE for that group. Nodes in the forwarding group are demoted to non-forwarding nodes if not refreshed before they timeout.

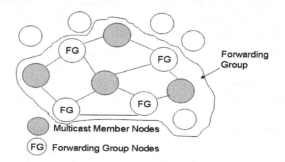

Fig. 2. On-Demand procedure for membership setup maintenance

3 Admission Control

To provide QoS requirements on a path, the admission control policy should guarantee the requested minimum bandwidth B_{min} for each flow. Bandwidth reservation by admission control is made at every node in the route setup phases, based on bandwidth calculations as described in the next section. To determine whether there is enough bandwidth for a new flow j, all we need to know is the available link capacity and the bandwidth to be consumed by the requesting flow. Because of the shared medium, a node can successfully use the channel only when all its neighbors do not transmit and receive packets at the same time. As illustrated in [6] the available channel bandwidth at node i can be given by:

$$B_{available}(i) = B - \sum_{j \in N(I)} B_{self}(j) \tag{1}$$

Where B is the raw data rate of the node i and B_{self} (j) is the total traffic between node j and its neighbors, i.e., the bandwidth consumed by the traffic transmitted or received by node j, and N(i) is the set of node i's neighbors. Given the requested bandwidth B_{min}, the bandwidth to be reserved for the flow j at node i is:

$$B_i(J) = \begin{cases} B_{min} & \text{If source or destination} \\ 2B_{mi} & \text{else} \end{cases} \qquad (2)$$

Since the intermediated nodes need to receive and then forward flow j traffic. The consumed bandwidth for flow j on node i's channel can be given by [6]:

$$B_{consumed}(i, j) = B_{uplink(i)}(j) + B_{downlink(i)}(j) \qquad (3)$$

Where $B_{uplink(i)}(j)$ and $B_{downlink(i)}(j)$ are the reserved bandwidth for flow j on upstream and downstream neighbor of node i, respectively. Note that $B_{uplink(i)}(j)$ and $B_{downlink(i)}(J)$ can either equal to B_{min} or $2B_{min}$ as shown in (2). In unicast protocols each node has only an upstream and a downstream node, therefore equation (3) can be used as an estimation of the consumed bandwidth. However, in multicast protocols each node may have more than one downstream node as a forwarding group. Therefore, equation (3) can not be used as an estimation of the consumed bandwidth. We propose a new technique for estimating consumed bandwidth in multicast protocols as shown below in pseudo code format.

```
if node i has already downstream node
{
    if the new downstream node is forwarder
        B_consumed(i,j)=B_consumed(i,j)+B_min;
    else
        Use the B_consumed(i,j) calculated by equation.3;
}
else
    B_consumed(i,j)=B_uplink(i)(j)+B_downlink(i)(j);
```

By comparing the value of $B_{available}(i)$ and $B_{consumed}(i, j)$, each node can now decide whether to accept the flow or not [6].

4   QoS-ODMRP Mechanism

- Neighborhood Maintenance: Neighborhood information is very important in QoS-ODMRP, since it provides the local topology and traffic information. To maintain the neighborhood information, every node in the network is required to periodically send out a HELLO packet, announcing its existence and traffic information to its neighbors. Each node i will include in the HELLO packet its self traffic (B_{self} (i)). The HELLO packets are sent at a default rate of one packet per two seconds, with time to live (TTL) set to 1. When a node receives a HELLO packet from its neighbor, it stores or updates the information of that neighbor in its neighbor list N(i). Failure to receive any packet from a neighbor for a certain period of time is taken as indication of broken link to the neighbor.

- Route Setup: Similar to ODMRP, when a source has data to send and no route information to the destinations is known; source appends the requested bandwidth for this session (B_{min}) to the JOIN-REQUEST packet and floods it. This process is shown below in pseudo code format:

```
if (!IsQuerySentFlag)
{
   IsQuerySentFlag=TRUE;
   Set timer to unset IsQuerySentFlag after
                                            T_UnSetIsQuerySendFlag;
   Source Calculates its AvailableBandiwdth and
                       ConsumedBandwidth for the new flow;
   if (AvailableBandwidth > ConsumedBandwidth AND
                            AvailableBandwidth >= B_min)
     Add B_min to the JOIN-QUERY Packet and send it;
   else
     There is not enough Bandwidth for this flow;
}
```

The IsQuerySentFlag is used to prevent sending JOIN-REQUEST for a certain time when there is not enough bandwidth. When the group is established or after $T_{UnSetIsQuerySendFlag}$, this flag will be reset for sending new JOIN-REQUEST. When a node receives a JOIN-REQUEST, it processes this packet as follow:

```
if This Packet is not duplicated
{
   Calculate the AvailableBandwidth and
                       consumedBandwidth for this request
   if (AvailableBandwidth > ConsumedBandwidth)
   {
       Change the reserved bandwidth for this flow to
                                            EXPLORED;
       Store upstream node and the bandwidth that it
                                   reserved for this flow;
       Hop_count++;
       if (hop_count < TTL)
         Rebroadcast JOIN-REQUEST packet;
       else
         Drop JOIN-REQUEST packet;
       if (this node is a receiver)
         Create a JOIN-REPLY packet and send it;
   }
   else
      There is not enough bandwidth for this flow, Thus
                                   Drop the packet;
}
else
   Drop JOIN-REQUEST Packet;
```

As described above, if the received packet is not duplicated, the node computes and compares the channel's $B_{Available}$ and $B_{Consume}$ of that flow. If $B_{Available}$ is greater than $B_{Consume}$ and also greater than the minimum bandwidth to be reserved (B_{min} or $2B_{min}$) this node will add a new entry in its reservation table and sets its status for that session to '*explored*' as shown in Figure 3. It also stores the last hop node information in its route table (i.e., backward route) and rebroadcasts the packet. The node will only remain in '*explored*' state for $T_{Explored}$. If no reply is received by the explored node during this time, the entry in reservation table will be discarded. Since the node which propagates JOIN-REQUEST is not aware of downstream node in the route to the destination, it uses Bw_{Uplink} as the estimation of $B_{Consume}$ for propagating JOIN-REQUEST packet. The destination replies back to the source via the selected route with JOIN-REQUEST. When a destination receives a JOIN-REQUEST packet, it builds the JOIN-REPLY packet and if it has enough bandwidth for this session, it broadcasts the JOIN-REPLY packet to the upstream node and updates its status to '*registered*' in the reservation table entry of this session (Fig 4). After registration, the nodes are ready to accept the real data packets. The node will only stay in *registered* state for $T_{Registered}$, if no data packet received by the registered node in that time; the entry in reservation table is discarded. When a node receives a JOIN-REPLY, it checks if the next node address of its entry matches its own address. If it is matched, the node realizes that it is on the path to the source. It checks its available bandwidth and compares it with $B_{Consume}$ of this flow. If $B_{Available}$ is greater than $B_{Consume}$, it then sets its forwarding flag for this session and broadcasts its own

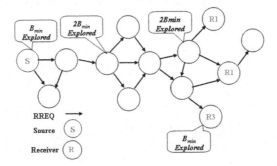

Fig. 3. On-Demand procedure for membership setup maintenance

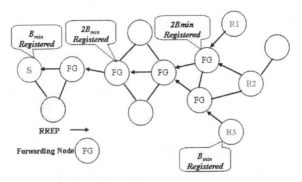

Fig. 4. On-Demand procedure for membership setup maintenance

JOIN-REPLY built upon the matched entry. The previous node address is obtained from the entry in route table which was recorded when JOIN-REQUEST was received. The JOIN-REPLY is propagated until it reaches the source. Intermediate nodes that relay JOIN-REPLY mark themselves as forwarding nodes and set their status to *'registered'* for this session. As explained before when a node handles JOIN-REQUEST, it is only aware of its upstream node, thus it used Bw_{Uplink} as an estimation of $B_{Consume}$, but after receiving JOIN-REPLY it will be aware of its downstream nodes and can calculate the precise $B_{Consume}$ for this session. This bandwidth availability rechecking is essential to reduce transient routes construction.

- Data Forwarding: After constructing the routes, if the source receives a certain number of JOIN-REPLY from destinations it can send packets to them via selected routes and forwarding nodes. After receiving a data packet, a node forwards it, only when it is not a duplicate packet and the forwarding flag for this session has not expired. After all, the node changes its *'registered'* status to *'reserved'* (Fig 5). The node will only stay in *'reserved'* state until receiving the next JOIN-REQUEST. This procedure minimizes the traffic overhead and prevents sending packets through the stale routes.
- Route Maintenance: QoS-ODMRP uses soft state for route maintenance. Each source periodically sends out JOIN-REQUEST packet, as well as route setup procedure, to refresh or repair break routes. JOIN-REPLY packets are backed through reversed routes and forwarding nodes are updated.
- Soft State: In QoS-ODMRP, no explicit control packets are required to be sent for joining or leaving the session. If a source wants to leave the session, it simply stops sending JOIN-REQUEST packet, since it does not have any data to send. If a destination no longer wants to receive from its source, it will not send JOIN-REPLY packet for that source. Forwarding nodes are demoted to non-forwarding if not refreshed (no JOIN-REPLY received). When a node leaves a session or is demoted to non-forwarding, it releases the requested bandwidth for that session and sends a HELLO packet to inform its neighbor about its new status.

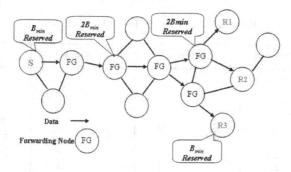

Fig. 5. On-Demand procedure for membership setup maintenance

5 Performance Evaluation

The simulation code was implemented within the Global Mobile Simulation (GloMoSim) library [4]. In the simulation, we modeled a network of 50 mobile hosts placed randomly within a 1000×1000m^2 area. Radio propagation range for each node was 250 meters and channel capacity was 2Mbit/sec. Each simulation runs for 300 seconds of simulation time. The IEEE 802.11 Distributed Coordination Function [5] was used as the medium access control protocol. We used Constant Bit Rate as our traffic. The size of data payload was 512 bytes. We used Random-Way point as mobility model.

To evaluate the performance of QoS-ODMRP, we simulated and compared the following schemes:

- ODMRP
- QOS-ODMRP

We evaluated both schemes as a function of mobility speed of nodes and the traffic of sessions. The number of groups was set to 3 and mobility speed of nodes was varied from 0, 9 and 18 Km/hr with pause-time equal to 3 seconds. Sessions were established with interval of 5 seconds and remained until end of simulation time. To evaluate the performance of our protocol, each source sent data as a rate of 128Kbit/s, 256Kbit/s and 512Kbit/s. we evaluated the following metrics for each session:

- Packet Delivery Ratio: The number of data packets received by destinations over the number of data packets sent by the source.
- Traffic admission ratio: ratio of the number of data packets sent to the network from the source and the number of data packets generated at the source.

6 Simulation Results

The packet delivery ratio for each group is shown in tables 1 and 2. As shown in these tables, we simulated three groups sending data packets with the rates of 128Kbit/s,

Table 1. Packet Delivery Ratio (First Scenario)

Groups	Data Rate (kb/s)	0 Km/h		9 Km/h		18 Km/h	
		ODMRP (%)	QoS-ODMRP (%)	ODMRP (%)	QoS-ODMRP (%)	ODMRP (%)	QoS-ODMRP (%)
Grp1	128	41.5	74.3	54.7	84.5	53	82.1
Grp2	512	7.23	0	10.2	0	9.92	0
Grp3	128	53.4	76.9	56.9	87.8	47.4	82.6

Table 2. Packet Delivery Ratio (Second Scenario)

Groups	Speed→ Data Rate (kb/s)	0 Km/h ODMRP (%)	0 Km/h QoS-ODMRP (%)	9 Km/h ODMRP (%)	9 Km/h QoS-ODMRP (%)	18 Km/h ODMRP (%)	18 Km/h QoS-ODMRP (%)
Grp1	128	50.8	64.8	61.6	76.3	62	79.2
Grp2	256	33	59.8	40.6	76.8	34.4	73.7
Grp3	128	57.7	75.8	62.8	79.9	55.7	71.5

Table 3. Traffic Admission Ratio (First Scenario)

Groups	Speed→ Data Rate (kb/s)	0 Km/h ODMRP (%)	0 Km/h QoS-ODMRP (%)	9 Km/h ODMRP (%)	9 Km/h QoS-ODMRP (%)	18 Km/h ODMRP (%)	18 Km/h QoS-ODMRP (%)
Grp1	128	100	80.5	100	90.8	100	85.5
Grp2	512	100	0	100	0	100	0
Grp3	128	100	83.5	100	83.6	100	82.9

512Kbit/s and 128Kbit/s. Because of low available bandwidth, the second group was not established. However, the first and the third groups in QoS-ODMRP had better performance than ODMRP. We can observe in these tables that as mobility speed of nodes increases, in both protocols, the packet delivery ratios decrease because of link broken. In QoS-ODMRP sources only send data when routes that are established have enough bandwidth, thus packet delivery ratio in QoS-ODMRP is higher than ODMRP. As shown in table 2, we simulated three groups sending data packets with the rates of 128Kbit/s, 256Kbit/s and 128Kbit/s. Although in this scenario all of the groups are accepted by QoS-ODMRP, packet delivery ratio for each group is about 70%, but in ODMRP it is about 40% and 60%.

The tables 3 and 4 show the traffic admission ratio for QoS-ODMRP. In our protocol, data packets will only be sent when there is enough bandwidth and groups are established, thus in some cases because of low bandwidth, routes are not established and sources will be prevented of sending data, thus admission ratios of our protocol are less than ODMRP. In scenarios that used the data rates of 128kb/s and 512kb/s traffic admission ratio is about 70%-80% and with the data rates of 128kb/s and 256kb/s, it is about 70%-80%.

Table 4. Traffic Admission Ratio (Second Scenario)

Groups	Data Rate (kb/s)	0 Km/h		9 Km/h		18 Km/h	
		ODMRP (%)	QoS-ODMRP (%)	ODMRP (%)	QoS-ODMRP (%)	ODMRP (%)	QoS-ODMRP (%)
Grp1	128	100	80.5	100	78.1	100	79.7
Grp2	256	100	73.4	100	80.8	100	62.6
Grp3	128	100	81.1	100	80.6	100	70.7

7 Conclusion

We introduced a new technique for estimating consumed bandwidth in multicast protocols, and also a new QoS routing algorithm with bandwidth reservation for multicast applications in ad hoc networks. This technique was applied to ODMRP and evaluated via several simulation scenarios. Simulation results show that our protocol will not accept new sessions when the required network resources are not available. QoS-ODMRP improves network performance in comparison of ODMRP by reducing overhead and efficient utilizing of bandwidth.

References

1. Chen, Y.-S., et al.: On-demand, link-state, multi-path QoS routing in a wireless mobile ad-hoc network. European Wireless (2002)
2. Lin, C.-R.: On-demand QoS routing in multi-hop mobile networks. IEEE INFOCOM, 1735–1744 (2001)
3. Chen, Y., Yu, Y.: Sprial-Multi-Path QoS Routing Protocol in a Wireless Mobile Ad-Hoc Network. IEICE Transaction on Communications 87-B (1), 104–116 (2004)
4. UCLA Parallel Computing Laboratory, GloMoSim: A scalable simulation environment for wireless and wired network system.

5. IEEE Computer Society LAN MAN Standards Committee, Wireless LAN Medium Access Protocol (MAC) and Physical Layer (PHY) Specification, IEEE Std 802.11-1997, IEEE, New York, NY (1997)

6. Xue, Q., Ganz, A.: ad hoc QOS on-demand routing (AQOR) in mobile ad hoc networks. Journal of parallel and distributed computing, 154–165 (2003)

7. Lee, S., Gerla, M., Chiang, C.: On-Demand Multicast Routing Protocol. In: IEEE WCNC (1999)

8. Darehshoorzadeh, A., Dehghan, M., Jahed Motlagh, M.R.: Quality of Service Support for ODMRP Unicast Routing in Ad hoc Networks. In: International Conference on Digital Telecommunications (ICDT 2006), Côte d'Azur, France (August 29-31, 2006)

9. Bür, K., Ersoy, C.: Multicast Routing for Ad Hoc Networks with a Multiclass Scheme for Quality of Service. In: Aykanat, C., Dayar, T., Körpeoğlu, İ. (eds.) ISCIS 2004. LNCS, vol. 3280, pp. 27–29. Springer, Heidelberg (2004)

10. Bur, K., Ersoy, C.: Multicast Routing for Ad Hoc Networks with a Multiclass Scheme or Quality of Service. In: Aykanat, C., Dayar, T., Körpeoğlu, İ. (eds.) ISCIS 2004. LNCS, vol. 3280, Springer, Heidelberg (2004)

11. Murphy, C.S.R., Manoj, B.S.: Ad Hoc Wireless Networks Architecture and Protocols, 1st edn., pp. 505–580. Prentice hall, Englewood Cliffs (2004)

12. Iiyas, M.: The Handbook of Ad Hoc Wireless Networks, pp. 1–3. CRC Press, Boca Raton, USA (2003)

The Analysis of Fault Tolerance in Triangular Topology Sensor Networks*

Diwen Wu and Dongqing Xie

School of Computer and Communication, Hunan University,
Changsha 410082, Hunan, China
diwenwu@sohu.com, dongqing_xie@hotmail.com

Abstract. In some specific applications, sensors can be deployed in a deterministic way to form regular network. This paper firstly analyzes the relationship between the probability of coverage and that of the node failure in the network with triangular topology. Then the fault tolerance performance in triangular topology network is analyzed by using the k-subnet. At last, the paper discusses the connectivity performance in triangular topology network and in grid topology network.

1 Introduction

Sensor nodes are often scattered by aircraft or other ways because they are mostly deployed in inaccessible or inhospitable areas in many applications [1, 2, 3, 4]. While, in some specific applications, networks area is secure so that sensors can be deployed in a deterministic way to form regular and efficient structures. Due to fragility of sensor nodes and bad environment, errors of nodes must be taken into account when we deploy such wireless sensor network [5].

Coverage and connectivity are two important issues in sensor networks. In some cases, sensor nodes will be out of order. So the probability that sensor node is out of order should be taken into account when computing the coverage and connectivity in sensor networks. This paper focuses on the relation between the probability of error node and the coverage as well as the connectivity of network in deterministic deployment.

Up to now, most researches on wireless sensor networks with regular topology is based on Grid Network [6]. However, the analysis in this paper mainly relies on Triangular topology and the performance comparison is also made with that of Grid network. Firstly, this paper defines a Triangular network model. Then the coverage of regular 2D wireless sensor networks in deterministic deployment is discussed. Moreover, by using the k-order Subnet the performance of fault-tolerance in regular wireless sensor network is discussed.

---

* This work is supported by the National Natural Science Foundation of China (60673156).

E. Kranakis and J. Opatrny (Eds.): ADHOC-NOW 2007, LNCS 4686, pp. 248–261, 2007.

2   Related Work

Coverage and connectivity are two important issues in study of wireless sensor networks. Coverage ensures that each position in target area is at least covered by one sensor node, which is a measure of QoS (Quality of Service) of sensing function. Each node needs to play a role of forwarding data to a remote sink for other nodes, due to limited energy and communication capability of individual node. Thus, network connectivity is of paramount importance to ensure successful data forwarding. It has been proved that a sufficient condition to ensure connectivity of active sensor nodes in a full coverage of a convex area is: Communication radius is at least twice of sensing radius [9, 10].

Many meaningful works have been done on coverage and connectivity of wireless networks, especially wireless sensor networks. Up to now, previous work mainly focus on the following two points. One is design of efficient algorithms and protocols concerning full coverage (connectivity), partial coverage (connectivity) or conditional coverage (connectivity) under given coverage and connectivity; the other point is theoretical analysis on interplay among coverage (connectivity), error probability, network size and node distribution etc.

Recent contributions on energy efficient coverage problems in the context of static wireless sensor networks are surveyed, including various coverage formulations, models, assumptions as well as an overview of the solutions proposed [11]. Problem of finding maximal number of covers in a sensor network is addressed, where a cover is defined as a set of nodes that can completely cover the target area [12]. It is proved a NP-completeness problem and a centralized solution was given. S. M. et al. How to use ILP (Integer Linear Programming) technique is also presented to solve several optimal 0/1 node scheduling problem in wireless sensor networks [13], viz how to schedule the minimum active nodes to completely cover a sensor field. In the context of providing network coverage, the relation between reduction in sensor duty cycle and required level of redundancy is examined for a fixed performance measure. And two mechanisms of Random Sleep and Coordinated Sleep were presented.

Apart from centralized processing above, alternative, suboptimal and distributed algorithm was provided to solve the same or similar problems [15, 16]. In [15], a subset of nodes is selected out initially and operates in active nodes until they run out of their energy or are destroyed. Other nodes fall asleep and wake up occasionally to probe their local neighbors. If there is no working node with its probing range, a sleeping node starts to work. Probing range can be adjusted to control sensing redundancy. However, original sensing coverage and connectivity may be reduced after node scheduling. Algorithm in [16] is mainly on node arithmetically calculating the union of all sectors covered by its neighbors and determining its working status according to calculation results.

The connected coverage problem with a given coverage ratio was discussed. The author firstly introduced partial coverage concept and analyzed the property for the first time to prolong network lifetime [17]. Then a heuristic algorithm was presented which takeed into account the partial coverage and sensor connectivity simultaneously. Further research on such topic was made in [18].

Sensing capabilities of network nodes are affected by environment factors in real deployment, which can not guarantee connectivity in all directions. Thus, studies in [19,20] explored probabilistic coverage for sensor networks from different aspects. Assuming probabilistic coverage for sensors, an error model targeting location estimation application was proposed [19]. A signal strength based approach is used to form a probabilistic function depending on distance between sensor and object. A probabilistic coverage algorithm is designed to evaluate area coverage in randomly deployed wireless sensor networks [20]. The algorithm takes into account the variation in sensing behavior of deployed nodes and adopts a probabilistic approach in contrast to widely used idealized unit disk model.

An important issue in study and deployment of wireless sensor networks is that to ensure given coverage and connectivity, how many nodes are needed, how large communication radius should be set and what range the probability of node error should be controlled in. necessary and sufficient conditions are derived for 1-coverage and 1-connectivity when n nodes are deployed in a $\sqrt{n} \times \sqrt{n}$ grid and each node is allowed to fail independently with probability p [21]. The sufficient and necessary condition for random grid network to cover the unit square area is firstly obtained in [22], as well as to ensure the connectivity of active nodes. Some work is extended to $\sqrt{n} \times \sqrt{n}$ grid, random uniform and Poisson sensor networks [23]. And sufficient conditions is got for k-coverage in three topologies above [24]. All methods above are based on asymptotic analysis, which requires number of nodes goes to infinity to achieve corresponding properties. When the number of nodes is finitethe coarse capacity of Wireless Mesh Networks is derived [25]. Geometry probability was firstly introduced into analysis of connectivity probability, which can be effectively derived in both infinite and finite cases [26].

In this paper, we mainly discuss the performance of fault-tolerance in triangular sensor networks.

3   Network Model

Network model in this paper mainly is applied to such scenario: in each deployed position, sensors are in charge of both collecting sensing data and forwarding packet so as to transmit collected data to sink.

Grid and triangular topology is widely used in wireless sensor networks with deterministic deployment. Assume that the sensing radius of nodes is the distance from one grid point to its adjacent point and the communication radius is large enough to reach the sink node, the performance of grid topology network is analyzed [7]. Besides excessive power consumption on long transmission, heavy load in sink will lead to low communication efficiency. Here, in case of full coverage, each sensor node only communicates with adjacent nodes and data transmission between sink node and nodes far away can take advantage of immediate nodes. The network model can be described as follows:points to be monitored are arranged in regular triangular topology (as shown in Fig. 1). Sensors can be accurately placed in corresponding points. Because sensors may

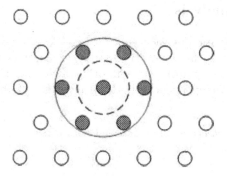

Fig. 1. Grid Network

be placed close to monitored point, sensing radius can be small enough only to cover object position. The communication radius should be equal to or a little larger than the distance between adjacent nodes.As shown in 1, each little circle represents position to be monitored, Shadowed circle means that sensors are placed in such position.the dotted circle is the sensing range and the solid circle is the communication range.

Because the communication range of each node is the distance between two adjacent nodes, if sensor is placed in each object position, the resulting topology comes out as Fig. 2. We call such network as Triangular Network.

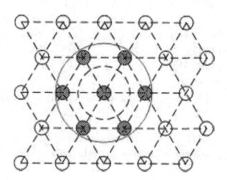

Fig. 2. Triangular Network

For convenience, we will use $Tri_{m\times n}$ denoting Triangular Networks with n rows and m nodes in each row, EP as error probability, and CP as connectivity probability. Moreover, assume that EP of node is p, the number of object positions is T and the number of sensor nodes is N. In real deployment, more than one sensor can be placed in each position to enhance the performance of fault-tolerance. In following analysis, we assume only one sensor is placed in each position. But the conclusion can be extended to the case where more than one sensors are in each position.

4   Coverage Analysis

In this section, the relation between full coverage of network and the probability of sensor node error is discussed. Because each sensor is placed close to the object position and is responsible of monitoring one point in Triangular network, coverage problem is independent of both arrangement of object points and the topology.

Lemma 1. *If each object point is detected by one sensor, the coverage rate of the network is* $(1 - p)^N$.

Proof omits.   ∎

Lemma 2. *If each object point is detected by* $n1, n2, \ldots, nT$ *sensor nodes respectively, the coverage rate of the network is* $(1 - p^{n1})(1 - p^{n2}) \ldots (1 - p^{nT})$.

Proof omits.   ∎

Theorem 1. *If each object point is detected by* $n1, n2, \ldots, nT$ $(n1 + n2 + nT = N)$ *sensor nodes, then the maximum coverage rate is* $(1 - p^{N/T})^T$, *when* $n1 = n2 = \cdots = nT = N/T$.

Proof: Assume that there are nI sensor nodes in i^{th} object point. The probability of ith point being covered is $1 - p^{nI}$. So the probability that all points in the network are covered is $P_{cov} = (1 - p^{n1})(1 - p^{n2})(1 - p^{n3}) \ldots (1 - p^{nT})$.
$\log(P_{cov}) = \log(1 - p^{n1}) + \log(1 - p^{n2}) + \log(1 - p^{n3}) + \cdots + \log(1 - p^{nT})$.
Let $f(x) = \log(1 - p^x)$, we can get $\log(P_{cov}) = f(n1) + f(n2) + \cdots + f(nT)$. Because $f''(x) = -p^x \log^2(p)/(1 - p^x)^2 < 0$, $f(x)$ is convex function. From the definition of convex function, $f(n1) + f(n2) + \cdots + f(nT) <= r \cdot f((n1 + n2 + \ldots + nr)/T) = r \cdot f(N/T)$. When $n1 = n2 = \cdots = nT = N/T$, $f(n1) + f(n2) + \cdots + f(nT)$ reaches its peak.
So we can say that the coverage rate is the maximum when $n1 = n2 = \cdots = nT = N/T = (1 - p^{N/T})T$.   ∎

Theorem 1 informs us that we should place the same number of nodes in each position. We can facilitate the conclusion to the case where more than one nodes are in one position.

5   Connectivity Analysis

According to the assumptions, when all object points are covered by sensor nodes in the network, the whole network must be connected. So in the following, when the network connectivity is analyzed, the network coverage problem is neglected. The following two important issues should be discussed in deployment of wireless sensor networks:

1. To ensure certain desirable probability of network connectivity, what range the probability of node error should be controlled in?

2. Given network size and the probability of node error, how much is the network connectivity?

The authors proved that when the size of grid increases, its CP may approach 0 asymptotically [8]. Such conclusion also may apply to regular sensor networks. We can start to analyze network connectivity when the probability of node error is given. Based on the definition of K-*Mesh Subnet* and the analysis of fault tolerance on *Mesh* network in [8], we are able to derive a lower bound on the connectivity probability of $Tri_{m \times n}$ in Triangular network.

5.1 Definition of K-Order Subnet

First of all the following difinition is given:

Definition 1 (k-Order Subnet). *Nodes in Triangular Sensor Network $Tri_{m \times n}$ are denoted as coordinate $(x, y), x = 1, 2, \ldots, m, y = 1, 2, \ldots, n$. For given positive integer b, d ($b < m/k, d < n/k$), the network which is composed of all nodes whose x coordinate locate in $bk + 1, bk + 2, \ldots, (b+1)k$ and y coordinate in $dk + 1, dk + 2, \ldots, (d+1)k$ is called k-Order Subnet in $Tri_{m \times n}$, which is dependent of integer b and d (For convenience, b and d can be divided by k). The following results can be easily applied to the general case.*

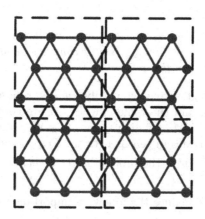

Fig. 3. k-Order Subnet

It can be seen from the Definition 1 that an regular sensor network $Tri_{m \times n}$ with size of $m \times n$ can be divided into $(m/k) \times (n/k)$ disjointed k-Order Subnets. The whole network is divided into m/k rows and n/k columns and there are m/k and n/k k-Order Subnets in each row and column respectively. We index each subnet as $Tri_{1,1}, Tri_{1,2}, \ldots, Tri_{(m/k),(n/k)}$. If there exists at least one node in one k-Order Subnet which can communicate with at least one node in another k-Order Subnet, then the two k-Order Subnet is adjacent. As shown in Fig. 3, Triangular Network $Tri_{6 \times 6}$ is divided into four 3-Order Subnets and the four 3-Order Subnets are adjacent each other.

Definition 2 (k-Order Subnet connected). *In Triangular sensor network* $Tri_{m \times n}$, *if the number of correct nodes in each of four sides of a k-Order Subnet is greater than that of error nodes, and all correct nodes in each k-Order Subnet form a connected graph, then the Triangular sensor network* $Tri_{m \times n}$ *is k-Order Subnet connected.*

We can draw a very important but obvious conclusion from the definition of k-Order Subnet connected: all correct nodes in a k-Order Subnet connected Triangular sensor network $Tri_{m \times n}$ form a connected graph.

5.2 Lower Bound on Connectivity Probability of K-Order Subnet Connected Sensor Network

For convenience, we denote the event that sensor network $Tri_{m \times n}$ is k-Order Subnet connected as $C(Tri_{m \times n})$. According to Definition 2, we can get Lemma 3.

Lemma 3. *Assuming that each node has independent error probability p, then the probability that sensor network* $Tri_{k \times k}$ *is k-Order Subnet connected is*

$$\Pr[C(Tri_{k \times k})] = \sum_{i=0}^{k^2} N_{k,i}(1-p)^{k^2-i}p^i,$$

where $N_{k,i}$ $(0 \le i \le k_2)$ *represents the number of method that when deleting i nodes from network* $Tri_{k \times k}$, $Tri_{k \times k}$ *is still connected and the remaining nodes in each of four sides exceed the half.*

Proof: Assuming that extracting arbitrarily i nodes from network $Tri_{k \times k}$, all of which are error nodes, and the remaining $k^2 - i$ are correct nodes, then such event probability is $(1-p)^{k^2-i}p^i$. Thus, $N_{k,i}(1-p)^{k^2-i}p^i$ is just the probability that there are i error nodes in sensor network $Tri_{k \times k}$ which is still connected. So, $\sum_{i=0}^{k^2} N_{k,i}(1-p)^{k^2-i}p^i$ is the probability that sensor network $Tri_{k \times k}$ keeps k-Order Subnet Connected. ∎

From Lemma 3, we can deduce the following important theorem.

Theorem 2. *Assuming that each node in Triangular sensor networks has independent error probability p, the probability that sensor network* $Tri_{m \times n}$ *is k-Order Subnet connected is* $Pr[C(Tri_{m \times n})] = \left(\sum_{i=0}^{k^2} N_{k,i}(1-p)^{k^2-i}p^i \right)^{(m/k) \cdot (n/k)}$, *where* $N_{k,i}$ *is a constant defined in Lemma 3.*

Proof: As before, we use $Tri_{1,1}$, $Tri_{1,2}$, ..., $Tri_{(m/k),(n/k)}$ to denote k-Order Subnet. According to Lemma 3, the probability that each k-Order Subnet is connected is $\sum_{i=0}^{k^2} N_{k,i}(1-p)^{k^2-i}p^i$. Triangular sensor network $Tri_{m \times n}$ contains mn/k^2 k-Order Subnets, and all mn/k^2 k-Order Subnets in $Tri_{m \times n}$ is disintersected, so the probability that sensor network $Tri_{m \times n}$ is k-Order Subnet Connected $Pr[C(Tri_{m \times n})]$ is:

$$\begin{aligned}
\Pr[C(Tri_{m\times n})] &= \Pr[C(Tri_{1,1}) \cap C(Tri_{1,2}) \cap \cdots \cap C(Tri_{m/k,n/k})] \\
&= \Pr[C(Tri_{1,1})] \cdot \Pr[C(Tri_{1,2})] \cdots \Pr[C(Tri_{m/k,n/k})] \\
&= \Pr[C(Tri_{k\times k})]^{(m/k)\cdot(n/k)} \\
&= \left(\sum_{i=0}^{k^2} N_{k,i}(1-p)^{k^2-i}p^i \right)^{(m/k)\cdot(n/k)}
\end{aligned}$$

∎

If Triangular sensor network $Tri_{m\times n}$ is k-Order Subnet Connected, all correct nodes in $Tri_{m\times n}$ must form a connected graph. So, CP of network is not less than probability that the Triangular sensor network $Tri_{m\times n}$ is k-Order Subnet Connected. Thus, we can deduce a lower bound on the connectivity probability of Triangular sensor network $Tri_{m\times n}$.

Corollary 1. *Assuming that each node in Triangular Sensor Networks has independent error probability p, then the probability that correct nodes in network $Tri_{m\times n}$ are connected is at least $\left(\sum_{i=0}^{k^2} N_{k,i}(1-p)^{k^2-i}p^i \right)^{(m/k)\cdot(n/k)}$, where $N_{k,i}$ is a constant defined in Lemma 3.*

5.3 Probabilistic Cpmputation on Fault Tolerance of Sensor Networks

In the following, we will discuss the application of Corollary 1 in analysis of fault tolerance of sensor networks. Firstly, the probability that a sensor network $Tri_{3\times 3}$ is a 3-Order Subnet is analyzed.

Corollary 2. *Assuming each node in a 3×3 Triangular sensor network has independent error probability p, then the probability that such network is a connected 3-Order Subnet is $\Pr[C(Tri_{3\times 3})] = (1-p)^9 + 9(1-p)^8p + 22(1-p)^7p^2 + 9(1-p)^6p^3$.*

Proof: When i nodes are deleted from $Tri_{3\times 3}$, there exist $N_{3,i}$ methods that $Tri_{3\times 3}$ maintains connectivity and there are at least two correct nodes in each of four sides. When there is no error node, $N_{3,0} = 1$. When there exists only one error node deleted from the Triangular sensor network, $N_{3,1} = 9$. When two error nodes are deleted from the Triangular sensor network $N_{3,2} = 22$. When three error nodes are deleted from $Tri_{3\times 3}$, $N_{3,3} = 9$. Finally, it can be seen that $N_{3,4} = N_{3,5} = \cdots = N_{3,9} = 0$. Therefore, according to Lemma 1, the probability that Triangular sensor network $Tri_{3\times 3}$ is a 3-Order Subnet Connected is

$$\Pr[C(Tri_{3\times 3})] = (1-p)^9 + 9(1-p)^8p + 22(1-p)^7p^2 + 9(1-p)^6p^3. \quad \blacksquare$$

By using Corollary 1 and Corollary 2, we can derive that the connectivity probability of sensor network $Tri_{m\times n}$ at least is

$(\Pr[C(Tri_{3\times 3})])^{mn/9}$, $m, n \geq 3$.

Assuming that it is required that the connectivity probability of Triangular Sensor Network $Tri_{m\times n}$ is at least 99%, we can use the above formula to compute the probability of node error in Triangular sensor network. Table 1 gives the computing results with various network size.

According to Corollary 1, we can obtain that if nodes can meet the demand listed in Table 1, the connectivity probability of sensor networks must be no less than 99%. It is worth noting that values in Table 1 are only an upper bound of error probability, and such upper bound is worked out based on the computation on connectivity of 3-Order Subnet.

When computing the upper bound probability of node error for given connection probability of Triangular sensor network, if a larger k-Order Subnet is taken into account, a better upper bound on connectivity probability of sensor network $Tri_{m \times n}$ can be obtained. In the following, 5-Order Subnet is used to analyzed the upper bound.

Firstly, $\Pr[C(Tri_{5 \times 5})]$ based on 5-Order Subnet is discussed. While computing $\Pr[C(Tri_{5 \times 5})]$, the value of $N_{5,i}$ should be calculated. According to the analysis on $N_{5,i}$, the value of $N_{5,i}$ is calculated by using computer programm to enumerate all possible methods. The results are $N_{5,0} = 1$, $N_{5,1} = 25$, $N_{5,2} = 298$, $N_{5,3} = 2207$, $N_{5,4} = 11180$, $N_{5,5} = 40350$, $N_{5,6} = 104734$, $N_{5,7} = 191796$, $N_{5,8} = 235295$, $N_{5,9} = 177107$, $N_{5,10} = 72534$, $N_5, 11 = 14233$, $N_{5,12} = 1238$, $N_{5,13} = 44$, $N_{5,14} = \cdots = N_{5,25} = 0$. Therefore, we can get the following corollary.

Corollary 3. *If each node has independent error probability p, the probability that Triangular Sensor Network is 5-Order Subnet Connected is*

$$
\begin{aligned}
\Pr[C(Tri_{5 \times 5})] =&(1-p)^{25} + 25p(1-p)^{24} + 298p^2(1-p)^{23} + 2207p^3(1-p)^{22} + \\
&11180p^4(1-p)^{21} + 40350p^5(1-p)^{20} + 104734p^6(1-p)^{19} + \\
&191796p^7(1-p)^{18} + 235295p^8(1-p)^{17} + 177107p^9(1-p)^{16} + \\
&72534p^{10}(1-p)^{15} + 14233p^{11}(1-p)^{14} + 1238p^{12}(1-p)^{13} + \\
&44p^{13}(1-p)^{12}.
\end{aligned}
$$

By using Corollary 1 and Corollary3, we can obtain that the connectivity probability of sensor network $Tri_{m \times n}$ at least is

$(\Pr[C(Tri_{5 \times 5})])^{mn/25}$, $m, n \geq 5$.

Assuming that it is required that the connectivity probability of Triangular Sensor Network $Tri_{m \times n}$ is at least 99%, we can use the above formula to compute the probability of node error in Triangular sensor network. Table 2 gives the computing results with various network size.

As mentioned above, a better lower bound can be obtained by using a larger k-Order Subnet. Although $N_{k,i}$ can be computed by computer program, it can not be obtained in feasible time when k is not small. When $k = 5$, it cost less than one minute to work out $N_{k,i}$ on a computer (CPU: 2.4GHz). While $k = 11$, we can not get the results in five days by using the same computer. In the following, we will discuss how to compute $N_{k,i}$ when k is not small.

When p is relatively small, for a large i, the value of p^i is very small and can be negligible when computing the connectivity probability of k-Order Subnet.

Table 1. The probability of node error for 99% CP in Triangular Networks (in 3-Order Subnet)

Network Size	Error Probability
12×12	≤ 0.6734
51×51	≤ 0.1578
81×81	≤ 0.0993
102×102	≤ 0.0788
201×201	≤ 0.0400

Table 2. The probability of node error for 99% CP in Triangular Networks (in 5-Order Subnet)

Network Size	Error Probability
10×10	2.7791%
50×50	0.6600%
80×80	0.4226%
100×100	0.3410%
200×200	0.1737%

In a 11×11 Triangular sensor network $Tri_{11 \times 11}$, each node has independent error probability p. Assuming that p is less than 1.5%, we will calculate the probability that there are more than 6 error nodes among the 121 nodes.

There are 121 nodes in $Tri_{11 \times 11}$. Thus the probability that there are less than h error nodes is $\sum_{i=0}^{h} \binom{121}{i} p^i (1-p)^{121-i}$, and the probability that there are more than 6 error nodes is $1 - \sum_{i=0}^{6} \binom{121}{i} p^i (1-p)^{121-i} < 0.0025$.

So it is very clear that we can ignore the case which there are many error nodes when p is small.

With the above result, for the connectivity probability of such 11-Order Subnet as $Tri_{11 \times 11}$, if $p \leq 1.5\%$, we will only take into account the case that there are 6 error nodes at most. The deviation from accurate value of $\Pr[C(Tri_{11 \times 11})]$ is about 0.25%, which has little impact on the lower bound of the connectivity probability. We conclude that for Triangular Sensor Network, $N_{11,0} = 1$, $N_{11,1} = 121$, $N_{11,2} = 7258$, $N_{11,3} = 287729$, $N_{11,4} = 8479820$, $N_{11,5} = 198153567$, $N_{11,6} = 3823873850$. So, we get the following Corollary.

Corollary 4. *Assuming that each node has independent error probability p, the probability that $Tri_{11 \times 11}$ is 11-Order Subnet Connected at least is*

$$\Pr{}^{11}[C(Tri_{11 \times 11})] = (1-p)^{121} + 121(1-p)^{120}p + 7258(1-p)^{119}p^2 +$$
$$287729(1-p)^{118}p^3 + 8479820(1-p)^{117}p^4 +$$
$$198153567(1-p)^{116}p^5 + 3823873850(1-p)^{115}p^6.$$

By using Corollary 1 and Corollary 4, we can conclude that the probability that Triangular sensor network $Tri_{m \times n}$ is connected is at least $(\mathrm{Pr}^{11}[C(Tri_{11 \times 11})])^{mn/121}$, $m, n \geq 11$.

In the same way, If the connectivity probability of sensor network $Tri_{m \times n}$ should be larger than 99%, the probability of error nodes should be no less than one value, which is shown in Table 3.

Table 3. The probability of node error for 99% CP in Triangular Networks (in 11-Order Subnet)

Network Size	Error Probability
55×55	$\leq 0.9588\%$
110×110	$\leq 0.6297\%$
209×209	$\leq 0.3635\%$
407×407	$\leq 0.1900\%$
506×506	$\leq 0.1532\%$

The above Results are helpful to arrange sensor node in Triangular Sensor Network. When the number of nodes number is quite large, the connectivity probability of network inevitably decreases.Thus, it is an important issue to determine the probability of node error for ensuring the connectivity probability. For evaluating the performance of sensor network, the connectivity probability of sensor network for given probability of node error is a very important metric. As shown in Table 3, we are inferred that even for large sensor network with more that 250000 nodes ($Tri_{506 \times 506}$), the connectivity probability can be guaranteed to above 99% as long as the probability of node error can be controlled within 0.15%.

6   Simulation and Analysis

We have analyzed the relation between CP and EP in sensor networks. In the following, we will evaluate the reliability of the lower bound of the connectivity probability and the upper bound of the probability of node error in Triangular sensor network. We will also do comparison between Triangular sensor network and Grid sensor network.

In Table 4, the connectivity probability of Triangular sensor network and Grid Sensor with different probability of node error and network size are given.It can be concluded from Table 4 that when p is relatively small ($p = 0.5\%$ and 1%), networks are almost connected. When p is large ($p = 5\%$), the connectivity probability of networks drops significantly. Comparing Triangular Network to Grid Network, we can see that the connectivity probability of the latter suffers from a faster drop than the former with the increasing of error probability.

From the above results, it can be seen that the fault-tolerance of Triangular network is higher than that of Grid network. But in areas with the same size, Triangular network needs more nodes than Grid network for full coverage. It is

Table 4. CP with Uniformly Distributed EP of node

Size	Number	EP	CP in Grid	CP in Triangular
10×10	100	0.5	99.98	99.98
50×50	2500	0.5	99.99	100
80×80	6400	0.5	99.99	100
100×100	10000	0.5	99.99	99.98
200×200	40000	0.5	99.95	99.98
10×10	100	1.0	99.91	99.95
50×50	2500	1.0	99.92	99.98
80×80	6400	1.0	99.87	99.96
100×100	10000	1.0	99.85	99.96
200×200	40000	1.0	99.58	99.90
10×10	100	5.0	97.43	98.94
50×50	2500	5.0	91.92	97.95
80×80	6400	5.0	85.12	97.03
100×100	10000	5.0	79.19	96.66
200×200	40000	5.0	49.46	94.50

very clear that Triangular network needs more nodes to cover the same size area than that of Grid network.

Assuming that the communication radius of each sensor is 1 unit , the probability of node error is is p,the expected CP is P_{conn} and the size of cover area is $(m-1) \times (n-1)$, we can obtain the following conclusion.

Theorem 3. *Assuming that for Grid topology network $Gri_{m \times n}$, to reach the desired connectivity probability P_{conn}, the probability of node error in each position should be less than P_g; for Triangular topology network $Tri_{m \times \left(\left\lfloor \frac{2\sqrt{3}}{3}n\right\rfloor+1\right)}$, to reach the desired connectivity probability P_{conn}, the probability of node error should be less than P_t. If such network is deployed in a $m \times n$ area with 1-unit communication radius and the probability of node error is p, there are at least $m \times n \times \lceil\log_p(P_g)\rceil$ nodes in Grid topology network and at least $m \times \left(\left\lfloor \frac{2\sqrt{3}}{3}n\right\rfloor+1\right) \times \lceil\log_p(P_t)\rceil$ nodes in Triangular topology network.*

Proof:

1. Grid topology

For $(m-1) \times (n-1)$ area, if node with communication radius 1 unit is used, then Grid network $Gri_{m \times n}$ with size $m \times n$ is needed to monitor the whole area. It is required that $m \times n$ Grid network must reach the connectivity probability P_{conn}, and the probability of node error in each position should less than P_g, thus nodes placed in each position should be less than $\lceil\log_p p_g\rceil$, then the minimum number of nodes in such area is $m \times n \times \lceil\log_p(P_g)\rceil$.

2. Triangular topology

It can be seen from Fig. 2 that if the communication radius is 1 unit, the distance between the adjacent nodes in the same row is 1 and the distance between the adjacent nodes in different rows is $\sqrt{3}/2$. Therefore, there are $\left\lfloor \frac{2\sqrt{3}}{3}n \right\rfloor + 1$ rows in the whole network. Also in Fig. 2, if there are m nodes in each row, then the width of network is $m - 1/2$. Assuming that m is large, for network with width $m - 1$, each row needs m nodes, so $m \times \left(\left\lfloor \frac{2\sqrt{3}}{3}n \right\rfloor + 1 \right)$ Triangular network is needed to monitor the whole area. In line with conditions in theorem, in the same way, the number of nodes should be more than $m \times \left(\left\lfloor \frac{2\sqrt{3}}{3}n \right\rfloor + 1 \right) \times$ $\left\lceil \log_p(P_t) \right\rceil$. ∎

7   Conclusion

We firstly described the fault-tolerance problem, then discussed the coverage and connectivity in Triangular sensor network. The results about coverage and connectivity are useful for further deployment and research of network.

Based on the analysis of k-Order Subnet, we derived the results from fault-tolerance on connectivity, which are just answering for the estimation on the upper bound of the probability of node error and the lower bound of the connectivity probability. Especially for Triangular network, which has higher connectedness than Grid network, such a method may underestimate the connectivity of such a kind of networks. Our following work will focused on further approximation to an accurate value. Moreover, a new analytic method is in great need for Triangular network to take full advantage of those features.

References

1. Akyildiz, I.F., Su, W., Sankarasubramaniam, Y.: Wireless Sensor Networks: A Survey. Computer Networks, 393–422 (March 2002)
2. Warneke, B., Last, M., Liebowitz, B., Pister, K.S.J.: Smart dust: Communicating with a cubic-millimeter computer. IEEE Computer Magazine 34(1), 44–51 (2001)
3. Tilak, S., Abu-Ghazaleh, N.B., Heinzelman, W.: A taxonomy of wireless micro-sensor network models. Mobile Computing and Communications Review 1(2), 1–8 (2002)
4. TinyOS, http://tinyos.millennium.berkeley.edu
5. Zhang, H., Hou, J.C.: Maintaining scheme coverage and connectivity in large sensor networks. Technical report, UIUC (2003)
6. Chakrabarty, K., Iyengar, S.S., Qi, H., Cho, E.: Grid coverage for surveillance and target location in distributed sensor networks. IEEE Transactions on Computers 51(12), 1448–1453 (2002)
7. Chen, B., Jamieson, K., Balakrishnan, H., Morris, R.: An Energy-Efficient Coordination Algorithm for Topology Maintenance in Ad Hoc Wireless Networks. In: ACM/IEEE International Conference on Mobile Computing and Networking (MobiCom 2001), Rome, Italy, (July 16-21, 2001)
8. Wang, G.: Probability analysis of mesh network fault [Doctor Degree Thesis]. ChangshaCentral South University (2004)
9. Wang, X., Xing, G., Zhang, Y.: Integrated coverage and connectivity configuration in wireless sensor networks, SenSys (2003)

10. Zhang, H., Hou, J.C.: Maintaining sensing coverage and connectivity in large sensor networks. In: NSF International Workshop on Theoretical and Algorithmic Aspect in Sensor, Ad Hoc Wireless and Peer-to-Peer Networks (2004)
11. Cardei, M., Wu, J.: Energy-efficient coverage problems in wireless Ad Hoc sensor networks. Computer Communications 29(4), 413–420 (2005)
12. Slijepcevic, S., Potkonjak, M.: Power efficient organization of wireless sensor networks. In ICC 2002, vol. 2, pp. 472–476 (2001)
13. Meguerdichian, S., Potkonjak, M.: Low power 0/1 coverage and scheduling techniques in sensor networks. UCLA Technical Reports 030001 (2003)
14. Hsin, C.F., Liu, M.: Network coverage using low Duty-Cycle sensors: random & coordinated sleep algorithm. Information Processing in Sensor Networks, pp. 433–442 (2004)
15. Ye, F., Zhong, G., Lu, S., Zhang, L.: Energy Efficient Robust Sensing Coverage in Large Sensor Networks. Technical Report (2002),
    http://www.cs.ucla.edu/~yefan/coveragetech-report.ps
16. Tian, D., Georganas, N.D.: A Coverage-preserving node scheduling scheme for large wireless sensor networks. In: processing of ACM wireless sensor network and application workshop
17. Liu, Y., Liang, W.: Approximate coverage in wireless sensor networks. In: Proceedings of IEEE Conference on Local Computer Networks (2005)
18. Wang, L., Kulkarni, S.S.: pCover: partial coverage for long-Lived surveillance Sensor Netowrks
19. Ren, S., Li, Q., Wang, H., Chen, X., Zhung, X.: A study on object tracking quality under probabilistic coverage in sensor networks. Mobile Computing and Communications Review (2005)
20. Wang, G., Cao, G., Porta, T.L.: Movement-assisted sensor deployment. In: INFOCOM 2004, pp. 2313–2324 (2004)
21. Shakkottai, S., Srikant, R., Shroff, N.: Unreliable Sensor Grids: Coverage, Connectivity and Diameter. In: Proceeding of INFOCOM 2003, pp. 1073–1083 (2003)
22. Gupta, H., Das, S.R., Gu, Q.: Connected sensor cover: Self-organization of Sensor Networks for Efficient Query Execution. In: Proc. of ACM MobiHoc, pp. 189–200 (2003)
23. Zhou, Z., Das, S., Gupta, H.: Connected k-coverage problem in sensor networks. In: Proc. of Intl. Conf. on Computer Communications and Networks (ICCCN), pp. 373–378 (2004)
24. Kumar, S., Lai, T., Balogh, J.: On k-coverage in mostly sleeping sensor networks. In: MobiCom 2004, pp. 144–158 (2004)
25. Jun, J., Sichitiu, M.L.: The nominal capacity of wireless mesh networks. IEEE Wireless Communications 10(5), 8–14 (2003)
26. Jia, W., Wang, J.: Analysis of Connectivity for Sensor Networks Using Geometrical Probability. IEE Proceedings-Communication, 305–312 (2006)

Performance Modeling of Mobile Sensor Networks

Jerzy Martyna

Institute of Computer Science, Jagiellonian University, ul. Nawojki 11,
30-072 Cracow, Poland
martyna@softlab.ii.uj.edu.pl

Abstract. In the paper a performance evaluation model of mobile wireless sensor network is presented. The model is based on renewal theory. The hierarchical structure of the network holds a clusters composed of ordinary sensors. Afterwards, all clusterhead sensors belong to a backbone network which is needed for data transmission to the sink of the sensor network. Some sensors are mobile and can send their data to the clusterhead sensors. In this paper, we address one of the fundamental problems, namely performance modeling of mobile wireless sensor networks. Moreover, our model can also be used for ordinary wireless sensor networks which are static and geometrically constrained. Specific models are suggested and analysed for different communication needs.

1 Introduction

Advances in the technologies of VLSI, RF and embedded processors provide the widespread use of wireless sensor networks to obtain some physical quantities, such as temperature, humidity, pressure, etc., from the physical environment. The wireless sensor networks are composed of hundreds or thousands of homogeneous devices (called sensor nodes) with the ability to collect the sensed data and to deliver them to a central server or sink [1, 5, 9, 15]. Sensor nodes may be deployed at random or installed at deliberately chosen spots. A sensor network can be composed of some nodes which are moving. For example, a military application where nodes are dropped from an aircraft onto land, and transmitters attached to soldiers. In the most general situation, the sensor nodes and sinks move - either they are in wind, currents, etc., or they have actuators for motion [7, 8, 16, 17].

The general research challenges for communication and coordination in sensor networks arise primarily due to the large number of constraints, such as small capacities in memory, limited CPU execution speeds, limited energy power, scare communication bandwidth, data delivery delay, location awareness, convertibility between mobility and capacity, etc. A limited transmission range of an individual sensor makes a given sensor unable to directly communicate with all the other sensors that might detect a common event. In succession, the limited energy of each sensor node cannot improve the quality and energy efficiency of

E. Kranakis and J. Opatrny (Eds.): ADHOC-NOW 2007, LNCS 4686, pp. 262–272, 2007.

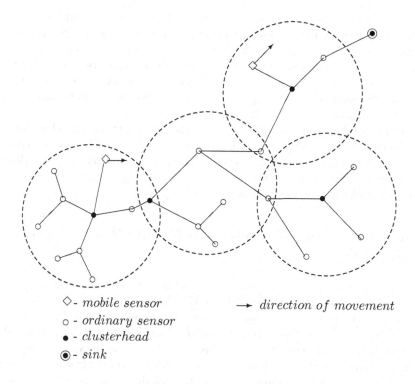

◇ - *mobile sensor*

○ - *ordinary sensor*

● - *clusterhead*

◉ - *sink*

→ *direction of movement*

Fig. 1. An example of mobile sensor network

the environmental system during the reconfiguration and customization in the future.

To reduce packet transmission time in wireless sensor network numerous mechanisms have been proposed. One of them is the clustering hierarchy presented by Heinzelman et al. [6]. Thus, the sensor network is partitioned into several clusters, each consisting of a cluster head, gateway (optional), and ordinary members. Clusterheads and gateways can be treated as nodes of the backbone of the network. They are responsible for relaying broadcast (or multicast) packets to all nodes and also for collecting data from environment and sending them to sink (see Fig. 1). The clustering hierarchy prevents large amounts of packet transmission and saves energy power.

The performance modeling of the sensor network was studied by many researchers. Among others, a Markov chain as a model of sensor network was used by Chiasserini [3]. In this model all sensor nodes are characterized by energy consumption, data delivery delay and typical states of sensor nodes, namely "active" and "sleep". A discrete Markov chain with a specialized iLTL analyser was proposed by Kwon [10, 11]. Nevertheless, none of the above-mentioned models takes into considerations the movement of sensors.

The main goal of this paper is to introduce a new model of a sensor network in which some sensors are moving in the sense given in [7, 14]. In this approach three possibilities are admissible. In the first, all static sensor nodes are seen as moving. In the second all the moving nodes are treated statically [2]. In the third possibility all the moving nodes are treated as moving. This representation is nearest to the presented solution. This model of wireless mobile sensor network requires new solutions, including a localization of the sensor [8], coordination [12], etc. However, the given approach provides the new dependencies which adequately way describe the behaviour of sensor network.

In the second section the details of our model are given. The third section gives some performance measures for our model of a mobile sensor network. Some numerical results obtained as a solution of our model are given in the fourth section. The paper is wrapped up by presenting a summary, conclusion and an outlook into the future work in section 5.

2   The Performance Model of a Mobile Wireless Sensor Network

Two layers of sensor hierarchy are considered. The first layer (primary) represents all the sensors which are responsible for the data transmission to the sink. It consists only of clusterhead sensors and sensors belonging to backbone network. The second layer (secondary) is composed of ordinary sensor nodes which collect the information and send it to the clusterheads. In our model, some sensor nodes are moving or the sink is in move in relation to the static sensor network. We assumed that all the data delivered by mobile sensor nodes are sent to clusterheads, while the ordinary static sensor nodes transmit their packets within their cluster.

For the mobile sensors handoffs are admitted between clusterheads. If the transmission channel is not available for the mobile sensor node, the handoff is broken. On the other hand, if the signal received by a mobile sensor node from the headcluster is getting smaller and it is impossible to open a new connection, the existed link is closed.

The submission process of a new connection is caused by one of three events: a) the mobile sensor node will send its data, b) the static sensor desires their data, c) the mobile sensor must execute the handoff. The liquidation of a connection is possible only in three situations: a) the termination of data transmission from the mobile sensor, b) the termination of data transmission from the static sensor, c) the termination of handoff for the mobile sensor node, d) the loss of connection by the mobile sensor node in the case of lack of radio range.

We assumed that the submission process of a new connection is the Poisson process with the parameter $\lambda^{(m)}$, in the case of mobile sensor node, and $\lambda^{(f)}$ for the static sensor node. The demand of handoff by a mobile sensor from the radio range of a new clusterhead to the radio range of another neighbouring clusterhead is the Poisson process with the parameter $\lambda_h^{(m)}$. We assumed that the transmission time is a random variable with the exponential distribution with the parameter μ. The time of connection between the mobile sensor node and the clusterhead sensor node is a random variable with the exponential distribution with parameter $\mu_h^{(m)}$. The time between two alternate breaks of connection for a mobile sensor node is given by $\mu_d^{(m)}$. This parameter is determined on the ground of experiment.

2.1   The Cluster Model of a Mobile Sensor Network

Let the call request for arrival process to the clusterhead sensor nodes be treated as a summarized process of three Poisson with the parameters: $\lambda^{(m)}$, $\lambda^{(f)}$, $\lambda_h^{(m)}$. The times of transmission processes are modeled by random variables with the exponential distribution whose mean value is equal to μ^{-1} for the static sensor and $(\mu_h^{(m)})^{-1}$ for the mobile sensor nodes.

Let $\pi(i)$ be the probability of an equilibrium state in which the channel number i is occupied in the cluster head node in the cluster. It is given by

$$\pi(i) = \frac{\rho^i}{i!}\pi(0) \quad \text{for} \ \ 1 \le i \le L^{(c)} \tag{1}$$

where $L^{(c)}$ is the number of transmission channels of clusterheads.

Thus, the service rate ρ is given by

$$\rho = \frac{\lambda^{(m)} + \lambda_h^{(m)} + \lambda^{(f)}}{\mu + \mu_h^{(m)}}, \quad \text{and} \ \ \pi(0) = \frac{1}{\sum_{i=1}^{L^{(c)}} \frac{\rho^i}{i!}} \tag{2}$$

Assuming that the system is in the steady-state, we obtain for the mobile sensors the number of handoffs which is equal to the mean number of closed connections. It is given by

$$\lambda_h^{(out)} = \sum_{i=1}^{L^{(c)}} i \cdot \mu_h^{(m)} \cdot \pi(i) \tag{3}$$

We suppose that any cluster can be treated independently from others. Then, we get

$$\lambda_h^{(m)} = \lambda_h^{(out)} \tag{4}$$

The probability that all transmission channels in the clusterhead node are busy can be computed with the help of the Erlang B formula. Thus, it is given by

$$Pr^{(c)}(busy) = \frac{\frac{\rho^i}{i!}}{\sum_{k=0}^{i} \frac{\rho^k}{k!}} \tag{5}$$

The above described process is a renewal process with the interval distribution whose Laplace transformation can be derived from the recursive formula

$$\phi_k(s) = \frac{\phi_{K-1}(s + \mu)}{1 - \phi_{K-1}(s) + \phi_{K-1}(s + \mu)} \tag{6}$$

where μ^{-1} is the mean service time, $\phi_0(s)$ is the characteristic function of the interarrival times at the queue. Let K be the number of the transmission channel which is equal to $L^{(c)}$ in the clusterhead of i-th cluster. Thus, we obtain the dependence

$$\phi_0(s) = \frac{\lambda^{(m)} + \lambda_h^{(m)} + \lambda^{(f)}}{s + \lambda^{(m)} + \lambda_h^{(m)} + \lambda^{(f)}} \tag{7}$$

Since the handoff times for the mobile sensors are independent and they have the exponential distribution time with the same parameter $\lambda_h^{(m)} > 0$, the renewal process is the Poisson process [4]. The renewal process that corresponds to the Poisson process belongs to to the special case of a renewal process of the Palm type or a pure birthday process [13]. By taking this into account we can compute the total number of all the transmission channels of the clusterhead sensor which are busy, namely

$$\overline{N}^{(c)}(busy) = (\lambda^{(m)} + \lambda_h^{(m)} + \lambda^{(f)}) \cdot Pr^{(c)}(busy) \tag{8}$$

Assuming that the number of cluster f_c is large, we can give a mean number of transmission requests that cause the occupation of the transmission channels in all clusterhead sensors. This number is equal to

$$\overline{N}^{(cluster)}(busy) = f_c \cdot \overline{N}^{(c)}(busy) = f_c \cdot (\lambda^{(m)} + \lambda_h^{(m)} + \lambda^{(f)}) \cdot Pr^{(c)}(busy) \tag{9}$$

Following Eq. (9) it is visible that together with the growth of the number of the cluster in the sensor field the number of transmission channels which are busy with the data transmission increases.

2.2   Model of the Sensor Field

We assume that in the sensor field the transmission requests in the sink arise from the mobile sensors and from the clusterhead sensors. The transmission request process is the superposition of three Poisson processes. The first of them corresponds to the transmission requests from the mobile sensors. The second Poisson process is caused by the handoff requests between the mobile sensors and the clusterhead sensors. The third Poisson process arises from the clusterhead sensors. All these Poisson processes possess the following parameters, $\lambda^{(s,m)}$, $\lambda_h^{(s,m)}$, $\lambda^{(s,c)}$, respectively. The time between transmission requests is assumed to be an exponentially distributed random variable with parameter $\mu^{(s,m)} + \mu^{(s,c)}$. The first component represents the mean transmission time between the mobile sensor and the sink and the mean transmission time between the clusterhead sensor and the sink, respectively. Therefore, the service rate of the sink for both types of data transmission is as follows:

$$\rho^{(s,m)} = \frac{\lambda^{(s,m)} + \lambda_h^{(s,m)}}{\mu^{(s,m)} + \mu^{(s,c)}} \tag{10}$$

$$\rho^{(s,c)} = \frac{\lambda^{(s,c)}}{\mu^{(s,m)} + \mu^{(s,c)}} \tag{11}$$

The first of them is associated with the mobile sensors and the second with the clusterhead sensors.

The probability that i channels of the sink are occupied by the data transmission from the mobile sensors and j channels are engaged by the data transmission from the clusterhead sensors is equal to

$$p(i,j) = \frac{(\rho^{(s,m)})^i \cdot \rho^{(s,c)})^j}{i! \cdot j!} \pi(0,0) \tag{12}$$

for all states $s = (i,j)$ belonging to S such that s,

$$S = \{(i,j) \mid i + j \leq L^{(c)} \cap j \leq L^{(c)} - L_m \tag{13}$$

where L_m is the number of channels reserved for the mobile sensors.

The probability $\pi(0,0)$ can be obtained from the normalization equation:

$$\pi(0,0) = \frac{1}{\sum_{(i,j)\in S} \frac{(\rho^{(s,m)})^i}{i!} \frac{(\rho^{(s,c)})^j}{j!}} \tag{14}$$

The number of the state is not too large here because its order is equal to $(L^{(s)})^2/2$, where $L^{(s)}$ is the number of channels in the sink.

The mean number of handoffs can be computed in the steady-state in which the mean number of handoffs incoming to the radio range of sink is equal to the mean number of handoffs leaving the radio range of the sink, namely

$$\lambda_h^{(out)} = \sum_{(i,j)\in S} i \cdot \mu_h^{(s,m)} \cdot \pi(i,j) \tag{15}$$

The blocking probabilities of free channels for the mobility sensor is given by

$$Pr^{(s,m)}(block) = \sum_{(i,j)\in S_m} \pi(i,j) \tag{16}$$

and for the clusterhead sensor it is in the form, namely

$$Pr^{(s,c)}(block) = \sum_{(i,j)\in S_f} \pi(i,j) \tag{17}$$

where

$$S_m = \{(i,j) \mid i + j = L^{(s)} \cup j = L^{(s)} - L_m\} \tag{18}$$

$$S_c = \{(i,j) \mid i + j = L^{(s)}\} \tag{19}$$

The blocking probabilities for new call and handoff requests coincide.

3   The Performance Measures of a Wireless Mobile Sensor Network

The usefulness of a wireless mobile sensor network can be defined by some performance measures. The probability that the data transmissions by the mobile sensors fail is defined by the $F_m^{(trans)}$. Thus, the probability requesting h handoffs of the data transmission is given by

$$1 - F_m^{(trans)} = \sum_{h=0}^{\infty} P_m(success \mid h\ handoffs) \cdot P_m(h\ handoffs) \qquad (20)$$

Let the probability of a release of transmission channels, which is caused by a successful handoff realization of mobile sensors be given by

$$B_m = \frac{\mu^{(s,m)}}{\mu^{(s,m)} + \mu^{(s,c)}} \qquad (21)$$

Thus, on the assumption that the different handoffs are independent, we can define the probability that the transmission is not realized by the mobile sensor. It is given by

$$F_m^{(trans)} = 1 - \sum_{h=0}^{\infty}(1 - Pr^{(s,m)})^{(h+1)}B_m^h(1 - B_m^h) = 1 - \frac{(1 - Pr^{(s,m)})(1 - B_m)}{1 - (1 - Pr^{(s,m)})B_m} \qquad (22)$$

Let G_f be the probability that the transmission by a static sensor is successfully completed. The successful end of this transmission is as result of a handoff of this connection to another sensor caused for instance by a change of route. The main reason is associated with the lack of energy of a given sensor. This probability is given by Q_f. Thus, the call blocking probability for a static sensor is given by

$$H_f = 1 - F_f = 1 - G_f Q_f \qquad (23)$$

where

$$Q_f = (1 - Pr^{(s,f)})^{d+1}F_f^d(1 - F_f) \qquad (24)$$

The transmission probability F_f is here defined as the transmission completion in a handoff caused by route change, d is the total number of route changes.

To recognize the handoff regions in terms of geographical locations criteria we use the following method. We supposed that a mobile sensor has a system which measures the signal power. For each mobile sensor l it is possible to compute requesting a handoff channel, its current power level $P_l^{(m)}$, and the rate of change of this power $\frac{dP_l^{(m)}}{dt}$. Thus, whenever a channel becomes free, the time $t_l^{(m)}$, that mobile sensor l exists the handoff region can be computed

$$t_l^{(m)} = \frac{\mid P_{min} - P_l^{(m)} \mid}{\frac{dP_l^{(m)}}{dt}}, \quad 1 \leq l \leq M^{(m)} \tag{25}$$

where P_{min} is the minimum value of power level, $M^{(m)}$ is the total number of mobile sensors.

4 The Numerical Results

The given analysis allows to find some dependencies for numerous load parameters in wireless mobile sensor network.

For the studied model of wireless mobile sensor network the main parameters which are given in Table 1 are assumed.

Table 1. Values of the parameters

Parameter	Value
$\lambda^{(s,c)}$	$0.4 \ s^{-1}$
$\lambda^{(s,m)}$	$[0.01 - 0.1] \ s^{-1}$
$\lambda_h^{(s,m)}$	$[0.005 - 0.012] \ s^{-1}$
$\mu^{(s,m)}$	$0.01 \ s^{-1}$
$\mu^{(s,c)}$	$0.04 \ s^{-1}$

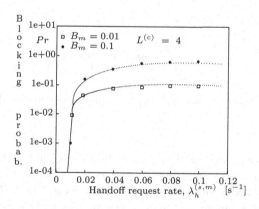

Fig. 2. Blocking probability for mobile sensor versus handoff request rate for given number of channels in clusterhead sensor, $L^{(c)} = 4$

One of the more important parameters of wireless, mobile networks is the call blocking probability between the mobile sensor and the clusterhead sensor. By means of the Eq. (22) we can give for wireless sensor network the call blocking probability for the mobile sensor. Next, we can obtain the graph of the call blocking probability which is given in Fig. 2.

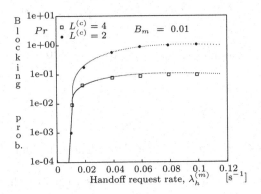

Fig. 3. Blocking probability for mobile sensor versus handoff request rate for given probability of release of transmission channels, $B_m = 0.01$

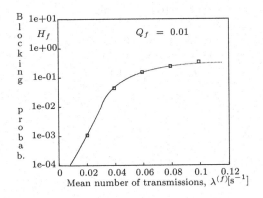

Fig. 4. Blocking probability for static sensor versus mean number of transmissions for given value of Q_f

For the supposed value of $\lambda^{(s,m)}$ from the interval $[0.01 - 0.1]$ $[\mathrm{s}^{-1}]$ and the reservation of two (four) transmission channels for the clusterhead sensors we obtain the call blocking probability which is given in Fig. 3. It is evident that the growth of the number of channels reserved for the clusterhead causes the decrease of the call blocking probability for mobile sensors. Analogously, for the static sensors in the wireless sensor networks with the static and mobile sensors, with the assumed values of the call (request) arrival rate for the static sensors the call blocking probability depends on the number of reserved channels of transmission for the transmission realized by the mobile sensors.

By enlarging the number of transmission channels for the mobile sensor nodes we cause the reduction of the blocking probability. Nevertheless, this influence is much smaller than in the case of the channel reservation for the transmission of static sensors.

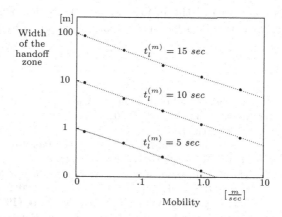

Fig. 5. Width of the handoff zone versus mobility of mobile sensor node for different values of $t_l^{(m)}$

The presented model can also be used for the wireless sensor network composed of only static sensor nodes. In the model the blocking probability of transmission is given by the graph presented in Fig. 4. The increase of the mean number of transmissions causes the growth of the number of blocking probability.

We are interested in the determining the width of the handoff zone. Since time $t_l^{(m)}$ is determined in Eq. (25), we obtained varying mobility for the mobile sensor node. Thus, we calculated the width of the handoff zone for varying mobility of the mobile sensor node (see Fig. 5).

5 Conclusion

In this paper, the performance model of wireless, mobile sensor networks with emphasis on reducing the blocking probability is proposed. To give priority to mobile sensor transmission, we introduce the idea of handoffs between the mobile sensor nodes and clusterhead sensor nodes. Nevertheless, our model is developed to analyze the performance of the whole sensor network.

The numerical results show that the model can reduce the blocking probability. This is achieved by the reservation of defined number of channels for possible transmission. Consequently, the probability of blocking probability is increased because the capacity of the wireless, mobile sensor network is limited. When a new call is coming, we should base our decision on the number of open connections of the local and adjacent clusterhead sensor nodes. Therefore their blocking probabilities are lower than the ones obtained from our model. The drawback of the conventional methods is that they might cause dropped calls in their adjacent cluster when they have reached their saturation points.

There are two main directions we are going to forward this work in the future. One direction is the settlement of the dependencies between the density of the sensor placement and the effectiveness of the use of mobile sensors for data collection. The second area of the future work is the assignment of the total lifetime of the sensor network to the function of the number of failed sensors.

References

1. Akyildiz, I.F., Su, W., Sankarasubramaniam, Y., Cayirci, E.: Wireless Sensor Networks. Computer Networks 38, 393–422 (2002)
2. Antol, J., Calhoum, Ph.,Flick, J., Gajos, G.A., Kolacinski, R., Minton, D., Owens, R., Parker, J.: Low Cost Mars Surface Exploration: The Mars Trumbleweed, NASA Langley Research Center, NASA/TM-2003-212411 (2003)
3. Chiasserini, C.-F., Gareto, M.: Modeling the Performance of Wireless Sensor Networks. In: Proceedings of the IEEE INFOCOM, vol. 4, pp. 220–231 (2004)
4. Cox, D.R.: Renewal Theory. John Wiley and Sons, New York (1994)
5. Hać, A.: Wireless Sensor Network Design. John Wiley and Sons, Hoboken (2003)
6. Heinzelman, W.B., Chandrakasan, A.P., Balakrishnan, H.: An Application-Specific Protocol Architecture for Wireless Microsensor Networks. IEEE Transactions on Wireless Networking 1(4), 660–670 (2002)
7. Howard, A., Mataric, M.J., Sukhatme, G.S.: Mobile Sensor Network Deployment Using Potential Fields: A Distributed, Scalable Solution to the Area Coverage Problem. In: 6th International Conference on Distributed Autonomous Robotic System (DSRS '02), Fukuoka, Japan, pp. 299–308 (2002)
8. Hu, L., Evans, D.: Localization for Mobile Sensor Networks (September26 - October 1) MobiCom '04 (2004)
9. Intel Research Oregon, Heterogeneous Sensor Networks, Technical Report, Intel Corporation, [URL] (2003),
   http://www.intel.com-research-exploratory-heterogenous.html
10. Kwon, Y., Agha, G.: Linear Inequality LTL (iLTL): A Model Checker for Discrete Time Markov Chains. In: Davies, J., Schulte, W., Barnett, M. (eds.) ICFEM 2004. LNCS, vol. 3308, pp. 194–208. Springer, Heidelberg (2004)
11. Kwon, Y., Agha, G.: Performance Evaluation of Sensor Networks: A Statistical Modeling and Probabilistic Model Checking Approach. ACM Transaction on Embedded Computing Systems (ACM TECS) (2006)
12. Nagpal, R., Shrobe, H., Bachrac, J.: Organizing a Global Coordination System from Local Information on Ad Hoc Sensor Network. In: Zhao, F., Guibas, L.J. (eds.) IPSN 2003. LNCS, vol. 2634, Springer, Heidelberg (2003)
13. Palm, C.: Intensity Variations in Telephone Traffic, North-Holland Publishing Co. (1987)
14. Poduri, S., Sukhame, G.S.: Constrained Coverage for Mobile Sensor Networks. IEEE International Conference on Robotics and Automation, 165–172 (2004)
15. Pottie, G.J., Kaiser, W.J.: Embedding the Internet: Wireless Integrated Network Sensors. Communication of the ACM 43(5), 51–58 (2000)
16. Shorey, R., Ananda, A., Chan, M.C., Ooi, W.T. (eds.): Mobile, Wireless, and Sensor Networks. IEEE Press, Los Alamitos (2006)
17. Wang, G., Cao, G., LaPorta, T., Zhang, W.: Sensor Relocation in Mobile Sensor Networks. In: Proceedings of the IEEE INFOCOM, vol. 4, pp. 2305–2312 (2005)

Electronic-Oriented IP Address Auto-Configuration Protocol for MANET

Jin-Ok Hwang, Hyo-Beom Lee, and Sung-Gi Min*

Dept. of Computer Science and Engineering, Korea University
Seoul, Korea
{withmind,eembryo,sgmin}@korea.ac.kr

Abstract. In an ad-hoc network, configuration of an IP address based on the current Internet protocol version, is an important task. Previous allocation methods for automatic IP configuration have several drawbacks, such as wasted address space, message implosion, and/or long delay (e.g., using the duplicate address detection [DAD] procedure). Such methods are unsuitable for connecting a mobile ad-hoc network directly to an external network, such as the Internet, because mobile nodes frequently move in a dynamic environment. Mobile devices need a quick and unique IP auto-configuration scheme in small-scale environments. Therefore, we propose the electronic-oriented IP address configuration scheme, which consists of electronic bit position (EBP) and electronic MAP (eMAP) schemes. The EBP allocates IP addresses, and the eMAP locates reusable IP addresses among crashed or abruptly abandoned mobile nodes. The protocol for message handling uses a random back-off mechanism that decreases delay time and response implosion. The protocol was tested with a network simulator (*ns-2*).

1 Introduction

Today, mobile computing and wireless technology are poised for greatness, particularly the mobile ad-hoc network (MANET), which is a dynamic network without supporting infrastructure. Many routing protocols have been used to develop the ad-hoc network, but are of no practical use, because the main ad-hoc routing protocols are based on IP addresses, with routing protocols that are tightly coupled with the IP address, in the network layer. Therefore, the assignment of IP addresses without infrastructure is an important task. However, the address auto-configuration in MANET has been left undeveloped. The IETF Zeroconf working group and MANET[1] are focusing on implementation of mobile ad hoc networks.

The IETF zeroconf working group deals with auto-configuration issue.

In a wired network, an allocation scheme such as the dynamic host configuration protocol is used to assign IP addresses, but this method is based on a centralized server. Several address allocation algorithms have been focused for MANETs. Previous studies [5]~[9] of automatic IP address configuration have

---

* Corresponding author.

E. Kranakis and J. Opatrny (Eds.): ADHOC-NOW 2007, LNCS 4686, pp. 273–284, 2007.

focused on allocating IP addresses without conflict, to compensate for the lack of central allocation servers.

However, these methods lead to the leakage of IP addresses, an issue that must be addressed for practical application of IP addresses to current systems. If mobile networks (MNs) have to be connected to the external network, which is currently based on Internet protocol version 4 (IPv4), it may not be practical for ad-hoc networks to require a public address space. To assign IP addresses using the current addressing method, the number of unused addresses must be minimized and the address-leak problem must be solved.

When a mobile node crashes, or there is an abrupt abandonment of a node, the IP address is lost. Therefore, IP address reallocation protocols are required, in order to reused IP addresses.

We summarize the contribution of this paper as follows:

- We ensure that IP address allocation is unique, without long delay. Mobile nodes should obtain unique addresses for correct routing and communication.
- We ensure that the IP address space is tightly-packed without unused IP addresses. In characteristic of current addressing method, an IP address is a valuable resource. Our protocol can economize the IP address spaces in the current IP stack.
- We eliminate plural IP allocation from the "Ack message"s which include the generated the IP address will be get back from several nodes. Mobile nodes need one IP address; the remaining IP addresses are discarded. So, we use the 4-way message protocol based on random-back off mechanism.

This paper is structured as follows. Section 2 discusses related work on automatic IP address assignment in ad-hoc networks. Section 3 describes the allocation protocol and address search for crashed mobile nodes or nodes that have abruptly left and Section 4 shows the results of a simulation with *ns-2* using our protocol. Section 5 compares the performance of EIA protocol with other methods. Section 6 provides concluding remarks.

2   Related Works

For IP address auto-configuration in MANET, allocation protocol should be provide the following requirements:

- If possible, allocation should occur as quickly as possible, because ad-hoc nodes are moving nodes in dynamic environments. Ad-hoc nodes require very short transmission delay times when they join the session.
- The IP address should be unique and without conflict. Two or more nodes in a network should not be assigned the same IP address.
- In current IP stack, the IP address is a valuable resource. So, IP address auto-configuration protocols should not create unused IP addresses or waste IP addresses.

When the node leaves the ad-hoc session, its IP address should become available for assignment to other nodes.

Auto-configuration[5], [6] produces unique IP addresses but takes a long period of time for allocation and requires temporary address space. The algorithm uses repeated DAD(Duplicate Address Detection) to test address duplication and uses a temporary address to send duplication detection messages. Because of the high mobility of MANET, repeated DADs do not guarantee uniqueness. In addition, the address de-allocation procedure is clearly defined, which means that address leakage is possible. When temporary address space is used only for DAD, this space is wasted. Therefore allocation time, address uniqueness, and address reuse are issues that have not been adequately addressed.

A dynamic configuration and distribution protocol (DCDP)[7] has been proposed as a fast allocation scheme for assigning unique addresses. DCDP involves macro auto-configuration of the address pool, and is a top-down modular protocol. The algorithm splits the currently available pool of addresses into two equal halves and does not maintain any information beyond that of its own configuration. In MANET, DCDP allocates IP addresses based on local information without the need to consult the MN. Unfortunately, there is no address maintenance functionality, so address leakage and uneven address allocation can occur.

Prophet[8] was proposed as an address allocation solution for large-scale MANETs without centralized agent servers. The method forecasts the IP address of the joining node using the assignment function. It does not require the DAD procedure. The uniqueness of the allocated IP address is guaranteed by the statistical properties of the assignment function. However, the assignment function requires a large address space, and creates address leakage.

In MANETconf[9], the basic algorithm uses a distributed mutual exclusion based on Ricart and Agrawala[10]. A configured node is proposed as a candidate IP address for allocation to a newly joining node. If the candidate IP address is accepted by all of the configured nodes that are part of the MANET, the candidate IP address is allocated to the newly joining node. This address allocation requires a positive response from all the configured nodes, indicating that the address is available for use. The unreliable multicasing in MANETs means that the algorithm uses a two-phase address allocation mechanism. To release the allocated address, the releasing node sends *"address clean_up"* messages. All nodes maintain the allocated address, and the state is flexible. The main disadvantage is that the allocation procedure takes a long period of time. MANETconf[9] uses distributed mutual exclusion[10] to manage IP address space, but it must wait for a large number of responses for allocation and reallocation.

We can resolve these problems with an electronic IP address configuration protocol, which will be suitable for small-scale MANET, prevent address leaks, unused IP address and plural IP allocation.

3   Electronic-Oriented IP Address(EIA) Auto-Configuration Protocol

Our electronic-oriented IP address(EIA) auto-configuration protocol consists of an electronic bit position(EBP) and an electronic MAP (eMAP). The EBP is used to allocate IP addresses quickly and eMAP is used to search for IP addresses that have been abruptly abandoned or crashed the mobile node.

For simplicity, we assume that there is only one initial mobile node in a MANET. This assumption can be easily satisfied by executing a distributed election procedure before starting the EIA protocol. The initial mobile node has a continuous address space before the initiation of the address allocation procedure, and the address space is called the "zone". A newly joining mobile node (MN_J) is called the "requestor" and members in a MANET that can allocate IP addresses to initiators are called "allocators". To request IP address allocation, a requestor must be one hop distance from an allocator.

3.1   Message Handling for EIA

The message handling of EIA consists of four messages: address_discovery, address_offer, address_request and address_response. The address_discovery message is sent via broadcast and the other messages are sent via unicast.

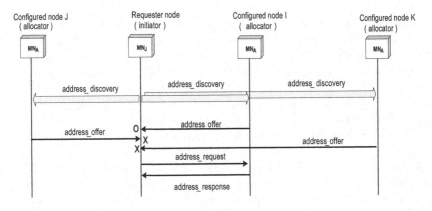

Fig. 1. The time-line of address allocation procedure

1. **address_discovery:** When a mobile node enters the MANET, it broadcasts the address_discovery message. The sending node is called the "requestor". The broadcast is repeated until the mobile node receives at least one address_offer message.

2. **address_offer:** The message sent in reply to the address_discovery message, by the allocators that received address_discovery message.

3. **address_request**: When a mobile node receives several "address_offer" messages from plural nodes, it must choose only one of them. The mobile node selects the first "address_offer" message that it receives, thus avoiding wasted IP addresses. The other "address_offer" messages are discarded. When the requestor has chosen one allocator("address_offer"), it sends the "address_request" message to the allocator. If the allocator does not have enough addresses, or no address left to assign, it may use the electronic MAP to find another allocator with sufficient addresses. The selection of the candidate allocator is based on the proxy protocol and uses the electronic MAP protocol. The detailed description of this protocol is presented in section 3.4.

4. **address_response**: The message from the allocator to the requestor containing the allocated address information.

To avoid a storm of address_offer messages (response storm), the requestor is prohibited from making a direct connection to allocators. Allocators perform a **random back-off mechanism** that sets an interval depending on the condition of the allocator. If an allocator has an allocatable IP address, the allocator transmits the *address_offer* message in response to the *address_discovery* message within $0 \sim 1$s. If the allocator does not have allocatable IP addresses, it waits several seconds ($1 \sim 3$s) before responding to the discovery message. If the allocator receives an *address_offer* message sent by another allocator, it does not send a response.

Note that the address_offer message does not yet contain any IP address. When the requestor receives an address_offer message, it sends the address_request message to the allocator. Upon receiving the allocation_request message, the allocator performs the IP address allocation procedure(see the next section 3.2) and then sends the allocation_response message.

3.2   Electronic Bit Position (EBP) Protocol

Fast and unique IP allocation must be performed without having to consult the member nodes of MANET. In our EIA scheme, each node has independent allocatable address space. When a new node wants an IP address, the AI message is sent based on the counter and level (see Fig.2). With this counter and level property, fast and unique IP address assignment is guaranteed.

For IP address assignment, our protocol uses the electronic bit position protocol with the **Address Information(AI)**.

The structure of AI is as follows:

$$AI = [\text{ zone}(\alpha, \text{ size, netmask}), \text{ counter, level }]$$

where α is the first IP address of a zone, and the zone represents the **size** of the continuous IP addresses starting from the α. The **netmask** represents the network mask of the zone. The **level** is the binary indicator for the new address, and **counter** creates the new address without conflict. At the initial mobile node, the counter and level are set to zero. The allocated IP address is derived from counter and level:

$$IPaddress = \alpha + counter \qquad (1)$$

Once an IP address is assigned to a node, the IP address of the node is not changed unless the lifetime of the IP address expires (see Section 3.3).

When an allocator receives an address_request message, it increases its level and calculates a new counter for the requestor as follows:

$$level = level + 1 \qquad (2)$$

$$counter = counter \| 2^{\,level-1} \qquad (3)$$

Then, the allocator makes $AI = [$ zone$(\alpha$, size, netmask), counter, level] and sends it to the requestor.

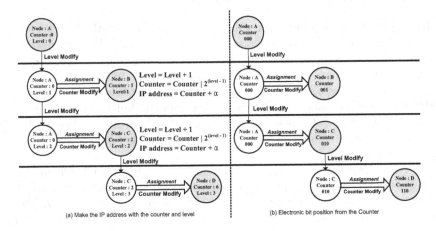

(a) Make the IP address with the counter and level   (b) Electronic bit position from the Counter

Fig. 2. The electronic bit positions(EBP) with AI

Figure 2 shows the electronic bit positions(EBP) with AI. The (a) procedure is making the AI with counter and level from expression(1)$\sim$(3). The (b) represents the electronic bit position from the counter.

Let the node A be the initial node, α be equal to 223.23.4.26 and the number of assignable IP addresses be seven(7). The AI information for node A is [(223.23.4.26, 7, 255.255.255.0), 0, 0].

When a mobile node B joins the MANET, the allocator A modifies its level to **1** and calculates the requestor counter to **1** (new counter), using expressions (2) and (3). The generated AI information for the requestor is [(223.23.4.26, 7, 255.255.255.0), **1, 1**]. Then node A sends the AI to the requestor. The IP address of the requestor becomes α(**223.23.4.26**)+**counter(1)** = **223.23.4.27**, using expression (1).

If node D joins the MANET via node C, the AI of node D is [(223.23.4.26, 7, 255.255.255.0), **6, 3**] and the allocated IP address is

$$\alpha(\textbf{223.23.4.26}) + \textbf{counter(6)} = \textbf{223.23.4.32}.$$

3.3   Electronic MAP(eMAP) Protocol

To search for IP address leak in abruptly left nodes or damaged nodes, we have introduced the electronic MAP protocol, which is based on the digital bit position.

The eMAP protocol seeks the IP address zone of crashed mobile nodes or nodes that have abruptly left. When a node leaves abruptly, the IP address space traced by the node is lost.

To recover the lost space, the member nodes of the MANET exchange their information periodically. The exchanged information is called electronic MAP and its structure is shown in Figure 3.

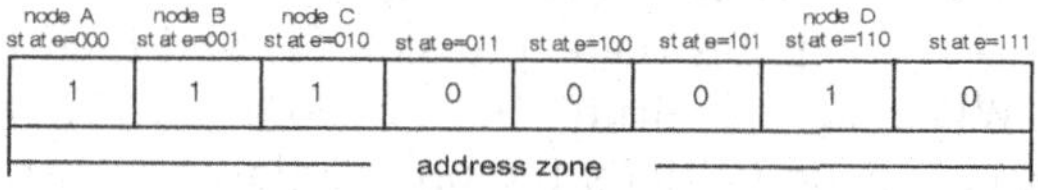

node A st at e=000	node B st at e=001	node C st at e=010	st at e=011	st at e=100	st at e=101	node D st at e=110	st at e=111
1	1	1	0	0	0	1	0

address zone

Fig. 3. An Example of a Electronic MAP

MAP consists of a bit(binary digit) array with a version number. Each binary digit bit represents a node. The initial electronic MAP values are set to zero.

If the node that has an IP address disappears from the electronic MAP, the node is considered to be departed from the MANET and its IP address is reclaimed by the address seek protocol(eMAP). This implies that if a node does not receive an electronic MAP message during a certain period, its IP address is no longer valid the node should re-join the MANET.

Tracing the MAP creates soft IP address allocation and no additional mechanism is required to seek soft allocation status. Each node has an eMAP. We have to converge the eMAP, when the MN does not have an available IP address and the beacon of neighbor node periodically. Then eMAP is broadcasted to member nodes using the *convergence message*. When the new electronic MAP arrives, it marks the bit position of the node in the new MAP, and the modified eMAP is forwarded. If the MN already receives the eMAP, it compares the electronic MAP that arrives with its new electronic MAP. If the two electronic MAPs are different, the node generates a new MAP by bitwise OR to the two electronic MAPs.

3.4   Proxy Allocator Protocol

If allocators do not have allocatable IP addresses, the configured node sends the *address_offer* message to a new requestor. After receiving the *address_request* message, the allocator becomes a **proxy allocator**. The proxy allocator forwards the *address_request* to another chosen allocator via unicast. When it receives the *address_response* message, it forwards the message to the requestor.

The proxy protocol uses the electronic MAP to determine which nodes have unused addresses. The electronic MAP represents the address allocation status

(Fig. 2). If the proxy allocator wants to forward an *address_request* message, it searches the electronic MAP for an unset bit. If an unset bit exists, the proxy allocator applies the expression (4), then " κ " to the notation (5).

$$\lfloor \log_2 unset\_counter \rfloor = \kappa \tag{4}$$

$$unset\_counter - 2^\kappa = Pcounter \tag{5}$$

Here, **Pcounter** represents the parent node of the unset bit position. Therefore, the Pcounter node can give the IP address to the new requestor through the proxy allocator.

For example, if the node D is the proxy allocator, the first unset bit is the bit position (011) from D's Electronic MAP in Fig 3.

The proxy allocator computes the expression (4) and then we get the result that "κ" is "*1*". When we substitute "κ" for notation (5), we find that the parent counter (001) is node B. Here, the proxy allocator D transmits the "*address_request*" to the chosen allocator (node B by expression (4),(5)) via unicast. After receiving the confirm message from node D, node D can allocate the IP address to the requestor. The unset bit position from B's Electronic MAP can obtain the IP address from expression (1).

Therefore, the requestor IP address becomes

(223.23.4.26) + counter(3) = 223.23.4.29.

4   EIA Protocol Simulation Experiments

The simulation was done on **ns-2** (version 2.29) with the CMU extensions to support ad hoc networks. The random way-point mobility model[14] was used. The mobile nodes have two different area sizes of $250m$ x $250m$ and $500m$ x $500m$.

In the simulation, the pause time of the mobile nodes was 0.2, 1.0, or 10.0 s, in two area sizes. The number of mobile nodes was fixed, with 60 in 64 available addresses. The wireless link speed was set to 2 Mbps. The routing protocol used was ad-hoc on-demand distance vector routing (AODV[3]).

Figure.4 shows the simulation results. The values are an average of 30 trials for each case. The electronic MAP takes 30s and electronic MAP convergence takes less than 1s. In Figure 4, several sparks are shown. These are caused by (forwarded) *allocation_request* message loss in the MANET, the spark values are multiples of 5s because the requestor retries the *address_discovery* message every 5s. The average allocation times are 0.91, 0.30, 0.29, 1.20, 0.52 and 0.70s, respectively.

When the mobile node participates in the MANET with a pause time of 0.2s, the EIA protocol consumes more time for the allocation (see Fig.4(a) and (d)). The reason for this is that the proxy allocator needs several attempts to find the right allocator because of the increasingly inaccurate electronic MAP.

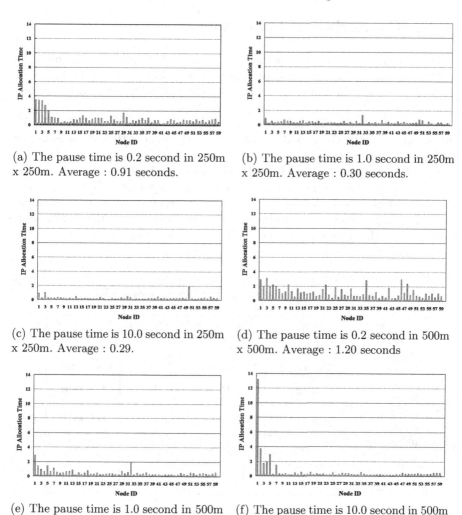

(a) The pause time is 0.2 second in 250m x 250m. Average : 0.91 seconds.

(b) The pause time is 1.0 second in 250m x 250m. Average : 0.30 seconds.

(c) The pause time is 10.0 second in 250m x 250m. Average : 0.29.

(d) The pause time is 0.2 second in 500m x 500m. Average : 1.20 seconds

(e) The pause time is 1.0 second in 500m x 500m. Average : 0.52 seconds

(f) The pause time is 10.0 second in 500m x 500m. Average : 0.70 seconds

Fig. 4. The IP Address allocation time

These results indicated that a 10.0s pause time and area size of 250m x 250m offered the most rapid allocation time. The worst case was wide area size(500m x 500m) and short pause time(0.2s).

The technical reason for the worst case was the coverage of the mobile node. If two nodes are far apart in area, one-hop transmission is impossible. In this case, the proxy node communicates to the requestor via neighbor nodes with multi-hop transmission, so it takes more time to get the IP address. Therefore, the allocation time is related to the pause time and area size.

5   The Comparison of Performance

Table 1 compares our strategy with AAA[5],[6] and Prophet[8]. The first five rows are feature briefs of the three allocation solutions, and the performance comparison focuses on unused IP addresses, plural IP addresses and IP address reuse.

The unused IP addresses in our system were managed very efficiently. The IP address zone was tightly packed with MNs, and the wasted IP address zone was eliminated.

In AAA[5],[6], the level of unused IP addresses is too high, due to this scheme's technique of splitting addresses into temporary and legal sets. Moreover, if networks consist of eight nodes, this scheme uses the eight addresses in the address zone of 256, and the remaining addresses (248) are unused. If all MNs (128 nodes) participate in the network, the unused IP zone is as much as 50% (temporary address pool).

Table 1. Characteristics and performance comparison

Evaluation	AAA[5][6]	MANETconf[9]	Prophet[8]	Our solution
Address conflict	Frequently Yes	Yes	No	No
Element	Need the DAD	Mutual exclusion	$f(n)$	EBP,eMAP
Unused IP address	High	High	High	Not exist
Plural IP address	Frequently Yes	Frequently Yes	Yes	Not exist
IP Address Reuse	Yes	Yes	No	Yes
Communication Overhead	O((n+1) x k)	O(r x n^2 x k)	O(2l / n)	O(4l / n)

Prophet[8] generates leak addresses (up to 37.5%) with the assignment function $f(n)$. For this reason, if many MNs participate in a MANET, a very broad address zone is needed. By contrast, if MNs are tightly packed, as they are in our solution, unused IP addresses cannot exist in the address space.

In addition, Prophet creates plural IP addresses which contain the available IP address from multiple nodes. Also, it can not reuse the IP addresses from crashed mobile nodes or nodes that left abruptly.

In fact, common network devices use the IPv4 to connect the other external network systems. It is premature to use IP version 6 (IPv6) in MANET. In current IP stack, public IP address space is a valuable resource, and locating reusable addresses is very important. Therefore, IP addresses have to be protected against loss until IPv6 is widely used. The problem of wasted IP address must be resolved now, and we need appropriate IP address allocation protocols for MANETs.

In the worst case, controlling the electronic MAP(eMAP) for seeking unused IP addresses has some overhead that depends on the convergence of the eMAP. However, the optimization of convergence overhead is left as future work. The maintenance of IP addresses induces the AODV routing protocol [3] to identify unused IP addresses.

In Table 1, communication overhead is represented by O during the allocation of the IP address, where n is the number of MNs, k is the retry time, and the number of links is 1.

Prophet[8] is the quickest in terms of communication overhead, but contains several drawbacks such as allowing unused IP addresses, plural IP addresses and Ack implosion.

To sum up the basic characteristics of the EIA protocol, it does not have unused IP addresses or plural IP addresses, and can reuse IP addresses for other nodes. This is very useful for MANETs.

6   Conclusion

An IP allocation and locating scheme was proposed for MANET. Prophet[8] has several drawbacks, such as allowing unused IP addresses and plural IP addresses, and can not reuse the IP addresses. But, our protocol provides high efficiency. To overcome the response storms(ack implosion), the EIA protocol uses a 4-way message protocol based on a random back-off mechanism.

In the worst case, the proposed system produces overhead over eMAP convergence periods, but locating such addresses with eMAP is important due to the limited number of addresses.

As mentioned above, we resolve the problems of IP allocation and locating in this paper. The simulation experiments show that the proposed solution has low latency and reasonable IP address space. The simulation results demonstrate that the scheme is well suited to small-scale MANETs.

However, the proposed scheme has some obvious analysis issues in terms of the worst case communication overhead and successful ratio relevant to MANET protocols. These issues will be the focus of future work.

References

1. Perkins, C.E., Royer, E.M., Das, S.R.: IP address autoconfiguration for ad hoc networks. Internet Draft, IETF Working Group MANET(Work in Progress) (July 2000)
2. Johnson, D.B., Maltz, D.A.: Dynamic Source Routing in Ad-Hoc Wireless Network, Mobile Computing. Kluwer Academic Publishers, Dordrecht (1996)
3. Perkins, C., Belding-Royer, E., Das, S.: Ad-hoc On-Demand Distance Vector(AODV) Routing. RFC 3626 (October 2003)
4. Droms, R.: Dynamic Host Configuration Protocol. Network Working Group RFC 2131 (March 1997)
5. Perkins, C.E., Elizabeth, M.: IP Address Autoconfiguration for Ad Hoc Networks, draft-ietf-manet-autoconf-01.txt (Work in Progress) (November 2001)

6. Sun, Y., Elizabeth, M., Perkins, C.E.: Internet Connectivity for Ad Hoc Mobile Networks, International Journal of Wireless Information Networks, Special Issue on "Mobile Ad Hoc Networks(MANETs); Standards, Research, applications (2002)
7. Misra, A., Das, S., Mcauley, A.: Autoconfiguration, Registration and Mobility Management for Pervasive Computing. IEEE Personal Communications (2001)
8. Zhou, H., Lionel, M.N.I.: Prophet Address Allocation for Large Scale MANETs. In: IEEE INFOCOM 2003 (2003)
9. Nesargi, S., Prakash, R.: MANETconf:Configuration of Hosts in a Mobile Ad Hoc Network. IEEE INFOCOM (2002)
10. Ricar, G., Agrawala, A.K.: An Optimal Algorithm for Mutual Exclusion in Computer Networks. Communications of the ACM 24(1), 9–17 (1981)
11. Chandra, R., Ramasubramajan, V., Birman, K.P.: Anonymous Gossip: Improving Multicast Reliability in Mobile Ad-Hoc networks. In: Proceedings of the 21st IEEE International Conference on Distributed Comuting Systems(ICDCS2001), Mesa, Arizona, pp. 275–283 (April 2001)
12. Jetcheva, J.G., Hu, Y.C., Maltz, D.A., Johnson, D.B.: A Simple Protocol for Multicast and Broadcast in Mobile Ad-Hoc networks(draft-ietf-manet-simple-mbcast-01.txt). Internet Engineering Task Force, MANET Working Group (July 2001)
13. Ozaki, T., Kim, J.B., Suda, T.: Bandwidth-Efficient Multicast Routing for Multihop, Ad-Hoc Wireless Networks. In: proceedings of IEEE INFOCOM, pp. 1182–1191 (2001)
14. Broch, J., Maltz, D.A., Johnson, D.B., Hu, y.-c., Jetcheva, J.: A Performance Comparison of Multi-Hop Wireless Ad Hoc Routing Protocols. In: Proceedings of the Fourth Annual ACM/IEEE International Conference on Mobile Computing and Networking, pp. 85–97 (October 1998)

Author Index

Lecture Notes in Computer Science

Sublibrary 5: Computer Communication Networks and Telecommunications

For information about Vols. 1– 4686
please contact your bookseller or Springer

Vol. 3976: F. Boavida, T. Plagemann, B. Stiller, C. West-phal, E. Monteiro (Eds.), NETWORKING 2006. Net-working Technologies, Services, and Protocols; Per-formance of Computer and Communication Networks; Mobile and Wireless Communications Systems. XXVI, 1276 pages. 2006.

Vol. 3970: T. Braun, G. Carle, S. Fahmy, Y. Koucheryavy (Eds.), Wired/Wireless Internet Communications. XIV, 350 pages. 2006.

Vol. 3964: M.Ü. Uyar, A.Y. Duale, M.A. Fecko (Eds.), Testing of Communicating Systems. XI, 373 pages. 2006.

Vol. 3961: I. Chong, K. Kawahara (Eds.), Information Networking. XV, 998 pages. 2006.

Vol. 3912: G.J. Minden, K.L. Calvert, M. Solarski, M. Yamamoto (Eds.), Active Networks. VIII, 217 pages. 2007.

Vol. 3883: M. Cesana, L. Fratta (Eds.), Wireless Systems and Network Architectures in Next Generation Internet. IX, 281 pages. 2006.

Vol. 3868: K. Römer, H. Karl, F. Mattern (Eds.), Wireless Sensor Networks. XI, 342 pages. 2006.

Vol. 3854: I. Stavrakakis, M. Smirnov (Eds.), Autonomic Communication. XIII, 303 pages. 2006.

Vol. 3813: R. Molva, G. Tsudik, D. Westhoff (Eds.), Se-curity and Privacy in Ad-hoc and Sensor Networks. VIII, 219 pages. 2005.

Vol. 3462: R. Boutaba, K.C. Almeroth, R. Puigjaner, S. Shen, J.P. Black (Eds.), NETWORKING 2005. XXX, 1483 pages. 2005.